METHODS IN MICROANALYSIS

Volume I

SIMULTANEOUS RAPID COMBUSTION

Methods in Microanalysis

Volume I

SIMULTANEOUS RAPID COMBUSTION

Translated Microchemical Research Papers
of the late Mirra Osipovna Korshun
Institute of Hetero-Organic Chemistry,
Academy of Sciences, USSR
and members of her school

Edited by J. A. KUCK, Ph.D., *Senior Research Microanalyst,*
American Cyanamid Company

Translated by PHYLLIS L. BOLTON,
formerly with
American Cyanamid Company

and

KURT GINGOLD, Ph.D. (ATA), *Research Literature Scientist,*
American Cyanamid Company

GORDON AND BREACH
Science Publishers
New York · London

Science Publishers, Inc.
150 Fifth Avenue, New York, N. Y. 10011

Library of Congress Catalog Card Number: 64-18800

For Great Britain and Europe:
Gordon and Breach, Science Publishers, Ltd.
171 Strand
London W. C. 2, England

PRINTED IN THE UNITED STATES OF AMERICA

The work of translating and editing this volume of Russian articles has been made possible by the cooperation of the AMERICAN CYANAMID COMPANY as a project in the public interest.

MIRRA OSIPOVNA KORSHUN (obituary)*

On December 1, 1958 there occurred the death of M. O. Korshun, founder of the Soviet school of quantitative organic microanalysis and director of the microanalytical laboratories of the Institute of Elemento-Organic compounds of the Academy of Sciences of the USSR.

Miss Korshun was born on August 21, 1900 in Vilna. After finishing the four-year junior high school, at the age of 13, she went to work as a pharmacist's apprentice. Working in pharmacies and factories, she continued her studies and in 1929 graduated from the Faculty of Chemistry of Moscow State University, majoring in Organic Chemistry. After her graduation, she continued to work as a plant chemist, and in 1933 started to work in the field of organic microanalysis as supervisor of an analytical group. In 1937 she organized and directed the Microanalytical Laboratory in the Institute of Organic Chemistry, and in 1954 she joined the laboratory in the Institute of Elemento-Organic Compounds of the Academy of Sciences of the USSR, which she led until the time of her death.

In striving to find the best solutions to the analytical problems posed by the development of organic chemistry, Miss Korshun turned away from the old, classical methods of microanalysis and found a more rational method for the elemental analysis of organic compounds. For this she put into practice a new principle for the decomposition of organic compounds —"pyrolytic combustion." On the basis of a profound and thorough study of this process, she established a whole [new] area of organic elemental analysis—rapid micro methods for the simultaneous determination of several elements in a single sample of a multi-element organic compound. Miss Korshun's

* Izvest. Akad. Nauk SSSR, Otdel, Khim, Nauk *1959*, 759

methods are characterized by great precision and speed, together with simplicity and elegance; because of their undoubted advantage over the classical methods, they very rapidly found widespread use in the Soviet Union and abroad. This was greatly aided by her important work in setting up standard types of microanalytical apparatus and balances and introducing these into industry. In 1958, her work on rapid methods of analysis was presented in the competition for the Lenin Prize.

Miss Korshun devoted much energy to the training of cadres of qualified microanalysts and to the promotion of new, progressive methods of analysis; she wrote two monographs and 50 scientific papers. She was an active member of the Analytical Committee of the Presidium of the Academy of Sciences of the USSR, and a short time before her death was elected as a member of the Committee on Microchemical Methods of the IUPAC.

The fruitfulness of her work was marked by the Soviet government with the Order of the Labor Red Banner and the Order of the Sign of Honor. She was not only a great scientist, but also a fine person. All who knew her will always remember her great love of people and her supreme devotion to Soviet Science and to the Motherland.

PREFACE

An unfortunate language barrier for many an American microanalyst makes inaccessible to him significant information about current developments in the microchemical field that have been appearing in the Russian scientific literature during the past two decades. Therefore on account of the novel methods often described by the Russians and their bold experimental approach in many instances to a difficult analytical problem, it seems worthwhile to bring together under one cover certain scientific papers of recognized value and make them available in English to a larger group of readers. It is hoped that by this means microanalysts in our own country and elsewhere may try out some of the procedures presented and thus acquire a new outlook upon old analytical difficulties.

Conspicuous among the few outstanding microanalysts the world over, Miss Korshun through both the number and quality of her publications on quantitative organic microanalysis merits special recognition. Consequently in any compilation of Russian contributions to our knowledge of microchemical analysis, it should be expected that her work would overshadow that of anyone else in her country. In fact, it was her fortune to be a pioneeer in the field and to found a flourishing institute of microanalysis at Moscow University. This institute, in addition to serving the needs of the organic research staff of the university there, has always maintained an ambitious development program of its own which is characterized by a high degree of excellence. From this center many students have emigrated to distant parts of the Soviet Union to start their own laboratories taking the knowledge of her analytical methods with them.

Therefore, the experimental directions laid down in the beginning by Miss Korshun had to be direct, reliable, and simple in execution. The possibility of a lack of even primitive microchemical facilities to an analyst in some far distant place had always to be reckoned with so that her procedures had to be designed for use under a variety of circumstances.

On the other hand, readers of this book will find that the research papers of Miss Korshun which we have selected for translation have been written from the specialist's point of view. They represent a high level of scientific treatment and their content is fundamental to practical quantitative organic microanalysis. The same comment can be made about the research papers of her students, a few of which we have also included.

The methods presented in this book stem from the basic philosophy of Miss Korshun about the art of microcombustion. As a proponent of the so-called "empty tube" method in preference to the classical procedure of Pregl, she published over the years a variety of determinations of which all had in common the principle of pyrolyzing the sample in a quartz capsule with an insufficiency of oxygen. It must be admitted that igniting the sample in this fashion, whenever scientifically permissible, offers one obvious advantage—rapidity.

Consequently, although Miss Korshun did not live to see automation take possession of the carbon-hydrogen determination, present developments clearly indicate that such achievement is now not far off and that her "rapid" method of sample ignition will be an indispensable feature toward making it possible.

Another contribution of Miss Korshun was her continuing preoccupation with the problem of simultaneous determination; i.e., analysis for several elements at the same time in one weighed sample of a complex substance. In the past this approach has not seemed attractive to some analysts on account of the increased variety of analytical interferences which can result from adding to the number of hetero elements that have to be dealt with. Nevertheless, there is increasing interest today in methods for simultaneous determination and everyone recognizes their economy in the average laboratory if practical procedures can be developed. Among the selected research papers contained in the present book the reader will note that twenty-five dealing with simultaneous determinations have been included. Many of these have practical value if one is to judge from the quality of the published analytical data.

The purpose of this book, therefore, is to acquaint the reader with the progress of quantitative organic microanalysis in Russia during the past fifteen years, as exemplified in the publications of Korshun and her successors. Several significant papers by other Russian or Czechoslovakian authors which happened to fit into the general theme of the collection have been grouped together in a supplementary section.

In the opinion of the Technical Editor, the microanalyst reader will find certain individual papers in the collection particularly rewarding in view of the scientific conclusions which the authors draw about the particular analytical determination under study. Among such might be mentioned: the oxygen determination, analysis for carbon-hydrogen in nitrogen-containing compounds, silicon carbide formation in the combustion of silicon-containing compounds, the mechanism of silver sulfate formation in the combustion of sulfur-containing compounds, the determination of carbon-hydrogen in organic phosphorus compounds, the pyrohydrolysis of magnesium fluoride in the fluorine determination, the use of an iron catalyst for ammonia in the reductive combustion for nitrogen, and the conductometric determination of carbon and hydrogen.

In preparing an English version of the scientific articles presented herein, accuracy of translation has been the goal and every effort has been made to find the corresponding English equivalent for the Russian technical term. In only a few instances was an extensive rewording of sentence structure found necessary in order to carry over into our own language the exact meaning of the original authors.

It is hoped that microanalysts will find the reading of this book profitable to them. Newcomers to the field as well as experienced specialists will gain from the valuable technical ideas which it can suggest. But what is more important is the hope that American microanalysts and those responsible for their employment will grasp its implications. The great value which Russian science places upon research in this field, the thoroughness with which the Russians train their microanalysts in their leading university, and the creativity which is evident in the articles herein presented are hard facts that warrant thoughtful consideration by the rest of us and which may call for a certain amount of introspection on the part of every American chemist.

J. A. Kuck

CONTENTS*
(In Summary Form)

*Papers comprising a series on the same subject bear a common repeating number.

CHAPTER 1

Microanalytical Determination of Oxygen in Organic Compounds Communication I*

M. O. Korshun

Of all the elements contained in organic compounds, oxygen is the most difficult to determine. In elementary analysis, oxygen is determined indirectly, that is, by subtracting the total percentage of all the other elements from 100.

In spite of the fact that the problem of direct determination of oxygen has been of interest to chemists for over 100 years,† it was completely solved only recently.

Most of the methods for the direct determination of oxygen suggested during the past 10 years are macroanalytical methods which do not meet the requirements of the present time. The increase in the importance of elementary microanalysis, due to the rapid development of synthetic and preparative organic chemistry, forces analysts to make macrochemical methods more precise by converting them to microchemical methods, or to work out new methods.

At present, we have three basic microanalytical methods for the determination of oxygen in organic compounds.

Kirner's [1] microchemical method, based on the complete catalytic oxidation of the substance in an oxygen

***Zavodskaya Lab.* **10**, 241-245 (1941).

†An excellent historical review of the development of the methods for determination of oxygen in organic compounds has been given by M. S. Skanavi-Grigor'eva in *Uspekhi Khim.* 7, 713 (1938).

stream, makes it possible to determine simultaneously carbon, hydrogen and oxygen in the presence of halides and nitrogen. It is regrettable that the apparatus necessary for this method is extremely difficult to construct under conditions in the ordinary analytical laboratory; besides, the method itself requires special techniques. Therefore, this method has not found wide use in laboratory practice.

The methods of Lindner and Wirth [2], Unterzaucher and Bürger [3] are micro modifications of the Ter-Meulen [4] method based on the principle of destructive catalytic hydrogenation of the substance with subsequent determination of the water formed. The analytical results obtained by Ter-Meulen are in satisfactory agreement with the theory. Lindner's and Unterzaucher's microchemical methods for the determination of oxygen were published simultaneously and do not differ essentially, the only difference being in the method of determining the water.

The next advance in methods for the direct determination of oxygen has been Schütze's [5] elegant method which was first worked out for the determination of oxygen in zinc dust. The principle of the method: the substance under investigation is reduced with carbon at 1000°C in a stream of practically pure nitrogen according to the reaction:

$$ZnO + C \rightarrow CO + Zn \tag{1}$$

Thus, the total oxygen (free or combined in the form of water) reacts with carbon quantitatively forming carbon monoxide; the latter passes over iodine pentoxide and is oxidized to carbon dioxide which is collected in a Pregl absorption apparatus, filled with Ascarite reagent and weighed. The oxidation of carbon monoxide into carbon dioxide occurs according to the equation:

$$I_2O_5 + 5CO \rightarrow 5CO_2 + I_2 \tag{2}$$

Schütze did not limit himself to the determination of oxygen in zinc dust but, on the basis of the above reactions, he worked out a semimicrochemical method for the determination of oxygen in organic compounds. Contemporaneously with Schütze, Zimmerman [6] worked out a micro modification of this method with automatic combustion and modified the apparatus correspondingly. Zimmerman subtracted 100 μg from the obtained amount of CO_2. This correction consists of 40 μg CO_2 obtained from the traces of oxygen in nitrogen, and 60 μg CO_2 obtained from the oxygen in the air which seeped into the combustion tube when the charge was introduced.

Later on, Unterzaucher [7] gave the second micro modification of Schütze's principle by modifying the apparatus somewhat and introducing a gravimetric completion of the analysis in which he absorbed the iodine, liberated according to reaction (2), in alkali and titrated it with thiosulfate after oxidation to iodate. Unterzaucher shows a faultless table with results for the determination of oxygen in various substances both organic and inorganic. He states that his apparatus makes it possible to work without the introduction of empirical corrections and therefore makes it possible to determine even traces of oxygen.

Schütze's idea is not new. In 1912, Walker and Patrick [8] gave a similar method for the determination of oxygen in steels and iron by melting the metal with graphite and determining the oxide of carbon with iodine pentoxide. The percentage of oxygen was calculated on the basis of the amount of liberated iodine which was absorbed by alkali and titrated with thiosulfate.

The reaction for oxidizing carbon monoxide into dioxide by means of iodine pentoxide was discovered by Ditte in 1870 [9] and apparently was first used for the quantitative determination of the traces of carbon monoxide in air by Nicloux [10] and then Gautier [11] in 1898. The value

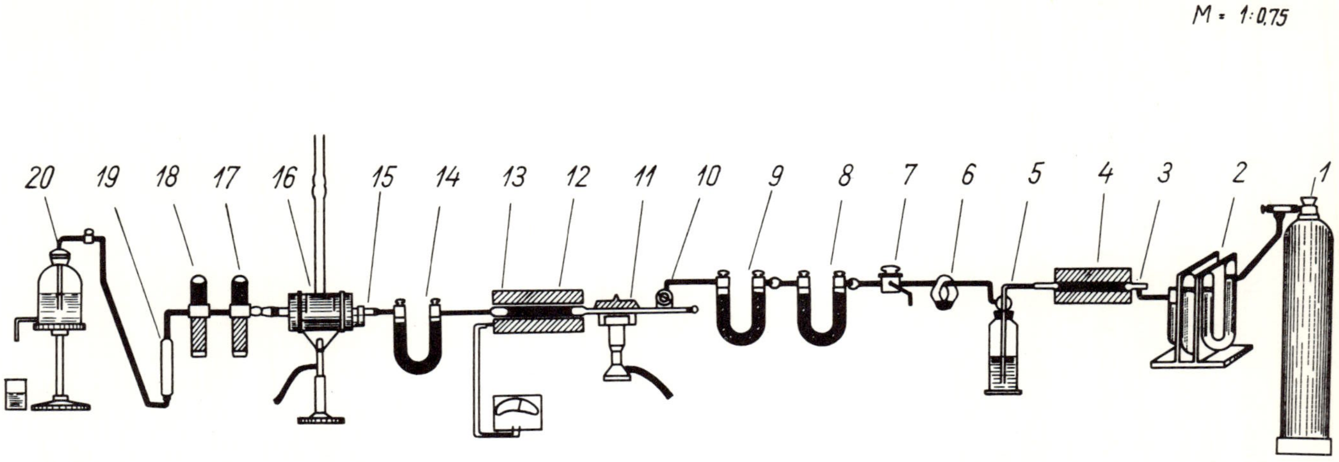

Fig. 1. Apparatus

of Schütze's work lies in the fact that he worked out a method for the quantitative determination of oxygen in organic compounds on the basis of the reduction of oxygen compounds with carbon at high temperature and a subsequent oxidation with iodine pentoxide of the carbon monoxide obtained. Thus, he solved the problem of the direct determination of oxygen, which had puzzled chemists for more than 100 years, and thus he filled in the gap in elementary microanalysis.

EXPERIMENTAL PART

Arrangement of the Apparatus and its Designation. We tried out Schütze's method with pure substances using almost the whole apparatus which was applied for determination of carbon and hydrogen [12]. The arrangement is shown in Fig. 1.

From the cylinder (1), with the aid of a precise reducing valve, nitrogen passes without preliminary drying and purification through three U-columns (2) connected in series; two of them are filled with calcium chloride and the third with activated carbon. Then the nitrogen goes into the tube (3) filled with activated copper. The length of the tube with copper is 250 mm.; its inner diameter, 10 mm. The copper is heated to 300°C with an electric furnace (4). It is necessary to note that the greatest difficulty in this work is the purification of nitrogen from the traces of oxygen. In the beginning, we purified nitrogen by passing it through a layer (60 cm.) of reduced copper heated to 800°C, but the blank tests always showed the presence of oxygen. The weight increase in the absorption tube was 200 μg. Lately we have used activated copper for the purification of nitrogen; this was suggested by Meyer and Ronge [13].

Preparation of Active Copper. 62.5 grams of $CuCl_2 \cdot 2H_2O$ is dissolved in 500 ml. of water, 62.5 g. of infusorial earth is added, and a solution of 60 g. of NaOH in 125 ml. of water added at 60°C with vigorous stirring. In about 15-20 min., the reaction mass is poured into a vessel containing 2.5 l. of distilled water. After the cupric hydroxide settles, the water is decanted, and the precipitate washed on a Büchner funnel. Then the paste-like mass is pressed through an iron sieve (1 $mm.^2$ mesh) and dried at 150-180°C until the preparation acquires a brownish tinge. Then it is cut into pieces of about 3-4 mm. in length, and the dust removed. The dried preparation is placed in a tube and reduced in a hydrogen stream at 300°C, the air having been previously displaced.

At the 200°C recommended by the authors the reduction proceeds very slowly. Due to its distribution over a large surface, this copper possesses a considerably greater activity than the usual reduced wire-form copper; but it still does not absorb the traces of oxygen, although it decreases the value of the blank.

After passing through the layer of activated copper, the nitrogen goes through a wash bottle (5) and bubble counter (6) filled with sulfuric acid, then enters a three-way stopcock (7) from which it can go either into the air or into a U-tube (8) filled with a mixture of equal parts of Ascarite reagent and soda-lime (slightly moistened) for the absorption of CO_2. For complete drying, the nitrogen passes through a second U-tube (9) filled with phosphoric anhydride deposited on granulated pumice†.

Thus, the purified and dried nitrogen is introduced into the combustion tube through a baffle with a three-way

† Pieces of pumice are ground to grains (1 $mm.^3$), sifted to remove the dust, boiled in dilute nitric acid, then in hydrochloric acid, and the process is repeated until the supernatant liquids become colorless. After this, the pumice is carefully washed with distilled water and with alcohol, dried and ignited in a muffle furnace. Pumice thus obtained is whitish gray.

stopcock (10). The combustion tube is made of transparent quartz (length, 500 mm.; inner diameter, 9-10 mm.); one end is drawn out into a thick-walled capillary (length, 100 mm.; outer diameter, 6 mm.; inner diameter, 1 mm.); the other end of the tube is closed with a ground-glass cap. The gas baffle available through use of the three-way stopcock is supposed to prevent the infiltration of air during the introduction and removal of the boat with its sample by passing nitrogen from A to B (Fig. 2) while the exit end of the combustion tube is closed.

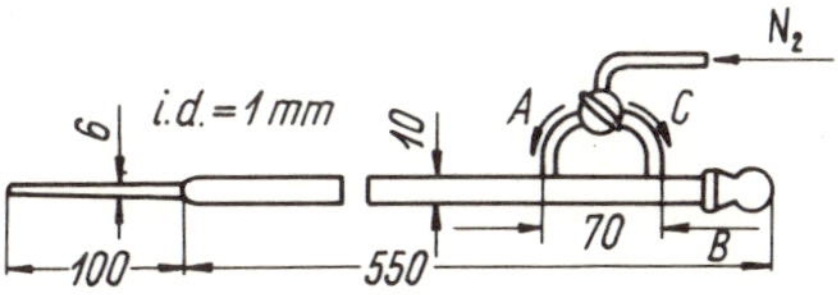

Fig. 2. Combustion tube

It is very simple to fill the tube. First, a layer of quartz wool is introduced (20 mm. in length), then a 15-mm. layer of ground quartz is dropped in, after which a layer of carbon granules follows (160 mm. in length). We used Baku gas black previously purified and prepared in the form of granules according to Unterzaucher [7]. The analysis of the carbon burned in a current of pure nitrogen at 1100°C gave the following results: sample = 2.594 mg.; CO_2 = 9.468 mg.; H_2O = 0.065 mg.; %C = 99.55; %H_2O = 0.28.

The electric furnace (13) serves for heating the contact layer (12) to 1100°C. The furnace consists of a metallic cylinder, 190 mm. in length and 100 mm. outer diameter. In the center of the cylinder is attached a porcelain tube with an inner diameter of 15 mm. around which is wound a coil of platinum wire. The furnace has a thermocouple well. A mixture of ground aluminum oxide and asbestos serves for thermal insulation.

A platinum—platinum-rhodium thermocouple connected with a millivoltmeter registers the temperature. The constancy of temperature at 1100°C is attained very easily by means of a rheostat. The furnace is heated to the required temperature in 30 min. It is evident that the temperature registered by the thermocouple in the well differs greatly from that in the reaction tube. Therefore, a new furnace ought to be calibrated before use. A U-tube (14) with ground KOH serves for the absorption of hydrogen halides and hydrogen sulfide in the case of combustion of compounds containing halogen and sulfur.

A tube with iodine pentoxide (oxidation tube) (15) has on one end a ground stopper with a vent tube similar to Pregl's apparatus and is connected with a U-tube by means of rubber; on the other end, the tube is ground to fit the iodine absorption tube (17) (Fig. 3). The total length of the tube is 170 mm.; inner diameter, 9-10 mm. Before introduction into the oxidation tube the fine crystalline iodine pentoxide is dried in vacuo for 3 to 4 hrs. at 180-190°C. The layer of I_2O_5 in the tube is held in place by a glass wool plug at each end.

Such a construction of the oxidation tube combined with a modified Abrahamczik [14] absorption tube makes it possible to collect quantitatively on reduced copper the iodine liberated during the reaction (2).

The ground stopper and the inner inlet tube of the absorption apparatus have an especially enlarged diameter in order to avoid clogging with iodide crystals during the analysis. The widened inner inlet tube should be directed upwards (Fig. 3).

The iodine absorption tube is filled as follows: a glass wool packing is placed in the upper part of the tube which is then half filled with fine reduced copper wire. The remaining space is filled with phosphoric anhydride on pumice with glass wool separators. The oxidation tube and the ground joint connecting it with the iodine absorption tube are placed in the heating mortar (16) which is heated to 130°C. This heating mortar can be moved horizontally as far as the ground joint of the absorption tube. The iodine liberated during the analysis is completely displaced into the absorption tube where it is fixed by the copper for weighing; the carbon dioxide formed is absorbed in the second absorption tube (18) which is filled with Ascarite and a layer of phosphoric anhydride. Next comes the guard tube with phosphoric anhydride (19), connected at one end to a Mariotte flask (20) and at the other to the second absorption tube (18).

All parts of the absorption train are connected by means of impregnated thick-walled rubber tubing connectors and glass to glass connections except for the one ground joint connection between the oxidation tube and the iodine absorption tube (Fig. 3).

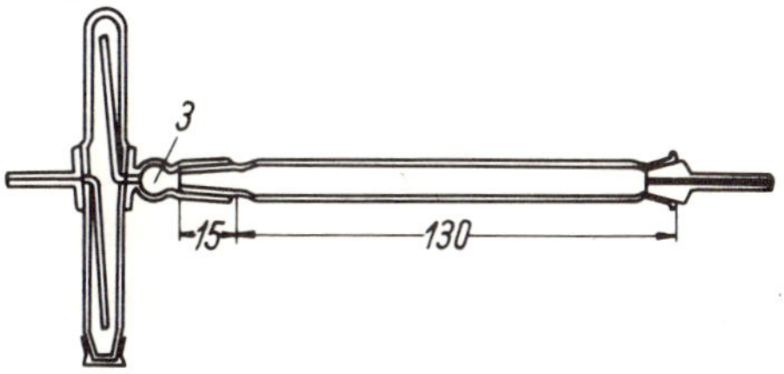

Fig. 3 The I_2O_5 oxidation tube and its ground joint connection to the iodine absorption tube.

Analysis. Before the start of the work both electric furnaces are switched on. The valve of the nitrogen cylinder is opened and nitrogen is let out through a three-way stopcock into the air with greater speed (7-8 ml.) than during the analysis. Meanwhile, the sample is weighed. When the furnaces are heated to the required temperature, the three-way stopcock (7) is turned to pass nitrogen into the combustion tube and the three-way stopcock (10) of the gas baffle is opened in the direction of from A to B (Fig. 2). Meanwhile, the free part of the tube is heated with a movable burner (11). When the heating mortar attains 130°C, the oxidation tube with iodine pentoxide and also the absorption tubes are connected to the system. Before the first analysis, a blank determination similar to that in the carbon-hydrogen determination is made. Nitrogen at a speed of 7-8 ml. per min. is passed through the whole apparatus in order to remove the traces of oxygen and water.

After the weighing of the absorption tubes, the analytical sample is introduced into the tube. The procedure is as follows: the stopcocks of the U-tubes (14) are closed and the three-way stopcock (10) is turned so that the nitrogen goes from A to B (Fig. 2). The combustion tube is opened, and the boat with the sample quickly inserted to a distance of 8-9 cm. from the furnace.

Meanwhile, nitrogen is passed at a speed of 7-8 ml. per min. However, in spite of all precautions, air diffuses into the tube; even after 30 min. of sweep, we obtained larger blank values than those in Zimmerman's data [6]. Even when the oxidation tube is attached immediately after introducing the sample, an abundant evolution of iodine vapors begins. Consequently, use of the gas baffle and the three-way stopcock is not completely justified. Unterzaucher, when introducing the boat, passed nitrogen in the reverse direction through the whole combustion tube for a 10-min. sweep and obtained excellent blank values; therefore, he does not apply any blank correction. The improvements inaugurated by Unterzaucher are in the checking stage with us now; therefore it is difficult to evaluate them yet.

With a 30 min. sweep we obtained about a 70-μg weight increase in CO_2. After the 30-min. sweep, the oxidation tube with iodine pentoxide is attached and also the absorption tubes (the I_2O_5 tube is disconnected each time during the weighing of the absorption tubes, the ground end of the tube being closed with a ground cap and the other end with a plug); the three-way stopcock of the gas baffle is turned to B (Fig. 2); the Mariotte bottle is connected to the last absorption tube, the gas flow is set at 20 bubbles per 14 seconds in the counter which corresponds to 5.5 ml. of nitrogen per min., and the combustion of the substance is started. The burner is placed at a distance of 2 cm. from the eye of the platinum boat. The combustion of the sample should be carried out at such a rate that the formation of the carbon on the walls of the tube is very slight, and the entrance of carbon monoxide into the oxidation tube is uniform. The beginning and end of the decomposition of the sample can be easily spotted by the color change in the oxidation tube. After completion of the combustion (usually consuming 15-20 min.), the sweep is begun; during the sweep the combustion tube is reignited. The sweep takes 10 min. In order to ensure complete displacement of all the iodine into the absorption tube, wire hooks which conduct heat from the heating mortar are placed against the concave part of the ground joint as in the carbon-hydrogen determination for driving water over into the $CaCl_2$-tube.

After the sweep has been completed, the stopcock of the Mariotte bottle is closed and then the carbon dioxide and iodine absorption tubes. Finally, the Mariotte bottle is disconnected (the P_2O_5 guard tube being closed with a plug), and the absorption tubes are taken off and weighed as usual [15]. The oxidation tube is removed from the heating mortar and closed until the beginning of the following analysis.

TABLE 1

Substance	Formula	Weight mg.	Weight[a] of CO_2, mg.	Determination on carbonic acid % oxygen: Calculated	Found	Deviation from the theory, in % abs.	Determination on iodine: Weight[b] of I_2 in mg.	Found % of O_2	Deviation from the theory, in % abs.
Benzoic acid	$C_7H_6O_2$	12 630	9.200	26.22	26.49	+0.27			
		13.612	9.900		26.45	+0.23			
		15.195	10.984		26.28	+0.06			
		11.808	8.465		26.07	−0.15			
		6.070	4.329		25.93	−0.29			
β-ethyldecalol	$C_{12}H_{18}O$	11.420	2.825	8.98	8.99	+0.01			
		8.910	2.280		9.02	+0.04			
α-aminobutyrolactone hydrochloride	$C_4H_8O_2NCl$	8.180	5.215	23.27	23.18	−0.09			
		8.230	5.245		23.17	−0.10			
Endo-9, 10-succinylanthracene	$C_{18}H_{13}O_3$	14.225	6.718	17.39	17.17	−0.22	7.768	17.21	−0.18
		11.777	5.587		17.25	−0.14	6.680	17.87	+0.48
2, 4-Dinitrophenylhydrazone-methylpropyl ketone	$C_{11}H_{14}O_4N_4$	10.000	6.549	24.05	23.81	−0.24	7.720	24.33	+0.28
		9.278	6.170		24.18	+0.13	7.230	24.56	+0.51

[a] Weight of CO with the correction of 125.
[b] Weight of I with the correction of 250.

During the sweep, the sample for the next analysis is weighed and introduced immediately after disconnecting the absorption train and the I_2O_5 oxidation tube.

The determination with the sweeping takes 30-40 minutes.

The results obtained are shown in the table. They leave no doubt that this method of direct determination of oxygen in organic compounds can become a working method in organic analytical laboratories. There is no doubt that by this method oxygen can be determined in one sample simultaneously by two chemical factors, CO_2 and I_2, which eliminates the necessity of carrying out two parallel determinations. At present, we are continuing work to improve details of the method as well as to broaden its analytical possibilities.

LITERATURE CITED

[1] W. R. Kirner. *Ind. Eng. Chem., Anal. Ed.* **6,** 358 (1934); **7,** 363 (1935); **8,** 57 (1936).

[2] J. Lindner and W. Wirth. *Ber.* **70,** 1025 (1937).

[3] J. Unterzaucher. *Ber.* **71,** 429 (1938).

[4] H. Ter Meulen. *New Methods for Organic Chemical Analysis.* Translated by ONTI (1931).

[5] M. Schütze. *Z. anal. Chem.* **118,** 241 (1939).

[6] W. Zimmermann. *Z. anal. Chem.* **118,** 258 (1939).

[7] J. Unterzaucher. *Ber.* **73,** 391 (1940).

[8] W. H. Walker and W. A. Patrick. *Ind. Eng. Chem., Anal. Ed.* **4,** 799 (1932).

[9] Ditte. *Bull. soc. chim.* **13,** 318 (1870).

[10] M. Nicloux. *Compt. rend.* **126,** 746 (1898).

[11] A. Gautier. *Compt. rend.* **126,** 931 (1898).

[12] M. O. Korshun. *Zavodskaya Lab.* **9,** No. 5-6, 615 (1940).

[13] F. R. Meyer and G. Ronge. *Angew. Chem.* **52,** 637 (1939).

[14] E. Abrahamczik. *Mikrochemie* **22,** 227 (1936).

[15] A. Friedrich. *Handbook for Quantitative Organic Microanalysis.* Translation with supplement by N. E. Zeligson, ONTI (1939).

CHAPTER 2

Apparatus for Direct Microdetermination of Oxygen Communication II[1]*

M. O. Korshun and N. E. Gel'man

The direct method for the microdetermination of oxygen in organic compounds [2] consists in the reaction of a strong reducing agent, such as red-hot carbon, with the products of thermal decomposition of the organic substance in an atmosphere of an inert gas.

The substance is decomposed in a quartz. tube in a stream of pure nitrogen; the decomposition products pass through a layer of granular carbon black heated to 1150°C. Thereby, the oxygen is quantitatively transformed into carbon monoxide. The carbon monoxide passes through a layer of iodine pentoxide heated to 115-120°C. and is quantitatively oxidized to carbon dioxide according to the equation:

$$5CO + I_2O_5 \rightarrow 5CO_2 + I_2.$$

The oxygen can be determined by the amount of CO_2, by the amount of I_2, and by the loss of weight of I_2O_5. The determination can be either gravimetric (by means of CO_2, I_2 and I_2O_5) or volumetric (I_2). The method requires a careful purification of nitrogen from traces of oxygen; this is achieved by passing nitrogen over carbon heated to 1150°C. in a special apparatus constructed by M. O. Korshun.

Zavodskaya Lab.* **12, 500-502 (1946).

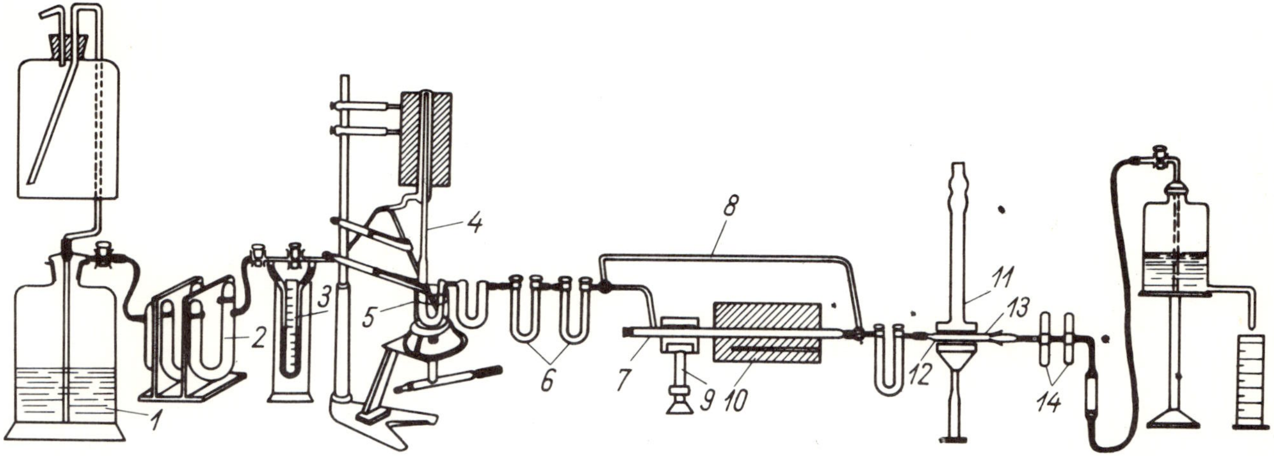

Fig. 1. Apparatus for Direct Microdetermination of Oxygen: 1—gasometer. 2—column for solid drying agents. 3—pressure regulator. 4—apparatus for purification of nitrogen. 5—glycerol bath. 6—U-shaped tubes. 7—tube for decomposition. 8—tube for reversed passage of gas. 9—burner. 10—electric furnace. 11—mortar. 12 and 13—oxidation units. 14—absorption apparatus.

The apparatus for the determination of oxygen (Fig. 1) consists of: the gasometer (1); two columns for solid drying agents (2); a pressure regulator (which is also a measuring device for the rate of gas flow) (3); a device for purification of nitrogen (4) and an electric furnace provided with a thermometer well and a rheostat; glycerol bath (5) with a microburner and a 200°C. thermometer; three U-tubes (6); a tube for decomposition (7) with a device for the reversed passage of gas (8) with two three-way stopcocks; a gas burner (9) with a wing top for a flat flame; an electric furnace (10) with wells for a

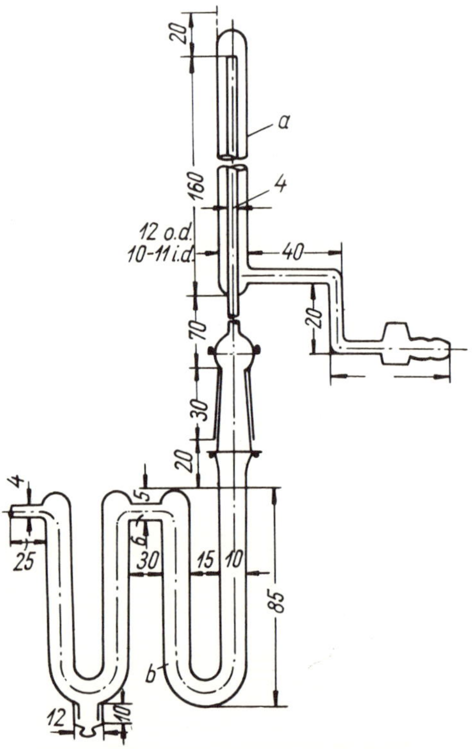

Fig. 2. Apparatus for Preliminary Purification of Nitrogen

thermocouple and a rheostat; a mortar (11) with a micro-burner, or electric winding with a rheostat; two tube-like devices (12) and (13) connected with each other by a ground-glass joint; two absorption tubes (14); a guard tube, a Mariotte bottle and a measuring cylinder.

Gasometer (1), pressure regulator (3), U-tubes (6), guard tube and Mariotte bottle have been described by us in the preceding communication [1].

Apparatus for preliminary purification of nitrogen (Fig. 2) is used for elimination of traces of oxygen from nitrogen. It consists of two parts: vertical quartz tube (a), and two U-tubes (b) of Pyrex glass sealed together. Both parts are connected by a ground-glass joint. The quartz part (a) is prepared from well-annealed quartz since any flaws in the quartz could lead to a rapid deterioration of the apparatus because of the high temperature used.

The tube (a) is prepared with an open upper end which is sealed off after the tube is filled with granular carbon black.

A tubular electric furnace [3] with a well for a thermocouple is used for heating the tube (a). The length of the furnace is 20 cm. The working temperature of 1150°C. is attained in 25 min., regulated by a rheostat and controlled by a thermocouple with a millivoltmeter.

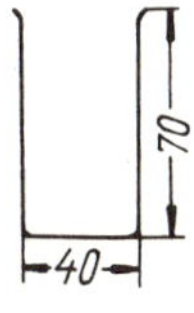

Fig. 3. Bath

The bath (Fig. 3) with glycerol, into which the U-tube (b) of the apparatus (4) (see Fig. 1) is immersed, is a thick-walled beaker made of Pyrex glass.

Decomposition Tube. The tube described in Communication I can be used as a decomposition tube by increasing the thickness of the walls to 1.5 mm. and by decreasing the length of the tube to 500 mm.

In selecting the decomposition tube, special attention should be given to careful treatment of the quartz (absence of bubbles and striations).

A more expedient form of tube is shown in Fig. 4. The wide opening of the tube is closed by a ground-glass cap. The cap has a stopcock for the outlet of gas during the blowing of nitrogen in the reverse direction through the ⊓-shaped tube which is equipped with two three-way stopcocks (see Fig. 1).

An electric furnace fastened horizontally is used for heating a layer of carbon in the decomposition tube.

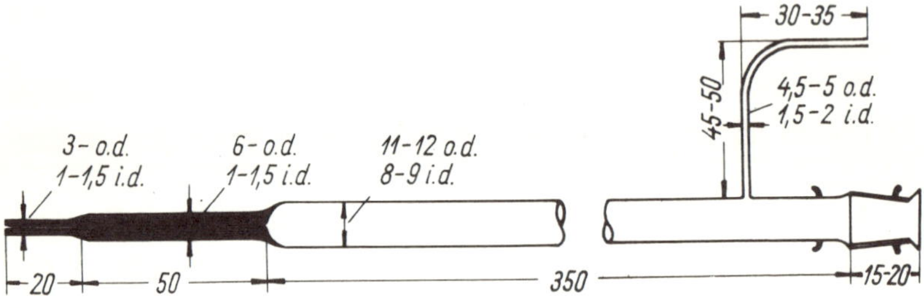

Fig. 4. Decomposition Tube

APPARATUS FOR GRAVIMETRIC DETERMINATION OF OXYGEN

In Fig. 1 is shown an apparatus used for the gravimetric determination of oxygen. In this variation are used absorption and oxidation units, a mortar, and absorption equipment described by Abrahamczik in the article, "Apparatus for the Microdetermination of Carbon and Hydrogen."

The oxidation and absorption units (1) and (2) (Fig. 5) are made of thin-walled glass. The joint (a) of the apparatus (1) is ground to the apparatus (2). Correspondingly, the cap (A) of apparatus (1) is ground to the stopper (B) of the apparatus (2). When the units are connected with the train, they are joined by ground parts (a) and (b). After being disconnected from the train, each unit is closed by a corresponding ground stopper or by a cap, and so weighed.

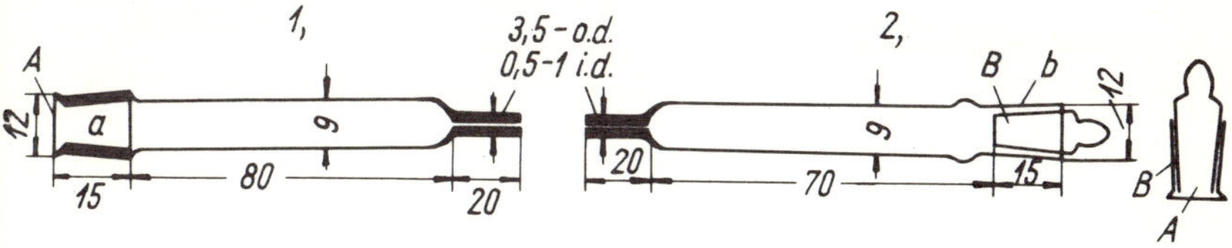

Fig. 5. Units for Gravimetric Determination of Oxygen: 1—oxidation unit; 2—absorption unit

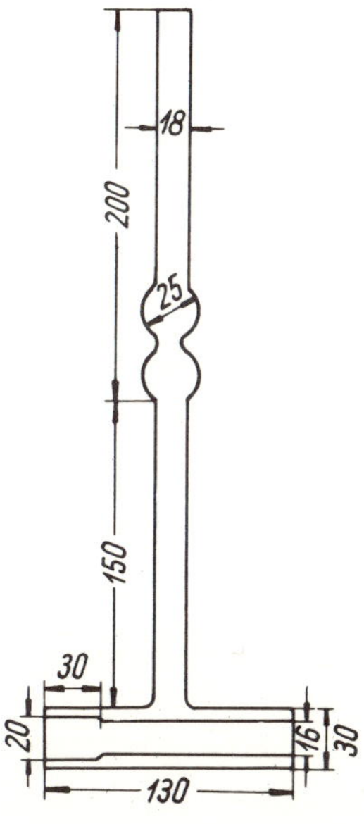

Fig. 6. Mortar for Gravimetric Determination of Oxygen

Mortar (Fig. 6). This variation of the mortar differs from the previously described mortar only by its length and by the inner diameter of the lower part. The heating is similar to that described above.

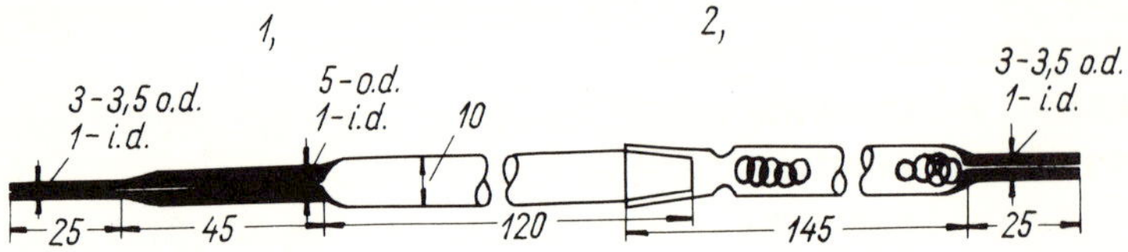

Fig. 7. Equipment for Gravimetric Determination of Oxygen: 1—oxidation unit; 2—absorption unit

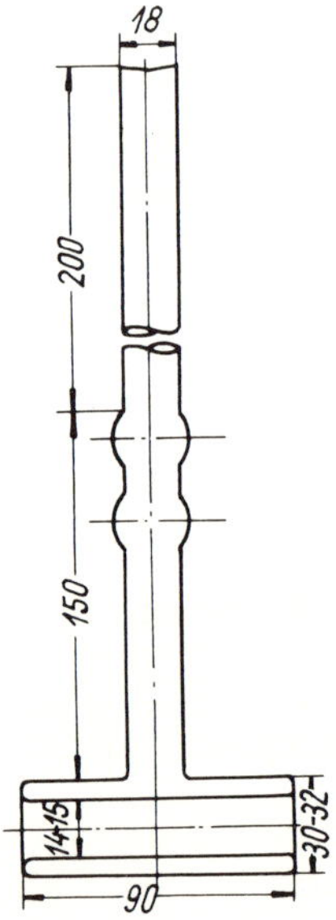

Fig. 8. Mortar for Volumetric Determination of Oxygen

APPARATUS FOR VOLUMETRIC DETERMINATION OF OXYGEN

In the volumetric determination of oxygen, oxidation and absorption units and a mortar are used.

Oxidation Unit (1) (Fig. 7). This unit is prepared from thin-walled Pyrex glass and fits into the mortar easily. At the wide end of the unit is an external ground joint corresponding to the cap in the apparatus (2). When the apparatus (1) is not in use, it is closed by a ground cap.

Absorption Unit (2) for Iodine (Fig. 7). The apparatus is prepared from alkali-resistant glass. The entire wide part of the apparatus is filled with open-end spiral rings made of solid glass rod, 1 mm. in diameter. The rings are oval in form with axes of 3 and 6-7 mm. The wide end with an inner ground joint fits easily into the widened part of the mortar. Because of the narrowing near the ground joint, it cannot drop out of the apparatus.

Mortar (Fig. 8). The third model of the mortar differs from those previously described only by the length of its lower horizontal part and by its inner diameter. The point of attachment of the units (1) and (2) (Fig. 7) is located in the widened part of the mortar.

LITERATURE CITED

[1] Communication I. *Zavodskaya Lab.* **10**, 241-245 (1941) (Chapter 1).

[2] M. O. Korshun. Direct Microanalytical Determination of Oxygen in Organic Substances. Dissertation, IOKh of the Academy of Sciences of the U.S.S.R., 1942

[3] Electric furnaces will be described in the article: "General and Auxiliary Equipment."

CHAPTER 3

The Determination of Carbon and Hydrogen in Substances Composed of Carbon, Hydrogen, and Oxygen*

M. O. Korshun and V. A. Klimova

A vast number of works [1] have been devoted to perfecting a method for the microanalytical determination of C and H in organic substances, but their principal aim has been merely a further development of the classical method of Pregl [2].

In the Pregl method there are three fundamental conditions:

1. The combustion of the substance proceeds at a constant oxygen rate of about 4 cc./min. with a constant elevated pressure 5-7 mm. above atmospheric in the system. This is achieved by means of a compressed asbestos plug placed in the combustion tube.

2. A so-called "universal" filling must provide complete oxidation of the organic substance and absorption of the acid gases†.

3. An efficient combination of the microequipment is used to carry out this method of analysis.

These three conditions have played an essential role in the nearly thirty years of the development and perfection of the microanalytical determination of C and H. There have been works directed toward maximum simplification

---

*Zhur. Anal. Khim. **2**, 274-280 (1947).

† Thus, lead chromate retains the gaseous sulfur compounds, silver the halides, and lead dioxide the nitrogen oxides.

of the "universal" combustion tube filling [3, 4, 5, 6], toward a change in the rate and pressure of the oxygen passed through, and finally toward the creation of simpler and more nearly ideal types of tube.

The literature data concerning the combustion rate are extremely varied.

Brodie [7] achieves complete combustion of the substance and displacement of the oxidation products in 25 minutes by increasing the oxygen rate and the pressure in the system.

Titov [8] spends 30 minutes burning a 50-60 mg. weighed sample by a similar technique.

Hallett [9] experimentally established that, by passing the oxygen through at the rate of about 20 cc./min., it is possible to burn 4-6 mg. of a substance and displace its oxidation products in 15 minutes.

Belcher and Spooner [10] effected the combustion of a 4-6 mg. weighed sample and the displacement in 10 minutes at an oxygen rate of 50 cc./min. In addition, these authors conducted the analysis in a combustion tube containing no filling, i.e., an empty combustion tube.

Belcher and Spooner [11] used what was essentially the rapid combustion method for determining C and H in coal, but they were relatively little concerned with the analysis of individual organic compounds and gave no attention at all to the analysis of liquids.

It must be noted that the rapid and uniform combustion of coal is an easier task than the combustion of organic compounds since the reaction $C + O_2 \rightarrow CO_2$ goes without an increase in volume and consequently requires no special precautions to prevent the passage of unburned material during combustion.

In view of the fact that rapid combustion in an empty tube had not been studied thoroughly with respect to its applicability to various organic compounds, such a study attracted our attention.

In the process of developing this method further we arrived at a totally different technique for the combustion of solids and also worked out a new method of burning liquids and volatile substances.

The essence of the method that we developed consists in taking a weighed sample of the substance to be tested in a quartz capsule, placing it in a quartz combustion tube (containing nothing except the weighed sample and oxygen) and burning it in a stream of oxygen passing through at the rate of 30-50 cc./min.

First the substance undergoes thermal decomposition; then the pyrolytic products pass through a zone heated to 850-900°‡ for a distance of 28-30 cm. Here they are completely oxidized to water and carbon dioxide, which are quantitatively absorbed by suitable absorbents.

Combustion of the substance and displacement of the oxidation products takes 10-15 minutes.

APPARATUS

The set-up for the determination of C and H in substances composed of C, H, and O is shown in Fig. 1.

A detailed description of the apparatus used has already been published [12].

CARRYING OUT THE ANALYSIS

1. *Combustion of solids.*

8-15 mg. samples of the substance are weighed in quartz capsules on a semimicrobalance with an accuracy of up to 0.01 mg.; 4-5 mg. samples are weighed with an accuracy of up to 0.002 mg. on a microanalytical balance.

‡ The optimum temperature conditions and oxygen rates are being studied and will be reported on later.

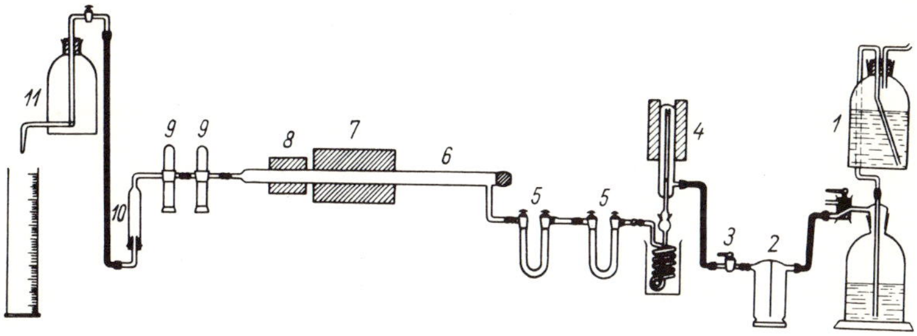

Fig. 1.

Simplified apparatus logically evolved, in all its parts, from the Pregl set-up: 1—gasometer with oxygen; 2—drying bottle containing calcium chloride and sulfate; 3—stopcock with grooves; 4—gas-scrubbing tube inside an electric furnace heated to 700°; 5—two U-tubes containing Ascarite reagent and calcium sulfate; 6—quartz combustion tube; 7—electric furnace 20 cm. long heated to 850-900°; 8—electric furnace 10-12 cm. long heated to 850-900°; 9—two absorption tubes for water and carbon dioxide; 10—guard tube; 11—Mariotte bottle over a 500 cc. measuring cylinder.

The capsule with the weighed sample is moved into the combustion tube to a location 10-12 cm. from the long furnace, which has been previously heated to 900°. The weighed absorption tubes are connected and the oxygen rate is established at 40-50 cc./min. by means of the stopcock, *3*, and the Mariotte bottle. The small furnace, heated to 850°, is located 6-7 cm. from the tip of the combustion tube and close against the large furnaces§.

A gas burner is placed at the open end of the capsule and gradually moved to the closed end, i.e., counter to the stream of oxygen. The speed with which the burner is moved depends on the nature of the substance being analyzed.

§ We consider it expedient to use two furnaces instead of one long one since these furnaces are standard items found in every microanalytical laboratory.

TABLE 1

Name and empirical formula of the substance and method of combustion	Amount found C in %	Amount found H in %	Amount calculated C in %	Amount calculated H in %	ΔC in %	ΔH in %
Levulose ($C_6H_{12}O_6$)						
Combustion in a boat	38.53	6.61	39.98	6.67	-1.45	-0.06
	38.63	6.67	39.98	6.67	-1.35	0.00
Combust. in a capsule	39.98	6.76	39.98	6.67	0.00	+0.09
	39.86	6.84	39.98	6.67	-0.12	+0.17
	40.01	6.71	39.98	6.67	+0.03	+0.04
Naphthalene ($C_{10}H_8$)						
Combustion in a boat	90.86	6.06	93.70	6.30	-2.84	-0.24
	88.06	6.54	93.70	6.30	-5.64	+0.24
Combust. in a capsule	93.75	6.25	93.70	6.30	+0.05	-0.05
	93.75	6.23	93.70	6.30	+0.05	-0.07
Bis(2,2,4,4-tetramethylbutyl)-napthalene ($C_{26}H_{40}$)						
Combustion in a boat	83.63	11.78	88.56	11.44	-4.93	+0.34
	85.20	11.83	88.56	11.44	-3.36	+0.37
Combust. in a capsule	88.30	11.30	88.56	11.44	-0.26	+0.14
	88.58	11.46	88.56	11.44	+0.02	+0.02

If the decomposition proceeds too rapidly, producing flashes, it is expedient to cool the tube with a moist pad at the point where the substance in the capsule is located and to move the burner to the open end of the capsule.

When all the substance is carbonized, the burner is moved to the end of the capsule and burning is discontinued. Then the combustion tube is brushed with the burner flame all the way up to the furnace.

In the event that water vapor condenses in the combustion tube tip, it must be entirely removed by slightly heating the tip with the short furnace for 1-2 minutes.

When the combustion is over, the absorption tubes are removed and weighed as usual.

The entire combustion is completed within 10 minutes. Only when highly volatile substances or substances that produce a large deposit of carbon black are burned, need the burning be prolonged to 13-15 minutes.

At the start of our investigations, the tests for the determination of C and H were conducted by burning the solid substance in a platinum microboat in strict accordance with the directions of Belcher and Spooner. According to our observations, this combustion technique, in contrast to ours (combustion in a capsule), produces vaporization of the substance but not decomposition; the vapors, passing into the heated zone of the large furnace, give a flash but are not completely oxidized. This results in the formation of a large quantity of carbon black and, obviously, carbon monoxide as well.

When combustion was effected in a microboat, we observed carbon black deposits not only in the tip of the combustion tube but also in the water absorption tube. In the majority of cases the analytical results obtained for carbon were very low when the substance was burned in a boat.

Table 1 presents the results of the analysis for C and H obtained by combustion in a microboat and in a quartz capsule. Burning and displacement lasted 10 minutes. Even in the case of the analysis of a standard substance—levulose by combustion in a microboat—the result obtained for carbon was about 1.5% lower than calculated; the result of the analysis for carbon in naphthalene and bis-(2, 2, 4, 4-tetramethylbutyl)naphthalene was 3-5% lower than the theoretical.

Analyses of these substances made by burning them in a quartz capsule give completely satisfactory results for carbon and hydrogen.

TABLE 2

Name and empirical formula of the substance	Amount found		Amount calculated			
	C in %	H in %	C in %	H in %	ΔC in %	ΔH in %
1,2,2-Trimethyl-3-isopropylidenylcyclopentan-4-one ($C_{11}H_{18}O$)	79.35	10.96	79.45	10.91	−0.10	+0.05
9,10-Dihydro-9,10-diethyl-1,2-benzanthracene($C_{22}H_{22}$)	92.29	7.84	92.28	7.72	+0.01	+0.12
	92.23	7.82	92.28	7.72	+0.05	+0.09
9,10-Dipropylanthracene ($C_{20}H_{24}$)	91.78	8.39	91.55	8.45	+0.23	+0.06
	91.80	8.27	91.55	8.45	+0.25	+0.18
9,10-Dihydro-9,10-dipropylanthracene ($C_{20}H_{24}$)	90.72	9.21	90.85	9.15	−0.13	+0.06
	90.67	9.24	90.85	9.15	−0.18	+0.09
3-Methyldecahydro-5,6-benzindan-1-one ($C_{14}H_{24}O$)	81.41	10.88	81.46	10.75	−0.05	+0.13
	81.49	10.96	81.46	10.75	+0.03	+0.21
Allyl-Δ^1-cyclopentenyl ketone ($C_9H_{12}O$)	79.55	9.15	79.41	8.82	+0.14	+0.34
	79.42	9.06	79.41	8.82	+0.01	+0.24
1,5-Dimethylindane ($C_{11}H_{14}$)	90.14	9.59	90.41	9.59	−0.27	0.00
	90.40	9.73	90.41	9.59	−0.01	+0.14
Dimesitylmethane ($C_{19}H_{24}$)	90.40	9.61	90.40	9.60	0.00	+0.01
	90.43	9.67	90.40	9.60	+0.03	+0.07

The data of Table 2 show the adequate degree of accuracy and the suitability of the rapid combustion method for solid substances difficult to analyze.

2. Combustion of liquids.

The combustion of high-boiling liquids presents no particular difficulty. Such liquids are analyzed by the combustion method that we have described above.

The rapid combustion of liquids with a high vapor pressure is more difficult.

In the Pregl method [2] a sealed capillary is placed in a combustion tube and while the capillary body is being cooled the capillary tip is softened by rapid heating; the internal pressure of the vapors, however, forces the capillary open. When this happens, the vapors that frequently escape drive the capillary up to the rubber stopper of the combustion tube and the analysis is ruined.

M. O. Korshun had previously proposed a combustion technique for volatile liquids [6] in which the liquid was evaporated from a spiral capillary into kieselguhr. Such a technique did not work in this instance, however, since the rapid stream of oxygen scattered the kieselguhr the entire length of the tube.

After many fruitless attempts we finally succeeded in producing several improvements and achieved accurate results in the combustion of volatile liquids by making a few small changes in the previous method.

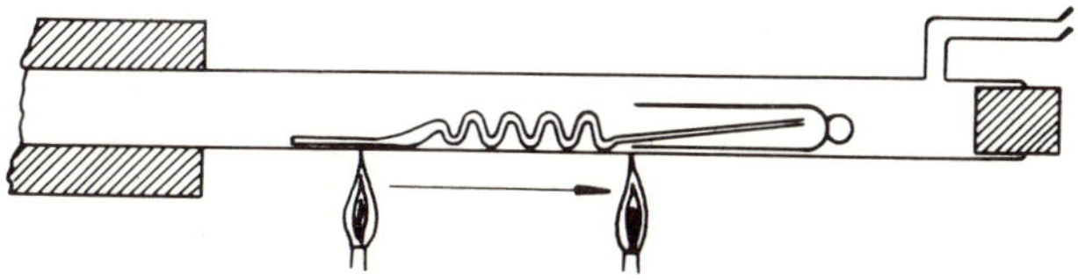

Fig. 2.

TABLE 3

Name and empirical formula of the substance	Amount found C in %	Amount found H in %	Amount calculated C in %	Amount calculated H in %	ΔC in %	ΔH in %
1,1,2,2-Tetramethylcyclopropane (C_7H_{14})	85.59	14.42	85.62	14.38	−0.03	+0.04
	85.60	14.47	85.62	14.38	−0.02	+0.09
α-Butoxyethyl isovalerate ($C_{11}H_{22}O_3$)	65.47	11.00	65.35	10.89	+0.12	+0.11
	65.16	10.84	65.35	10.89	−0.19	−0.05
α-Butoxyethyl isobutyrate ($C_{10}H_1\ O_3$)	63.70	10.67	63.83	10.64	−0.13	+0.03
	63.55	10.76	63.83	10.64	−0.28	+0.12
α-Phenoxyethyl acetate ($C_{10}H_{14}O_3$)	66.79	6.71	66.67	6.67	+0.12	+0.04
	66.77	6.71	66.67	6.67	+0.10	+0.04
Diphenylacetal ($C_{14}H_{14}O_2$)	78.43	6.60	78.51	6.54	−0.08	+0.06
	78.44	6.59	78.51	6.54	−0.07	+0.05
Vinylethynyldiisopropylcarbinol ($C_{11}H_{18}O$)	79.42	10.87	79.45	10.91	−0.03	+0.04
2-Methyl-3-isopropyl-2,6-hepten-4-yne ($C_{11}H_{16}$)	89.11	10.97	89.11	10.89	0.00	+0.08
	88.99	11.14	89.11	10.89	−0.12	+0.25
1-Vinylethynyl-3-methylcyclohexane ($C_{11}H_{14}$)	90.35	9.59	90.41	9.59	−0.06	0.00
Benzine (appr. C_7H_{16})	84.27	15.53	–	–	–	–
	84.23	15.71	–	–	–	–

The weighed sample is taken in a spiral capillary in the usual manner. Just before the capillary is introduced into the combustion tube its tip is broken off and the open end is quickly inserted in a quartz capsule (all this manipulation is done with forceps).

Fig. 2 shows the position of the capsule with the capillary in the combustion tube. The closed end is 4-5 cm. from the long furnace.

By careful heating of the tube at the place where the long handle of the capillary is located the liquid is driven into the capsule; then the burner flame is increased and moved to the open end of the capsule.

The vapors of the substance at the open end of the capsule, mixing with the oxygen, produce uniform flashes and, passing through the short section of the tube heated by the gas burner, burn completely. Drops of water condensing on the cool parts of the combustion tube up to the point where it enters the long furnace testify to this. This phenomenon may be explained by the fact that the quantitative ratio between the oxygen and the vapors of the volatile organic substance inside the combustion tube are favorable to complete burning.

Thus we were always able to make an accurate analysis of highly unsaturated and volatile compounds, acetals, ethers, aliphatic hydrocarbons, etc., spending no more than 10-15 minutes on their combustion (Table 3).

When analyzing volatile liquids with a high carbon content, in particular hydrocarbons, we often obtained a C and H total greater than 100%. In the previous method we considered the tube filling to be the source of error, but when these substances were burned in an empty tube the error was repeated and the figures obtained in the analysis of liquid hydrocarbons were 1-1.5% higher than calculated.

Subsequently we established that, in taking weighed samples of liquids with a high vapor pressure, error crept in unless the weight of the air displaced by the vapors of the substance was taken into account. It was proved that the weight of the air displaced was about 50 μg when a spiral capillary was used and about 25 μg when a Pregl capillary was used. This weight must be added to the weighed sample.

Only after the corrections were made, were good results obtained for the analysis of 1, 1, 2, 2-tetramethylcyclopropane, which has a boiling point of 75°, and benzine, which boils within the limits of 70-100° (Table 3).

The recommended rapid method for the determination of C and H differs sharply from contemporary methods based on the classical method of Pregl.

First of all, it is obvious that the use of an empty combustion tube eliminates the sources of error associated with the use of oxidizing mixtures, lead dioxide, and layers of asbestos [13, 14, 15]; next, the analyst is saved time and effort by no longer having to maintain constant ratios between the rate and pressure of the gas in the system. In addition the assembly of the set-up is considerably simpler (absence of a pressure regulator, mortar, etc.). Finally, and very important, the productivity of the analyst is multiplied 2-3 times.

In addition, the rapid method of determining C and H demolishes the opinion widely held by microanalysts that the combustion and displacement rates cited in many earlier works are critical [16].

At first glance it seems impracticable to effect combustion within 10 minutes. Such a tempo creates a certain lack of confidence in the analyst who has just changed to the rapid method.

We recommend that those starting to work with this method first conduct a few combustions of a standard; some easily analyzed substance such as levulose.

CONCLUSIONS

1. A rapid semimicro- and microanalytical method for determining C and H in organic substances composed of C, H, and O has been described. This method consists in burning the substance at 900° in an empty tube in a stream of oxygen passing at the rate of 30-50 cc./minute. The combustion time is 10-15 minutes.

2. Solids and liquids, including volatile substances (ethers, acetals, highly unsaturated compounds, etc.) were analyzed. Accuracy of the analysis is ±0.3%.

3. The technique developed for the rapid combustion of liquids and solids has greatly extended the field of application of this method.

LITERATURE CITED

[1] J. B. Niederl and V. Niederl, *Micromethods of Quantitative Organic Analysis*, New York, 1942, p. 146.

[2] F. Pregl, *Kolichestvennyi Organicheskii Mikroanaliz* [Quantitative Organic Microanalysis]. United Scientific-Technical Publishers, Moscow, 1934.

[3] J. B. Niederl and B. Whitman, *Mikrochemie*, 11, 274 (1932).

[4] M. O. Korshun and N. E. Gel'man, *Zavodskaya Lab.* 14, 114 (1935).

[5] A. Friedrich, *Praktika Kolichestvennogo Organicheskogo Mikroanaliza* [Practice of Quantitative Organic Microanalysis] United Scientific-Technical Publishers, Moscow, 1939.

[6] M. O. Korshun, *Zavodskaya Lab.* 9, 615 (1940).

[7] S. S. Brodie, *Ind. Eng. Chem. Anal. Ed.* 11, 226 (1939).

[8] E. M. Titov, *Zavodskaya Lab.* 9, 853 (1940).

[9] L. T. Hallett, *Ind. Eng. Chem. Anal. Ed.* 14, 956 (1942).

[10] R. Belcher and C. Spooner, *J. Chem. Soc.* 1943, 313.

[11] R. Belcher, *J. Inst. Fuel* **17**, 160 (1944).

[12] M. O. Korshun and N. E. Gel'man, *Zavodskaya Lab.* **12**, 109, 500, 617, 754, 871, 968 (1946).

[13] M. Boetius, *Ueber die Fehlerquellen bei der Mikroanalytischen Bestimmung des Kohlenstoffes und Wasserstoffes nach der Methode F. Pregl*, Berlin, 1931.

[14] J. Lindner, *Mikromassanalytische Bestimmung des Kohlenstoffes und Wasserstoffes mit grundlegender Behandlung der Fehlerquellen in der Elementaranalyse*, Berlin, 1935.

[15] C. Weygand, *Chem. Tech.* **16**, 15 (1943).

[16] N. E. Gel'man, *Zavodskaya Lab.* **9**, 38 (1940).

CHAPTER 4

The Formation of Nitrogen Oxides During the Determination of Carbon and Hydrogen in Substances Containing Carbon, Hydrogen, Oxygen, and Nitrogen*

M. O. Korshun and V. A. Klimova

In our first report [1] we described a rapid method for the semimicro- and microanalytical determination of C and H in liquid and solid organic compounds composed of C, H, and O.

The aim of the present work is to extend this rapid method of determining C and H to organic compounds containing C, H, O, and N.

The presence of nitrogen in the molecule of an organic substance always complicates the determination of C and H, since it necessitates filling the combustion tube with reagents to collect the nitrogen oxides or to reduce them to elementary nitrogen.

In the combustion of nitrogen-containing organic compounds one may expect to find ammonia, hydrocyanic acid, and cyanogen as well as nitrogen oxides and elementary nitrogen among the thermal decomposition products. Further oxidation of these decomposition products, however, in a large excess of oxygen at a temperature of 900° in the combustion tube must in turn lead to the formation of nitrogen oxides and elementary nitrogen alone.

**Zhur. Anal. Khim.* 3, 176-180 (1948).

TABLE 1—The Determination of Nitrogen Oxides

No.	Name and empirical formula of the substance	Structural formula	% N in the substance	Weighed sample of the substance in mg.	0.01 N $Na_2S_2O_3$ in cc.	Found % N as NO_2	Found NO_2 in γ	Possible error in % C.
1	Blank test				None			
2	Levulose $C_6H_{12}O_6$	$CH_2OH(CHOH)_3COCH_2OH$		8.03	None			
3	Benzidine $C_{12}H_{12}N_2$		15.22	7.30 9.34	0.037 0.039	0.47 0.38	17 18	0.06 0.05
4	p-Phenylene-diamine $C_6H_8N_2$		25.93	8.68 7.57	0.070 0.073	0.41 0.52	32 34	0.10 0.12
5	p-Tolyl-β-naphthyl-amine $C_{17}H_{15}N$	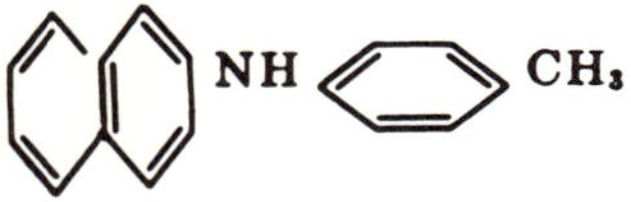	6.01	8.30 7.80	0.015 0.016	0.42 0.48	7 7	0.02 0.02
6	m-Amino-benzoic acid $C_7H_7O_2N$		10.22	7.38 7.82	0.022 0.045	0.40 0.79	10 21	0.04 0.07
7	Glycocoll $C_2H_5O_2N$	$CH_2(NH_2)COOH$	18.67	7.01 9.28	0.033 0.020	0.36 0.16	15 9	0.06 0.03
8	Benzamide C_7H_7NO		11.57	6.68 7.10	0.130 0.000	0.24 0.00	6 0	0.03 0.00
9	Succinimide $C_4H_7O_3N$	$C_2H_4(CO)_2NH.H_2O$	11.96	8.04 9.12	0.040 0.022	0.58 0.28	18 10	0.07 0.03

The proportions of nitrogen oxides and elementary nitrogen obtained as a result of combustion under these analytical conditions depend first on the nature of the nitrogen bond in the molecule and second on the method of burning the substance. It is quite obvious that the formation of nitrogen oxides is more likely in the analysis of nitro and nitroso compounds and less likely in the combustion of amino and azo compounds.

The remaining classes of nitrogen-containing compounds assume an intermediate position.

As to the method of burning the weighed sample, Friedrich [3] found twice the amount of nitrogen oxides when the nitrogen-containing substance was burned in a platinum boat as when it was thermally decomposed in a capsule sealed at one end.

According to our observations, the thermal decomposition of a substance can be effected very easily, almost without any admission of oxygen, if the capsule used is no shorter than 5-6 cm. It is then possible to guarantee a minimum formation of nitrogen oxides.

Starting from the above hypotheses, we undertook to investigate the amount of nitrogen oxides formed during the rapid combustion of several classes of nitrogen-containing compounds, expecting that nitrogen oxides generally would not form in compounds where the nitrogen was bound directly to carbon and hydrogen (amines, amides, imides, etc.). Thus it would be possible to analyze such compounds just like compounds composed of C, H, and O, without any additional changes.

THE FORMATION OF NITROGEN OXIDES DURING RAPID COMBUSTION

Dennstedt [2] has already shown that not only nitrogen oxides but also elementary nitrogen is formed under the conditions of the catalytic oxidation of nitro and nitroso compounds. Here only 50% of total nitrogen is converted to oxides; when other nitrogen-containing compounds are burned, only 20% of the total nitrogen is converted to oxides.

After a number of experiments devoted to the question of the formation of nitrogen oxides during the microanalytical determination of C and H by the Dennstedt method, Friedrich [3] came to the same conclusion.

As a result of the experimental data obtained under the conditions of our rapid combustion method, we have come to completely different conclusions concerning the amount of nitrogen oxides formed.

We conducted our experiments on the determination of nitrogen oxides according to the method recommended by Friedrich [3].

To collect the nitrogen oxides we used a tube developed earlier by M. O. Korshun and filled with Fenske rings. Before the combustion is begun, the tube is filled with a 20% potassium iodide solution; after the rings have been moistened, the excess liquid is drained off and the tube is connected to the tip of the combustion tube by means of a rubber connector. The combustion is effected as previously described [1]. Combustion time of the weighed sample, 10 minutes; rate of the oxygen passed through, 40-50 cc./min.; temperature of the two furnaces, 900°.

After the combustion is finished, the tube with the rings is disconnected and the contents are quantitatively washed out, first with 10 cc. of a 20% potassium iodide solution and then with water, into an Erlenmeyer flask. The iodine that separates out during the combustion is titrated with 0.01N thiosulfate solution; 1 cc. of 0.01N $Na_2S_2O_3$ corresponds to 0.140 mg. of nitrogen or 0.460 mg. of NO_2.

The results of our tests are presented in Table 1. These results clearly show that, under the conditions of rapid burning in an empty combustion tube, amino, amido, and imido compounds form no more than 0.5% nitrogen oxides. Furthermore, such amounts of NO_2 can lead to a carbon increase of only a few hundredths of a percent (permissible error ±0.3%).

TABLE 2
The Determination of C and H without the Collection of Nitrogen Oxides

No.	Name and empirical formula of the substance	Found % C	Found % H	Calculated % C	Calculated % H	Δ % C	Δ % H
1	Benzidine $C_{12}H_{12}N_2$	78.47	6.45	78.26	6.52	−0.21	+0.23
		78.22	6.60	78.26	6.52	−0.04	+0.08
2	p-Phenylenediamine $C_6H_8N_2$	66.91	7.52	66.66	7.41	+0.25	+0.11
		66.76	7.54	66.66	7.41	+0.10	+0.13
3	p-Tolyl-β-naphthylamine $C_{17}H_{15}N$	87.70	6.64	87.55	6.44	+0.15	+0.24
		87.65	6.60	87.55	6.44	+0.11	+0.20
4	Diphenylacetamidine $C_{14}H_{14}N_2$	80.24	6.86	80.00	6.67	+0.24	+0.19
		80.17	6.87	80.00	6.67	+0.17	+0.20
5	p-Aminoacetophenone C_8H_9ON	70.85	6.72	71.11	6.66	−0.26	+0.06
		71.14	6.67	71.11	6.66	+0.03	+0.01
6	m-Aminobenzoic acid $C_7H_7O_2N$	61.19	5.20	61.37	5.11	+0.12	+0.09
		61.37	5.28	61.37	5.11	+0.06	+0.17
7	p-Aminobenzoic acid $C_7H_7O_2N$	61.39	5.25	61.37	5.11	+0.02	+0.14
		61.63	5.20	61.37	5.11	+0.26	+0.09
8	Glycocoll $C_2H_5O_2N$	31.75	6.65	32.00	6.66	−0.25	−0.01
		31.99	6.67	32.00	6.66	−0.01	+0.01

This supports the hypothesis that the above-mentioned classes of nitrogen-containing compounds can be analyzed for C and H without collection of the nitrogen oxides.

For illustration we conducted several analyses of nitrogen-containing substances in which the determination of C and H was effected without collecting the NO_2. As is evident from the data of Table 2, the results obtained for the percentage content of carbon and hydrogen lie within the limits of admissible error.

As far as we know, we were the first to note this possibility of analyzing several classes of nitrogen-containing organic compounds for carbon and hydrogen in the same manner as compounds containing only C, H, and O.

It must be mentioned that, according to our observations, when the substance decomposes in the capsule with a flash there is always the danger that larger amounts of NO_2 will form.

Special experiments showed that when a weighed sample of benzidine is decomposed in a capsule up to complete carbonization, without flash, 0.5-0.4% NO_2 is found; decomposition with weak flash gives up to 1.2% NO_2, and combustion of the weighed sample with strong flash gives about 4% nitrogen oxides, but no more.

CONCLUSIONS

1. Research was conducted to study the amount of nitrogen oxides formed when certain classes of nitrogen-containing compounds are burned in a rapid stream of oxygen in an empty combustion tube.
2. It was found that, under the conditions of the method and combustion technique described, amino, amido, and imido compounds form up to 0.5% nitrogen oxides.
3. It was established that these classes of nitrogen-containing compounds can be analyzed by this method in exactly the same manner as compounds comprising C, H, and O, without any further changes.

LITERATURE CITED

[1] M. O. Korshun and V. A. Klimova, *Zhur. Anal. Khim.* **2**, 274 (1947).

[2] M. Dennstedt, *Z. anal. Chem.* **42**, 417 (1903).

[3] A. Friedrich, *Mikrochemie* **23**, 129 (1937).

[4] M. O. Korshun and N. E. Gel'man, *Zavodskaya Lab.* **12**, 500 (1946).

CHAPTER 5

The Microanalytical Determination of Mercury in Organic Compounds*

M. O. Korshun and E. V. Lavrovskaya

Increased study in the field of the synthesis of organic mercury compounds and the extensive use of these compounds in medicine, agriculture, etc. have made it necessary to find a simple method for the quantitative determination of mercury.

The methods hitherto described in the literature do not give good results when halogen, sulfur, or nitrogen is present in the organic mercury compound. Furthermore, each of these cases requires a change in the course of the analysis, which complicates the method considerably and deprives it of universality.

The majority of contemporary textbooks on quantitative organic microanalysis fail to discuss this subject.

Only in the works of Pregl [1] and Friedrich [2] is there a description of the methods of Verdino and of Meixner and Krocker for the determination of mercury.

Verdino decomposes the substance under analysis with strong nitric acid (Carius method). Metallic mercury is electrolytically precipitated at a gold electrode and then weighed. The accuracy of this method lies within the limits of $\pm 0.4\%$; the analysis lasts about four hours.

Meixner and Krocker burn a weighed sample of the substance in a stream of oxygen or air. The combustion products pass over a heated layer of CaO, which retains chlor-

Zhur. Anal. Khim. **3**, 322-328 (1948).

ine, sulfur dioxide, carbon dioxide, and water. The mercury vapors condensing in the tip of the combustion tube are driven into a gold-filled absorption tube, which is then weighed.

This method is not suitable for the analysis of substances containing mercury and nitrogen together.

Boetius [3] based his investigations on the method of Meixner and Krocker, after having first submitted it to thorough study. He demonstrated experimentally that calcium oxide retains chlorine easily, bromine considerably less readily, and iodine hardly at all, since the latter combines with metallic mercury to form mercury iodide and leads to high results in the analysis of iodine-containing compounds.

As a result of many experiments, Boetius settled on two types of combustion tube filling.

1) When determining mercury in organic compounds that do not contain nitrogen, he conducts the combustion in oxygen. The combustion products pass over a heated layer of lead oxide, which retains sulfur, chlorine, and bromine. The iodine is retained by a layer of silver-impregnated clay plate.

2) When the organic mercury compounds contain nitrogen, they are burned in a stream of carbon dioxide in a combustion tube filled with lead chromate, copper, and silver-impregnated clay plate.

The mercury is weighed, as usual, in the form of a gold amalgam.

It must be admitted that at present the most accurate microanalytical method for the determination of mercury is that of Boetius. Nonetheless, the two types of filling proposed by this author and the different methods of conducting the analysis associated with them deprive the method of simplicity and universality.

It is interesting to note that the work of Boetius has met no response in contemporary literature on organic microanalysis. Thus, in their textbook *Micromethods of Quantitative Organic Analysis*, published in 1942 [4], V. Niederl and J. Niederl only mention the incomplete method of Meixner and Krocker.

The aim of our research was the creation of a simple, universal micro method that would permit determining mercury in organic compounds of any composition with the use of one type of combustion tube filling.

We felt it likely that we could take the second type of filling proposed by Boetius for this purpose, not only using it for the determination of mercury in organic compounds containing nitrogen but also extending it to organic mercury compounds of any type.

ESSENCE OF THE METHOD

The substance being analyzed is subjected to thermal decomposition in a stream of nitrogen. The decomposition products pass over a layer of lead chromate, metallic copper, and silver-impregnated clay plate heated to 750°. The acid gases are completely retained by this filling, while the mercury vapors are driven over into a gold-filled absorption tube, which is dried and weighed.

The method that we have developed for the microdetermination of mercury permits analyzing different organic mercury compounds that contain chlorine, bromine, iodine, nitrogen, and sulfur in various combinations.

Although in this work Boetius proposed a filling with which nitrogen-containing mercury compounds could be analyzed, he tested this filling on only two substances.

We first determined mercury in organic compounds containing chlorine, iodine, and nitrogen; iodine and nitrogen; or nitrogen and bromine together with the mercury.

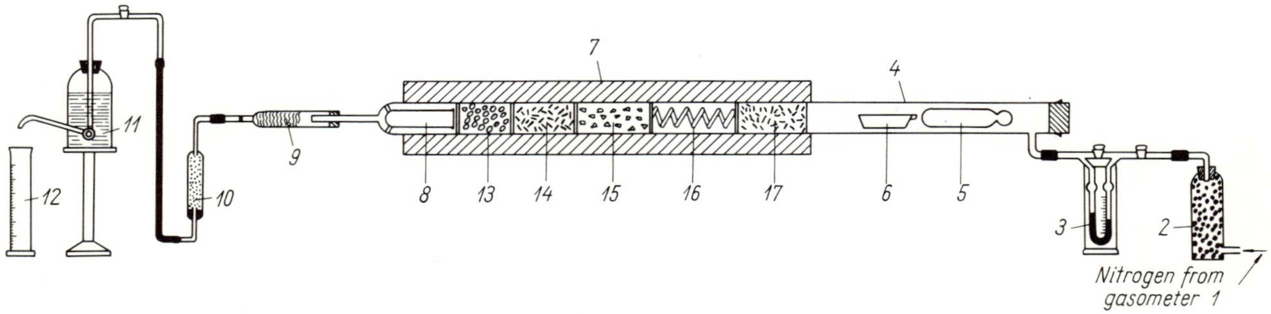

Fig. 1.

Diagram of the Set-up for the Determination of Mercury. 2—Fresenius drying tower with $CaCl_2$; 3—flowmeter; 4—combustion tube; 5—baffle; 6—boat; 7—electric furnace; 8—baffle; 9—absorption tube; 10—$CaCl_2$ tube; 11—Mariotte bottle; 12—measuring cylinder; 13—silver-impregnated clay plate; 14—metallic copper; 15—lead chromate; 16—copper spiral; 17—metallic copper.

As an inert gas, we used nitrogen instead of carbon dioxide.

A supplementary layer of metallic copper in the form of a spiral [16] and of wire [17] (Fig. 1) was introduced into the combustion tube. It was placed next to the layer of lead chromate and served to protect the first layer of metallic copper [14] from rapid oxidation as well as to retain the elementary oxygen contained in the nitrogen. This layer was easily replaced as it became oxidized.

In addition, we determined the influence of individual factors on the accuracy of the analysis and found that the filling must be heated to 700-750°. Under these conditions, however, the combustion tube must be so located that the layer of silver-impregnated clay plate has a temperature higher than the boiling point of mercury.

It was also established that during combustion the ratio of the amount of air to the amount of nitrogen must be no less than 7:4.

DESCRIPTION AND FUNCTIONS OF THE EQUIPMENT

The assembly diagram shown in Fig. 1 consists of the following pieces.

1. Gasometer with nitrogen (not illustrated).
2. Fresenius drying tower containing fused calcium chloride.
3. Flowmeter for regulating the rate of the nitrogen being passed through.
4. Transparent quartz combustion tube having a side arm for the gas inlet and ending in a tip 30 mm. long and 2.5 mm. in outside diameter. Over-all length of the tube, 650 mm.; inside diameter, 10 mm.
5. Baffle made of a quartz tube; it is hermetically sealed and serves to prevent flashing back of the vapors of the substance during the decomposition. Length, 80 mm.; outside diameter, 8 mm.

6. Quartz boat 23 mm. long, 5 mm. high, and 7 mm. wide.

7. Electric furnace at 750°; 250 mm. long.

8. Baffle of transparent quartz; 60 mm. long and 7 mm. in outside diameter. It is hermetically sealed. One end is flattened to form a disc 8.5-9.0 mm. in diameter; three quartz beads are fused to the other end so that the baffle does not close off the opening in the combustion tube tip. The purpose of this baffle is to keep the tube filling some distance from the end and to decrease the cross section in this part of the tube.

9. The absorption tube is made of transparent quartz and has an over-all length of 80 mm. It consists of a wide part 50 mm. long, 5 mm. in outside diameter, and 3.5 mm. in inside diameter, and a narrow part 20 mm. long, 3 mm. in outside diameter, and 1 mm. in inside diameter. The wide part is filled with narrow strips of gold foil in a layer 20 mm. long. A strip of foil is also placed along the entire length of the narrow end.

10. Calcium chloride tube.

11. Mariotte bottle.

12. Measuring cylinder.

REAGENTS

Silver-impregnated clay plate. Pieces of an unglazed, white clay plate are crushed and sifted through a sieve with 2 mm. meshes. A solution of 2 g. of silver nitrate in 3 ml. of water is added to 20 g. of plate and the whole is thoroughly mixed. After the solution has been completely absorbed, the mixture is submitted to preliminary drying in a desiccator and then to heating in a crucible over a burner until the evolution of nitrogen oxide ceases. In this operation the plate assumes a grey-green color and contains 5-6% silver.

Lead chromate. This is crushed, sifted, and heated in a crucible furnace for 1 to 1½ hours at 800°.

Metallic copper. Wire-form cupric oxide is reduced in a stream of hydrogen or in methyl alcohol.

Gold in the form of foil, wire, or gauze.

COMBUSTION TUBE FILLING

1. The baffle (8) (Fig. 1), flattened end up, is placed in the tube and covered with a thin layer of calcined asbestos.

2. Silver-impregnated clay plate is poured in to a depth of 40 mm., and covered with a thin layer of asbestos.

3. A 25 mm. layer of wire-form metallic copper is inserted, followed by another layer of asbestos.

4. A layer of crushed lead chromate is added, again followed by one of asbestos.

5. The copper spiral, 30 mm. long, is inserted and covered with asbestos.

6. Another layer of wire-form metallic copper, 50 mm. deep, is poured in, covered by a final layer of asbestos.

A thickness of 2 mm. is sufficient for each of the layers of asbestos.

THE COURSE OF THE ANALYSIS

The electric furnace (7) is turned on before the start of the analysis and within 15-20 minutes heats to 700-750°. During this time a 3-6 mg. sample is weighed out and the absorption tube is weighed. The tube is first wiped with a wet chamois, then with a dry one, and is quickly passed through the oxidizing flame of a gas burner to remove the charge. It is then left near the balance for 10 minutes and on the balance for 5 minutes. The reading is taken at the fifteenth minute after wiping.

The narrow end of the weighed absorption tube is connected by rubber tubing to the calcium chloride tube (10). The wide part is drawn over the combustion tube tip so that the end of the tip is 5 mm. away from the layer of gold. After this, the boat with the weighed sample is placed in the combustion tube 50 mm. from the layer of metallic copper (17). The baffle (5) is moved up against the boat and nitrogen is passed through at the rate of 4 ml./min.

Since the absorption tube is not connected tightly to the combustion tube, air suction must be created at this point in order to avoid a loss of mercury. This is done by setting the side arm of the Mariotte bottle at such a level that the air is sucked through at the rate of 7-8 ml./min. while the nitrogen rate is 4 ml./min.

The absorption tube is wrapped around several times with a moist flannel rag, 3-4 mm. of gold layer being left exposed next to the tip.

A movable burner with a small flame is placed under the baffle (5) and moved gradually toward the boat. After a few minutes, flaming volatilization, and carbonization of the substance are observed. The burner is left stationary under the boat until these phenomena cease; then it is slowly moved forward up to the layer of metallic copper. The burner flame is increased and the combustion tube is again brushed along its entire length.

When the combustion of the substance is over, the electric furnace is moved 20-30 mm. in the direction of the boat, and the combustion tube is carefully heated from the furnace to the tip with the small flame of a microburner. This operation is repeated several times until all the mercury collects, first at the beginning of the tip and then in the layer of gold foil. Finally, the mercury is driven over completely into the layer of gold foil by careful heating of the front end of the absorption tube.

After 3 minutes' cooling, the flannel rag is removed from the absorption tube, the tube is connected to a drying arrangement (Fig. 2), and 300 ml. of dry air is sucked through at the rate of 7 ml./min. The dried absorption tube is then weighed as described above.

It is advisable to use two absorption tubes in order to save time.

Depending on the percentage content of mercury in the substance being analyzed, it is possible to carry out from 3 to 5 determinations with the same gold layer in the absorption tube.

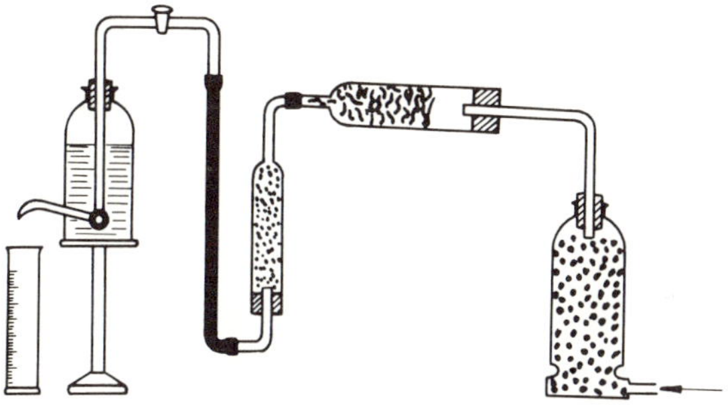

Fig. 2

REGENERATION OF THE GOLD AMALGAM

The mercury is removed from the absorption tube as follows. First the empty front end of the tube is heated and, after cooling, is immersed in a solution of nitric acid (1:1). After the mercury has dissolved, the tube is rinsed several times with water, wiped dry, and connected to a water-jet pump. The gold layer is heated to redness; then heating is discontinued and the mercury residue is removed by turning on the pump.

Such a method of cleaning the absorption tube makes it impossible for the laboratory air to be contaminated with mercury vapors.

DISCUSSION OF THE METHOD

When organic mercury compounds are analyzed, some water condenses in the absorption tube along with the mercury. In order to make a correct determination of the weight of the mercury collected, the absorption tube must be dried. We found that, when 300 ml. of dry air is sucked through the absorption tube at the rate of 7 ml./min. by means of the arrangement illustrated in Fig. 2, moisture is entirely removed and correct results are obtained. In the analysis of inorganic salts it is sufficient to pass through 200 ml. of dry air. It is not recommended that the tube be dried in a desiccator.

In avoiding a loss of mercury special attention must be accorded to establishing the correct ratio between the rate of air suction and of the nitrogen being passed through. The optimum ratio (7:4) is achieved by appropriately adjusting the side arm of the Mariotte bottle and measuring the outflow of water at a previously established nitrogen rate.

Correct temperature control of the tube filling is essential if accurate analytical results are to be obtained.

As is shown by the tests made on the double salt of mercuric chloride and m-nitrophenylchlorovinyliodonium chloride (Table 1), the silver-impregnated clay plate should not be subjected to intense heating during the analysis of iodine-containing mercury compounds, since the silver iodide then partially decomposes and sublimes, thus leading to high analytical figures. Overheating of the clay plate is not reflected, however, in the analytical results for mercury in chlorine- and bromine-containing substances. This is explained by the fact that, although silver iodide has a higher melting point than silver chloride and bromide, it is much more volatile than they.

Placing the capillary end of the combustion tube 20-30 mm. from the furnace is enough to prevent overheating of the clay plate.

It is not recommended that work be conducted at a furnace temperature below 700° since account should be

TABLE 1 — The Analysis of Organic Mercury Compounds

Name and empirical formula of the substance	Weighed sample in mg.	Increase in weight of the Hg in mg.	Found % Hg.	Calculated % Hg.	Δ in %
Metallic mercury	10.372	10.354	99.84	100.00	−0.16
	4.600	4.590	99.78	100.00	−0.22
Mercuric acetate ($C_4H_6O_4Hg$)	4.700	2.994	63.70	62.96	+0.74[a]
	5.556	3.490	62.82	62.96	−0.14
	6.401	4.026	62.90	62.96	−0.06[b]
Mercuric chloride ($HgCl_2$)	6.295	4.643	73.76	73.88	−0.12
	8.925	6.591	73.85	73.88	−0.03
p-Tolylmercury iodide (C_7H_7IHg)	4.790	2.288	47.77	47.92	−0.15
	5.364	2.566	47.83	47.92	−0.09
α-Dithienyl mercury ($C_8H_6S_2Hg$)	6.300	3.427	54.40	54.68	−0.28
	4.981	2.714	54.49	54.68	−0.19
Phenylmercury chloride (C_6H_5ClHg)	5.530	3.544	64.08	64.06	+0.02
	4.499	2.243	64.10	64.06	+0.04
N-(bromomercuriformyl) piperidine ($C_6H_{10}ONBrHg$)	3.884	1.950	49.95	51.09	−1.14
	3.711	1.853	49.93	51.09	−1.16
	4.580	2.325	50.76	51.09	−0.33[c]
	6.828	3.478	50.94	51.09	−0.15
Binary compound of piperidine and N-(chloromercuriformyl)piperidine ($C_{11}H_{21}ON_2ClHg$)	6.248	2.912	46.61	46.29	+0.32
	5.525	2.556	46.26	46.29	−0.03
Double salt of mercuric chloride and m-nitrophenylchlorovinyliodonium chloride ($C_8H_6O_2NCl_6IHg_2$)	5.089	2.304	45.27	45.14	+0.13
	3.630	1.654	45.55	45.14	+0.44[d]
	5.697	2.563	44.99	45.14	−0.15
N-(chloromercuriformyl) piperidine ($C_6H_{10}ONClHg$)	4.461	2.580	57.83	57.61	+0.22
	5.352	3.086	57.66	57.61	+0.05
Binary compound of mercuric chloride and formylpiperidine ($C_6H_{11}ONCl_2Hg$)	6.058	3.159	52.14	52.15	+0.01
	5.920	3.088	52.17	52.15	+0.02
Binary compound of mercuric chloride and o-tolyl-β-chlorovinyliodonium chloride ($C_9H_9Cl_4IHg$)	4.890	1.675	34.25	34.20	+0.05
	4.508	1.542	34.21	34.20	+0.01

[a] 200 ml. of dry air passed through during drying.
[b] 300 ml. of dry air.
[c] Substance recrystallized again.
[d] Silver-impregnated clay plate overheated.

taken that in electric furnaces the temperature is lower at the ends than at the center, and hence there is always the danger that near the ends the mercury will clog the filling.

We always obtained low analytical results when we worked at temperatures of 600-650°.

A study of the optimum conditions for collecting mercury by the vaporization of a weighed sample of metallic mercury showed that driving the mercury from the tip into the absorption tube must be done very carefully. Even when the tip is heated with the very small flame of a Teclu burner, overheating sometimes occurs, the mercury boils up and, despite the air suction, partially vaporizes outside.

We established that only when the tip is carefully heated with a microburner are completely satisfactory results obtained. The mercury losses are then of the order of 0.16% (Table 1).

The speed with which the absorption tube is passed through the oxidizing flame of a burner in order to remove the charge after wiping with a chamois also affects the accuracy of the analysis.

The tube must be passed through instantaneously, and not held in the flame. Otherwise the analytical results may be greatly lowered.

All the organic mercury compounds studied were supplied us by Dr. R. Kh. Freidlina, to whom we express our sincere thanks.

CONCLUSIONS

1. The method proposed for the microdetermination of mercury in organic compounds is based on the thermal decomposition of the substance in a stream of nitrogen. The decomposition products pass through a layer of "universal filling" heated to 750°, which completely retains the

acid gases. The mercury vapors condensing in the tip of the combustion tube are driven over into a layer of gold located in the absorption tube, which is weighed upon drying.

2. It has been found that a filling composed of lead chromate, two layers of metallic copper, and silver-impregnated clay plate may be used for the combustion of organic mercury compounds, independent of whether or not they contain nitrogen. This eliminates the necessity of using different types of tube filling.

3. Analyses were made of various organic mercury compounds containing nitrogen, halogen, and sulfur together. The accuracy of the method varies from ±0.01 to ±0.3%. The determination takes from 1½-2 hours.

LITERATURE CITED

[1] F. Pregl, *Kolichestvennyi organicheskii mikroanaliz* [Quantitative Organic Microanalysis], translated by ONTI [United Scientific and Technical Publishers] Moscow, 1934, p. 148.

[2] A. Friedrich, *Praktika kolichestvennogo organicheskogo mikroanaliza* [Practice of Quantitative Organic Microanalysis], translated by GONTI [State United Publishing House of Science and Technology] 1939, p. 162.

[3] M. Boetius, *J. prakt. Chem.* **151**, 279 (1938).

[4] J. B. Niederl and V. Niederl, *Micromethods of Quantitative Organic Elementary Analysis*, New York, 1942.

CHAPTER 6

The Determination of Carbon and Hydrogen in Nitro Compounds Without the Use of Lead Dioxide*

M. O. Korshun and V. A. Klimova

In our second report [1] we showed that under the conditions of the rapid combustion of nitrogen-containing compounds much smaller amounts of nitrogen oxides form than when these compounds are burned by the classical method. This has also proved true for nitro compounds. In our method the combustion is carried out with prepyrolysis of the substance, and we assume that the carbon separating out during pyrolysis reduces the nitrogen oxides formed from nitro compounds to elementary nitrogen at high temperatures.

Moreover, the hydrogen and carbon, as strong reducing agents, are oxidized at the moment of decomposition by combining with the oxygen already bound to the nitrogen. Should pyrolysis occur under conditions in which the parts of the molecule that split off could not leave the sphere of the reaction before completion of the process, then, most probably, the nitrogen oxides would disappear entirely from the decomposition products. Therefore, in order to decrease the amount of nitrogen oxides it is necessary to proceed with complete thermal decomposition of the substance without the admission of oxygen.

This task is best accomplished by rapid decomposition of the substance in a weighing capsule. In our work we

Zhur. Anal. Khim. **4**, 292-297 (1949).

TABLE 1

Amount of Nitrogen Oxides Obtained During Combustion by the Friedrich Method

Substance	Length of the capsule in cm.	Combustion time in minutes	Weighed sample in mg.	Consumption of 0.01N $Na_2S_2O_3$ in ml.	Nitrogen oxides collected: NO_2 in γ	Nitrogen oxides collected: Upon conversion to Nitrogen in γ	Nitrogen oxides collected: Nitrogen separated as NO_2 in %
Picric acid	5	40	14.59	6.160	3336	862	32.20
Picric acid	5	25	12.33	4.304	1980	603	25.50
Picric acid	3	40	13.94	6.900	3174	966	37.76
Picric acid	3	30	12.57	6.000	2760	840	34.90
Benzidine	5	40	12.25	0.904	416	127	6.71
Benzidine	5	30	16.68	0.702	323	98	3.87
Benzidine	3	40	8.13	0.950	437	133	10.75
Benzidine	3	50	8.33	1.600	736	224	17.67

demonstrated experimentally that fewer nitrogen oxides form upon rapid than upon slow combustion, since in the former case a larger amount of substance undergoes pyrolysis per unit of time and the larger amount of pyrolytic products formed displaces oxygen more effectively and prevents its future access to the capsule, i.e., to the substance being analyzed. Indeed, when we attempted to carry out the combustion exactly according to Friedrich, the slower the combustion, the more nitrogen oxides we obtained.

Furthermore, the nitrogen oxide yield also increased when we used a shorter capsule for taking the weighed samples. (See Table 1).†

In 1947, when our work on the determination of C and H in amino compounds and on the formation of nitrogen oxides during the combustion of these compounds was already in press, Heron [2] published his study on the formation of nitrogen oxides during the determination of

† The collection and determination of the nitrogen oxides were effected as described in the second report.

C and H in coals and nitrogen-containing compounds by the rapid method of Belcher and Spooner.

The experimental data obtained by Heron generally confirm the fact that the amount of nitrogen oxides formed during combustion depends chiefly on the nature of the nitrogen bond in the molecule and that it is considerably smaller for coals and amino compounds than for nitro compounds.

Nonetheless, this method of burning a substance without first subjecting it to pyrolysis (according to Belcher's directions) is, as we have shown, entirely unsatisfactory for solving the problem presented by Heron, since it does not permit obtaining reproducible results for the percent of nitrogen converted to oxides in comparative experiments. Indeed, Heron obtained non-reproducible data in the analysis of both coals and individual organic compounds. In coal the amount of nitrogen oxides varied from 10 to 31%, in amino compounds from 8 to 17% of the total nitrogen contained in the molecule.

In our first report we showed that Belcher and Spooner's method of combustion in an open boat without pyrolysis is not suitable for the analysis of organic compounds [5]. As we have pointed out, this method cannot be used to study the problem of the formation of nitrogen oxides during the combustion of various classes of nitrogen-containing compounds, since it does not provide reproducible figures for the percent of oxidized nitrogen.

Combustion with pyrolysis, on the other hand, makes it possible to obtain completely reproducible data concerning the oxides formed during combustion.

Despite the fact that the conditions for lowering the nitrogen oxide yield are considerably more favorable in our method than in any previous method, we were still unable to completely eliminate the formation of nitrogen oxides during the combustion of nitro compounds. Sometimes these oxides amounted to 15% (see Table 2) of the total nitrogen contained in the molecule; this introduced substantial error into the carbon determination and hence, in the analysis of nitro compounds, it was necessary to collect the nitrogen oxides before the combustion products passed into the carbon dioxide absorption tube.

In the classical methods for the microdetermination of carbon and hydrogen the generally accepted procedure was to collect the nitrogen oxides with lead dioxide. Owing to the transience of its absorbing properties, however, lead dioxide is one of the major sources of error in the determination of carbon and hydrogen. When heated, PbO_2 always liberates a certain amount of carbon dioxide and water since it can absorb not only halogens and acid gases formed from nitrogen and sulfur, but also organic substances, CO_2, and moisture. The carbon dioxide and water slowly separate from the lead dioxide during its absorption of acid gases, and the organic materials are gradually oxidized. They, in turn, also produce a certain amount of carbon dioxide and water. The properties of PbO_2 enumerated above lead to the conclusion that one unsuccessful combustion in which vapors of unburned substance reach the PbO_2 would be enough to put the combustion tube filling out of commission entirely or for a long time. Thus, even in work done by the classical methods, i.e., at strictly regulated gas rates, lead dioxide produces a series of errors in the determination of C and H.

Naturally, the number of these errors will not decrease but rather will quickly increase when work is carried out in a rapid stream of oxygen where the gas rate is 30-50 cc./min. Moreover, we wanted to find an absorbent that could be used outside the combustion tube so the bulk of the work could be kept in the empty tube. Therefore we completely abandoned the use of PbO_2.

TABLE 2

The Formation of Nitrogen Oxides During the Combustion of Nitro Compounds

	Substance	Calculated amount N%	Calculated amount C %	Weighed sample in mg.	Consumption of 0.01 N $Na_2S_2O_3$ in ml.	Nitrogen content of the weighed sample in mg.	Nitrogen oxides collected: NO_2 in γ	Nitrogen oxides collected: Conversion to nitrogen in γ	Nitrogen separated as NO_2 in %	Possible error in the C determination due to NO_2 in %
1	Picric acid	18.35	31.44	7.29	1.372	1.338	631	192	14.36	2.52
				8.30	1.440	1.615	663	201	12.48	2.18
2	p-Nitrotoluene	10.22	61.31	12.51	1.180	1.279	543	165	12.92	1.19
				8.65	0.890	0.884	409	125	14.14	1.29
3	p-Nitrophenol	10.07	51.80	9.09	0.430	0.915	198	60	6.56	0.59
				8.92	0.410	0.871	189	57	6.54	0.58
4	2,4-Dinitrophenol	15.22	39.13	8.37	0.490	1.274	225	69	5.41	0.74
				9.07	0.510	1.380	235	71	5.00	0.67
5	p-Nitrobenzoic acid	8.38	50.30	7.53	0.590	0.631	271	83	13.15	0.97
				6.09	0.590	0.516	271	83	16.08	1.21
6	Trimethylolnitromethane	9.27	31.79	10.13	0.350	0.939	161	'9	5.22	0.43
				6.58	0.280	0.610	129	39	6.72	0.55
7	m-Dinitrobenzene	16.66	42.85	8.49	2.260	1.414	1040	316	22.28	3.34
				8.39	2.540	1.398	1168	356	25.46	3.80
8	2,4-Dinitrotoluene	15.38	46.15	9.03	0.110	1.389	51	15	1.11	0.15
				9.04	0.090	1.390	42	13	0.91	0.13
9	p-Nitrobenzaldehyde	9.27	55.63	9.44	0.130	0.875	60	18	2.08	0.17
				9.12	0.140	0.845	64	20	2.36	0.19

We effected the absorption of the nitrogen oxides with a 0.02M solution of $KMnO_4$ or a saturated solution of $K_2Cr_2O_7$ in H_2SO_4. We were convinced that either H_2SO_4 or solutions of oxidizing agents therein would absorb the nitrogen oxides quantitatively [3, 4] and used the latter only because the change in color makes it easy to determine the extent to which the absorbing solution has become exhausted. Such an indication is, of course, extremely helpful.

APPARATUS

The set-up for the determination of carbon and hydrogen has already been described in our first report [5].

When nitrogen-containing compounds are burned with collection of the nitrogen oxides a small change need be made only in the part of the set-up where the combustion products are absorbed. This part is illustrated in Fig. 1. A small bubble counter (2), and another absorption tube (3), just like the water absorption tube (1), are placed between the customarily used absorption tubes. Tube (4) is filled with Ascarite and serves to collect the carbon dioxide. There are no other changes in the apparatus. Anhydrous magnesium perchlorate or calcium sulfate is used as the water absorbent. The use of calcium chloride

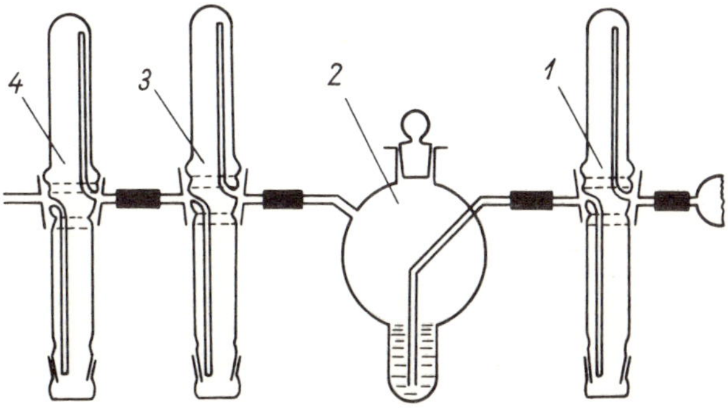

Fig. 1

TABLE 3 — Analysis of Nitrogen-Containing Compounds With Removal of the Nitrogen Oxides

No.	Name of the substance	N in the substance in %	Weighed sample in mg.	Weight of H_2O in mg.	Weight of CO_2 in mg.	Amount found H %	Amount found C %	Amount calculated H %	Amount calculated C %	Δ H %	Δ C %
1	Picric acid	18.34	9.41	1.11	10.79	1.32	31.27	1.32	31.44	0.00	-0.17
			11.21	1.51	12.88	1.51	31.34	1.32	31.44	+0.19	-0.10
2	p-Nitrotoluene	10.22	7.90	3.68	17.84	5.21	61.59	5.11	61.31	+0.10	+0.28
			9.33	4.41	21.00	5.29	61.39	5.11	61.31	+0.18	+0.08
3	Dinitrotoluene	15.38	10.28	3.12	17.42	3.39	46.21	3.30	46.16	+0.09	+0.05
			8.61	2.39	14.61	3.11	46.28	3.30	46.16	-0.19	+0.12
4	Trimethylamine picrate	19.44	8.84	3.18	12.24	4.00	37.76	4.16	37.50	-0.16	+0.26
			9.17	3.33	12.63	4.06	37.56	4.16	37.50	-0.10	+0.06
5	3-Methyl-6-tert.butyl-2, 5-dinitrophenetol	9.93	15.78	9.02	32.14	6.39	55.54	6.38	55.32	+0.01	+0.22
		9.93	11.58	6.48	23.52	6.26	55.40	6.38	55.32	-0.12	+0.08
6	Trimethylammonium (2-amino-pyridyl-3-butenyl) dipicrate	19.03	8.81	2.96	14.01	3.76	43.37	3.65	43.51	+0.11	-0.14
		19.03	7.54	2.68	11.99	3.98	43.37	3.65	43.51	+0.33	-0.14
7	N-Methyl-α-pyridonimine-8-(4'-methylacetylamino-1'-butenyl) picrate	18.18	6.305	2.740	11.412	4.86	49.40	4.76	49.35	+0.10	+0.05
		18.18	7.185	3.120	13.047	4.86	49.55	4.76	49.35	+0.10	+0.20
8	α-Amino-β-piperidylaldehyde phenylhydrazone	26.43	5.770	2.980	14.370	5.78	67.96	5.65	67.92	+0.13	+0.04
		26.43	5.965	3.060	14.847	5.74	67.93	5.65	67.92	+0.09	+0.01
9	α-Amino-β-pyridylaldehyde semicarbazone	39.10	4.830	2.232	8.305	5.17	46.92	5.03	46.86	+0.14	+0.06
		39.10	4.400	1.985	7.525	5.05	46.67	5.03	46.86	+0.02	-0.19
10	α-Amino-β-pyridylaldehyde	22.95	4.190	1.850	9.054	4.94	58.97	4.91	59.01	+0.03	-0.04
		22.95	4.155	1.820	8.988	4.90	59.03	4.91	59.01	-0.01	+0.02
11	α-Amino-β-pyridylaldehyde phenylhydrazone	26.41	4.780	2.385	11.865	5.58	67.74	5.65	67.92	-0.07	-0.18
		26.41	6.040	3.060	15.065	5.67	68.06	5.65	67.92	+0.02	+0.14

is not recommended, since this material can retain a certain amount of nitrogen oxides.

CARRYING OUT THE ANALYSIS

Before work is begun, an approximately 0.02M solution of potassium permanganate or a saturated solution of potassium dichromate in chemically pure sulfuric acid, sp. gr. 1.84, is prepared.

This solution should not be made up in advance since it is difficult to keep it completely anhydrous over a long period of time. It is preferable to prepare the solution in the counter itself. Sulfuric acid is poured by graduated pipette into the completely dry counter until the narrow part is entirely filled (as shown in the drawing). Then an oxidizing agent of 4 mg. of $KMnO_4$ or 25 mg. of $K_2Cr_2O_7$ per ml. of H_2SO_4 is added. The capacity of the narrow section of the counter generally varies from 1 to 1.5 ml. and must be filled completely, since, if too little sulfuric acid solution is used, the liquid foams easily. When prepared, the absorbing solution must be transparent and dark green (in work with $KMnO_4$) or orange-red (in work with $K_2Cr_2O_7$). The sulfuric acid must contain no volatile impurities. Hence, when the counter is ready for work, 1-1.5 liters of oxygen is passed through while the counter is connected directly to the tip of the combustion tube for 20-30 minutes.

Absorption tube (3), which is placed after the counter, is not weighed. It is filled exactly like the water absorption tube, (1), and serves to maintain a uniform moisture content in the gas leaving the sulfuric acid bubble counter.

After the weighed sample has been introduced into the combustion tube, the whole absorbing part of the apparatus is connected and burning is effected as usual. The entire course of the analysis is the same as for substances composed of C, H, and O, which was described in detail in our first report [5].

The few points that are especially important for the analysis of nitrogen-containing compounds should be emphasized here: 1) The combustion time must be 10-12 minutes. It should not be prolonged, since this may lead to an undesirable increase in the amount of nitrogen oxides, which means a more rapid exhaustion of the sulfuric acid solution. 2) Care must be taken that the end of the combustion tube be located 5-7 cm. from the small furnace (as specified in our method). This is particularly important in the analysis of nitro compounds. It is known that nitrogen dioxide completely dissociates into NO and O_2 at 650°, i.e., at a temperature lower than that of our furnaces. The reverse reaction proceeds at 130°. Thus, by moving the tube away from the furnace, we guarantee the complete oxidation of NO to NO_2. One filling of the bubble counter is enough to absorb the nitrogen oxides formed from the burning of 50-80 mg. of picric acid, i.e., for a whole day of work.

As the absorbent is exhausted, its color changes from green to brown or brownish yellow and from orange to greenish brown. The absorbent can be used as long as the original color remains. The precipitate does not impede the work if the solution above it keeps its starting color.

When work is interrupted, the apparatus must be carefully closed to prevent the entry of moist laboratory air into the sulfuric acid solution.

We have analyzed a large number of different nitro compounds by this method but only those difficult to analyze by the classical methods are presented in Table 3. The results shown were obtained from two determinations.

CONCLUSIONS

1. It has been determined that in the rapid combustion of nitro compounds quick pyrolysis of the substance (10-12 minutes) reduces the nitrogen oxide yield to 15% as compared with the 50% of the Friedrich method.

2. A method has been presented for the analysis of nitro compounds with collection of the nitrogen oxides outside the combustion tube by means of H_2SO_4 or a solution of strong oxidizing agents in H_2SO_4. Accuracy of the method, ±0.3%.

LITERATURE CITED

[1] M. O. Korshun and V. A. Klimova, *Zhur. Anal. Khim.* 3, 176 (1948).

[2] A. E. Heron, *Analyst* 72, 142 (1947).

[3] P. J. Elving and W. R. McElroy, *Ind. Eng. Chem., Anal. Ed.* 13, 660 (1941).

[4] K. Bürger, *Die Chemie* 55, 260 (1942).

[5] M. O. Korshun and V. A. Klimova, *Zhur. Anal. Khim.* 2, 274 (1947).

CHAPTER 7

The Microanalytical Determination of Nitrogen by the Hydrogenation of Organic Substances*

N. E. Gel'man and M. O. Korshun

There are two fundamental methods of determining nitrogen in the elementary analysis of organic substances. One is based on oxidation of the molecule of organic substance to CO_2 and H_2O and isolation of all the nitrogen in the form of elementary nitrogen, which is measured gasometrically (Dumas). The other consists in oxidizing the organic substance with sulfuric acid in the presence of catalysts. In this case the nitrogen (reduced by the carbon) is converted to ammonia, which, after removal from the reaction mixture, is determined by titration (Kjeldahl). Suitable micro modifications of both methods are used in organic microanalysis.

The application to the microanalytical determination of nitrogen of the principle of decomposing the organic substance by reduction—by heating it in a hydrogen atmosphere in the presence of a catalyst [1]—rather than by oxidation has proved of interest. As a result of this decomposition, i.e. destructive hydrogenation of the organic substance, all the nitrogen is converted to ammonia which is absorbed by acid and determined by titration. In many cases the decomposition is accompanied by the separation of carbon and resin on the catalyst. This results in a

Doklady Akad. Nauk S.S.S.R. **72** 895-998 (1950).

rapid loss of catalyst activity and necessitates frequent cleaning of the apparatus from carbon and resin.

As we have already established [2, 3], in the decomposition of an organic substance in a stream of hydrogen it is preferable first to subject the substance to thermal decomposition in an atmosphere of hydrogen and then to pass the decomposition products over the catalyst in mixture with excess hydrogen. According to our observations, preliminary thermal decomposition of the organic substance is a factor extremely favorable to the quantitative conversion into ammonia of the nitrogen contained therein. Moreover, with such a decomposition procedure it is possible to achieve a more nearly complete hydrogenation of the substance to gaseous products, to sharply decrease the amount of carbon and resin separated, and to avoid deposition of the carbon on the catalyst. Even if carbon deposits form, the major portion condenses on the walls of the weighing capsule, while the apparatus and catalyst remain clean and the latter can be used for a large number of analyses without regeneration. Thus, for example, 130 tests were made with the same nickel catalyst, 60 with the same iron catalyst, and in neither case was any decrease in the activity of the catalyst observed.

Our experiments showed that destructive hydrogenation without a catalyst does not lead to a quantitative conversion into ammonia of the nitrogen present in the molecule of organic substance. Low results are obtained even in those cases where the ammonium salt of an organic acid (e.g., ammonium oxalate) is hydrogenated. Reduced nickel in mixture with thorium oxide [1] can be used as the catalyst. We also studied iron, metallic platinum, and palladium reduced on crushed quartz, as specific catalysts for making ammonia. Low results were obtained with platinum and palladium catalysts, good results with nickel and iron catalysts (see Tables 1 and 2).

The iron catalyst was prepared from the nitrate of ferrous oxide, which was first decomposed by heating and then reduced in a stream of hydrogen at 475-500°.

The decomposition of the substance was effected rapidly at a hydrogen rate of 30-50 ml./min. The average dura-

TABLE 1

Determination of Nitrogen by Hydrogenation with a Nickel Catalyst (Sample 3-5 mg.; Temperature to Which the Catalyst Was Heated, 300-340°; Rate at Which the Hydrogen Was Passed, 30-50 ml./min.)

Substance and empirical formula	% N Calculated	% N Found	Difference in %
Ammonium oxalate $C_2H_8O_4N_2 \cdot H_2O$	19.71	19.56 19.55	−0.15 −0.16
p-Aminoacetophenone C_8H_9ON	10.36	10.32 10.22	−0.04 −0.14
Methylbenzamide C_8H_9ON	10.36	10.31 10.28	−0.05 −0.08
Oxamide $C_2H_4O_2N_2$	31.82	31.98 31.82	+0.16 0.00
Ethyl oxamate $C_4H_7O_3N$	11.97	12.13 12.07	+0.16 +0.10
Diphenylurea $C_{13}H_{12}ON_2$	13.20	13.30 13.26	+0.10 +0.06
Triphenylguanidine $C_{19}H_{17}N_3$	14.62	14.63 14.65	+0.01 +0.03
Benzoinoxime $C_{14}H_{13}O_2N$	6.16	6.18	+0.02
Azoxyanisole $C_{14}H_{16}O_3N_2$	10.77	10.74 10.93	−0.03 +0.16
α-Aminopyridine $C_5H_6N_2$	29.77	29.88 29.75	+0.11 −0.02
Dipyridylurea $C_{11}H_{10}ON_4$	26.16	26.21	+0.05

TABLE 2

Determination of Nitrogen by Hydrogenation with an Iron Catalyst (Sample 3-5 mg.; Rate at Which the Hydrogen Was Passed, 30-50 ml./min.)

Substance and empirical formula	Temp. of Catalyst in °C.	% N Calc.	% N Found	Difference in %
Ammonium oxalate $C_2H_8O_4N_2.H_2O$	310	19.71	19.67	-0.04
	320	19.71	19.69	-0.02
p-Aminoacetophenone C_8H_9ON	315	10.36	10.37	+0.01
	400	10.36	10.39	+0.03
Methylbenzamide C_8H_9ON	280	10.36	10.44	+0.08
	280	10.36	10.50	+0.14
Ethyl oxamate $C_4H_7O_3N$	405	11.97	12.02	+0.05
p-Phenylenediamine $C_6H_8N_2$	375	25.91	25.91	0.00
	400	25.91	25.79	-0.12
Triphenylguanidine $C_{19}H_{17}N_3$	405	14.62	14.51	-0.11
	420	14.62	14.48	-0.14
α-Aminopyridine $C_5H_6N_2$	380	29.77	29.62	-0.15
	400	29.77	29.54	-0.23
Dipyridylurea $C_{11}H_{10}OH_4$	400	26.16	26.20	+0.04
	420	26.16	26.11	-0.05
Carbazole $C_{12}H_9N$	400	8.38	8.38	0.00
	400	8.38	8.40	+0.02

tion of the decomposition† for nitrogen determination was 8 min., but in a number of cases it was considerably less. Owing to the rapidity of the decomposition, the time spent for one determination, including the sample weighing and titration, is 30-35 minutes.

In those cases where decomposition of the substance is accompanied by abundant separation of carbon the results obtained are low. Correct results can be obtained for

† By the duration of the decomposition we mean the exact time consumed for decomposition of the weighed sample, beginning with the first signs of change in the substance (volatilization, flaming, carbonization) and ending with its complete disintegration.

these substances by mixing the weighed sample with reduced nickel dust [1].

For titrating the ammonia obtained in the hydrogenation of the weighed sample we developed a method of determining the excess acid iodometrically. This is based on the reaction:

$$IO_3^- + 5I^- + 6H^+ \rightarrow 3I_2 + 3H_2O$$

We used a standard 0.02N solution of potassium biiodate to absorb the ammonia. The excess $KH(IO_3)_2$ was determined by adding a potassium iodide solution, which reacted with the potassium biiodate according to the equation:

$$6KH(IO_3)_2 + 5KI \rightarrow 11KIO_3 + 3H_2O + 3I_2$$

The evolved iodine was titrated with sodium thiosulfate.

A comparison of the reproducibility of the results obtained in acidimetric with that for iodometric titrations shows that the end point for the iodometric method can be determined more than twice as accurately as that for the acidimetric titration (see Table 3).

TABLE 3

Reproducibility of the Results Obtained upon Acidimetric and Iodometric Titration of 0.02 N Solutions

Acidimetric Titration		
H_2SO_4 taken, ml.	5.000	5.000
Average consumption of KOH, ml.	4.958	4.903
Average deviation, ml.	0.005[a]	0.004[a]
Number of tests	6	5
Iodometric Titration		
$KH(IO_3)_2$ taken, ml.	5.000	5.000
Average consumption of $Na_2S_2O_3$, ml.	5.369	5.344
Average deviation ml.	0.002[a]	0.002[a]
Number of tests	4	8

[a] These figures represent the average deviation of the alkali or sodium thiosulfate consumption from the average consumption of these solutions found for this series.

The same apparatus was used for the determination of nitrogen as was used in the determination of sulfur by hydrogenation [3, 4].

LITERATURE CITED

[1] G. Ter-Meulen and I. Hesling, *Novye Metody Organicheskogo Khimicheskogo Analiza.* [New Methods of Organic Chemical Analysis], 1931.

[2] N. E. Gel'man, Dissertation, Institute of Organic Chemistry, Academy of Science, U.S.S.R., 1948.

[3] M. O. Korshun and N. E. Gel'man, *Novye metody elementarnogo mikroanaliza,* [New Methods of Elementary Microanalysis], 1949.

[4] N. E. Gel'man, *Zavodskaya Lab.* 8, 673 (1939).

CHAPTER 8

A New Dry Absorbent for Collecting Nitrogen Oxides in the Determination of Carbon and Hydrogen in Nitrogen-Containing Organic Compounds

V. A. Klimova and M. O. Korshun

In our third report [1] we demonstrated the possibility of determining carbon and hydrogen in nitrogen-containing organic substances by a method that we had developed earlier [2, 3]. This comprised collecting the nitrogen oxides without the use of lead dioxide. The nitrogen oxides were absorbed outside the combustion tube in a special tube filled with a solution of $KMnO_4$ or $K_2Cr_2O_7$ in concentrated sulfuric acid.

We consider the sulfuric acid absorbent unquestionably superior to lead dioxide since it is not a source of error in the determination of carbon and hydrogen. Nonetheless, the use of a liquid sulfuric acid absorbent creates some inconveniences in the work. These include foaming of the solution at the oxygen rate of 35-50 ml./min., the necessity of frequently changing the solution, and the use of an additional tube to keep the moisture content constant in the gases passing through. Furthermore, when the pressure in the system is changed and, particularly, when the absorption tubes are disconnected, the liquid may be drawn into one of them.

Meanwhile, the formation of nitrosylsulfuric acid took place so smoothly under our operating conditions that it

Zhur. Anal. Khim.* **6, 230-233 (1951).

did not seem expedient to stop absorbing the nitrogen oxides with sulfuric acid, especially since no solid absorbent that could quantitatively collect nitrogen oxides in the cold in a stream of oxygen was known at the time. We then undertook to prepare a solid absorbent for nitrogen oxides based on this sulfuric acid solution. For this, we made use of the ability of silica gel to absorb large quantities of liquid while remaining dry outside. Such silica gel, impregnated with the liquid absorbent for nitrogen oxides, preserves all the advantages of solid absorbents and possesses the ability to take up nitrogen oxides over a longer period of time than the previously used liquid absorbent.

The proposed absorbent is silica gel impregnated with a 0.02 M solution of $K_2Cr_2O_7$ in sulfuric acid, specific gravity 1.84.

When various brands of silica gel are moistened with the sulfuric acid solution it becomes clear that coarsely porous silica gel takes up more sulfuric acid solution than silica gel with small pores and that the finished preparation obtained from coarsely porous silica gel is consequently able to absorb more nitrogen oxides than that obtained from silica gel with small pores. Thus, although silica gel of any porosity may be used if necessary, it is preferable to use coarsely porous silica gel since the absorbent made from it can be used three or four times as long as that made from the finely porous type. However, even if only finely porous silica gel is available, a solid absorbent must be preferred to a liquid one since it is more convenient to work with and absorbs a larger quantity of nitrogen oxides. The dry absorbent does not retain carbon dioxide and, as it is used up, sharply changes color from orange-yellow to green. Consequently it is easy to judge to what extent the absorbent has been exhausted. An absorption tube of any construction may be used to hold the absorbent.

For the past year and a half the dry absorbent described here has been used with unvarying success in our laboratories in the analysis of nitrogen-containing compounds, and its practicality can be considered completely confirmed. Similar oxidizing agents may, of course, be used instead of $K_2Cr_2O_7$.

PREPARATION OF THE ABSORBENT

Silica gel ground to 0.5-1 mm. granules is dried in a stream of dry oxygen in a combustion tube at 200° for an hour or in a drying oven at the same temperature for three hours. The dried silica gel is placed in a flask with a ground glass stopper and a freshly prepared 0.02 M solution of $K_2Cr_2O_7$ in H_2SO_4, sp. gr. 1.84, is gradually added. Two or three drops of the sulfuric acid solution are run in; then the flask is closed and shaken vigorously until all the solution has been absorbed. The addition of sulfuric acid solution is continued, two or three drops at a time, until the silica gel begins to adhere to the sides of the flask—a sign that it is completely saturated. Care must be taken that neither the sulfuric acid solution nor the absorbent

TABLE 1

Silica gel	Amount of silica gel taken in g.	Amount of sulfuric acid consumed for complete saturation of the silica gel		
		g.	%	Average %
Coarsely porous	3.4	3.7	109	123
Coarsely porous	5.9	8.1	137	123
Finely porous	4.8	1.5	31	27
Finely porous	3.5	0.8	23	27

remain exposed to the atmosphere for a long time since this weakens the acid and consequently impairs the absorbent.

The following four preparations listed in Table 1 were made up as described above.

As is evident from the data of Table 1, coarsely porous silica gel absorbs four to five times as much sulfuric acid with respect to its weight as finely porous.

STUDY OF THE ABSORBING CAPACITY OF THE NEW PREPARATION FOR THE COLLECTION OF NITROGEN OXIDES

The four preparations described above (see Table 1) were used to test the activity of the absorbent. The test was conducted as follows: a certain amount of absorbent, about 4 cc., was poured into a tubular absorbing apparatus between two layers of anhydrous magnesium perchlorate. The filled tube was preweighed and then connected to the apparatus used to determine the carbon and hydrogen.

A quartz capsule containing a 15-30 mg. weighed sample of silver nitrate, which served as the source of the nitrogen oxides, was placed in the combustion tube. The weighed

TABLE 2

Silica gel	H_2SO_4 solution impregnating the silica gel in %	Nitrogen oxides absorbed[a]: Up to complete saturation of the preparation in mg.	Nitrogen oxides absorbed[a]: Up to the beginning of non-quantitave absorption in mg.
Coarsely	137	125	85
porous	109	152	144
	109	146	119
Finely	23	80	25
porous	31	87	40
	38	62	42

[a]Volume of absorbent 4 cc.

sample was decomposed by heating with a gas burner for 7-10 minutes while a stream of oxygen was passed through at the rate of 35-50 ml./min. When the weighed sample had been decomposed and the resulting nitrogen oxides completely expelled, the tube was weighed.

The amount of nitrogen oxides absorbed in a series of tests with one filling of the absorption tube is shown in Table 2.

The absorbing capacity of the preparations tested is presented graphically in Figs. 1 and 2.

The total amount of nitrogen oxides (in mg.) absorbed from the start of the operation is plotted along the abscissa. The absorption of the nitrogen oxides (in %) separating out in each individual experiment is plotted along the ordinate.

As Fig. 1 shows, the absorbent prepared by impregnating coarsely porous silica gel does not lose its activity for a long time. It stops collecting the nitrogen oxides quantitatively only after it has taken up about 100 mg. At this point, a wide scatter in the absorption results begins to be observed. This scatter becomes understandable if account is taken that we are working at a high oxygen rate

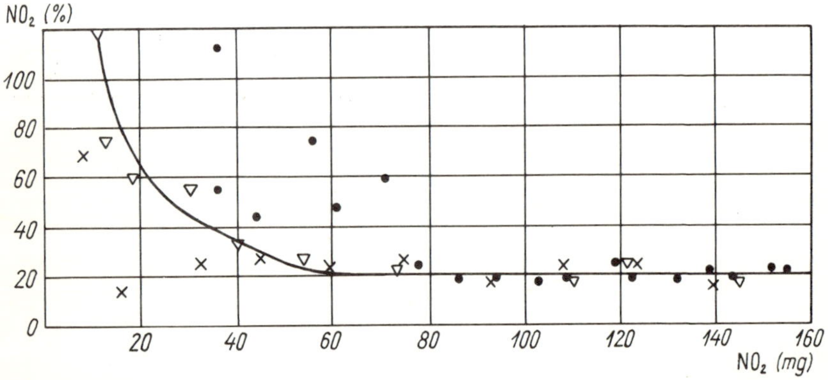

Fig. 1

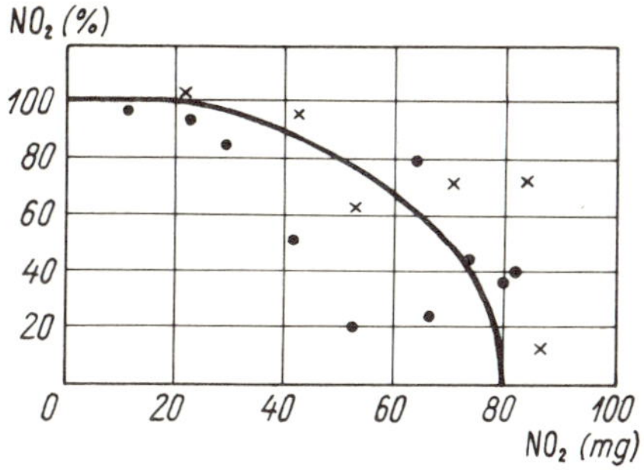

Fig. 2

and that when the absorbent is substantially saturated with nitrogen oxides part of these oxides may be carried away by the stream of oxygen. The blank tests that we conducted confirm this hypothesis.

The absorbent prepared by impregnating finely porous silica gel (Fig. 2) absorbs only 25-40 mg. of nitrogen oxides quantitatively. Thereafter, absorption is highly irregular.

Inasmuch as in the combustion of nitrogen-containing organic compounds by the rapid method only a small quantity of nitrogen oxides separates out [1, 4], i.e., no more than 1 mg. in weighed samples of up to 20 mg., one filling of the absorption tube, even with an absorbent prepared by impregnating finely porous silica gel, should be sufficient for 20-30 determinations. An absorbent prepared by impregnating coarsely porous silica gel is effective considerably longer, and one filling of the absorption tube should serve for 100-150 determinations, i.e., for burning, for example, 1000 mg. of picric acid. The work of our laboratory confirmed this conclusion: not until after 150 combustions of various nitrogen-containing compounds had been made was it necessary to change the filling.

It must be noted that the change in color from orange-yellow to green, along the layer of absorbent, which takes place gradually as the absorbent becomes exhausted, is particularly sharp in absorbents prepared by impregnating coarsely porous silica gel. In this case the inflection point of the curve coincides exactly with the moment at which the color of the absorbent changes to green along almost the entire layer and only 5-7 mm. remains the original orange-yellow.

CONCLUSIONS

1. A new solid absorbent for collecting the nitrogen oxides formed upon the rapid combustion of nitrogen-containing organic compounds has been proposed and put into use in organic microanalysis. This absorbent is made of silica gel impregnated with a solution of $K_2Cr_2O_7$ in sulfuric acid.

2. It has been established that one filling of the absorption tube is sufficient for 100-150 analyses of any nitrogen-containing compounds. The change in color of the absorbent permits judging the extent to which it has been exhausted.

3. It has been shown that coarsely porous silica gel gives the best results; finely porous silica gel can also be used for the preparation of the absorbent but in this case the active life is appreciably shorter.

LITERATURE CITED

[1] M. O. Korshun and V. A. Klimova, *Zhur. Anal. Khim.* **4**, 292 (1949).

[2] M. O. Korshun and V. A. Klimova, *Zhur. Anal. Khim.* **2**, 274 (1947).

[3] M. O. Korshun and N. E. Gel'man, *New Methods of Elementary Microanalysis*, Moscow-Leningrad 1949, p. 46.

[4] M. O. Korshun and V. A. Klimova, *Zhur. Anal. Khim.* **3**, 176 (1948).

CHAPTER 9

The Simultaneous Determination of Carbon, Hydrogen, and Silicon*

V. A. Klimova, M. O. Korshun, and E. G. Bereznitskaya

The determination of carbon and silicon in organic silicon compounds is extremely difficult, since silicon carbide forms upon their thermal decomposition. In rapid combustion, however, the formation of carbide does not occur with the same degree of ease in the case of all organic silicon compounds. Thus it has been determined that silanes containing one alkoxy or naphthyl radical decompose without the formation of silicon carbide when burned rapidly. Silanes that contain only alkyl and (especially) alkenyl radicals pyrolyze with the formation of carbide. Organic silicon derivatives of benzene apparently decompose without the formation of carbide, but we still do not possess sufficient material for complete confirmation.

Since, in the rapid combustion process, pyrolysis of the substance being analyzed is effected in a quartz capsule, it should be possible in the analysis of organosilicon compounds that do not form carbide to determine not only carbon and hydrogen but also silicon, by weighing the capsule containing the silicon dioxide formed; thus carbon, hydrogen, and silicon could be determined simultaneously from one 3-15-mg. weighed sample. We have actually succeeded in carrying out such determinations. The analysis of the tetraalkylsilanes, on the other hand, requires a

---

Doklady Akad. Nauk S.S.S.R.* **84, 1175-8 (1952).

TABLE 1

Compounds Containing Oxygen

Substance	C %			H %			Si %		
	Found	Calc.	Δ	Found	Calc.	Δ	Found	Calc.	Δ
Cyclohexyltriethoxysilane	58.44	58.50	0.06	10.72	10.64	0.08	11.32	11.40	0.08
	58.68	58.50	0.18	10.54	10.64	0.10	11.57	11.40	0.17
Dinaphthyldiethoxysilane	77.48	77.38	0.10	6.20	6.49	0.29	7.62	7.52	0.10
	77.49	77.38	0.11	6.23	6.49	0.26	7.72	7.52	0.20
Trinaphthylethoxysilane	84.42	84.54	0.12	5.92	5.77	0.15	6.61	6.17	0.44
	84.26	84.45	0.28	5.87	5.77	0.10	6.39	6.17	0.22
Naphthyloctyldiethoxysilane	73.85	73.69	0.16	9.68	9.56	0.12	7.63	7.82	0.19
	73.66	73.69	0.03	9.47	9.56	0.09	7.86	7.28	0.04
Hexahexyldisiloxane	73.95	74.23	0.28	13.29	13.40	0.11	9.39	9.62	0.23

special combustion technique in which pyrolysis is slow and incomplete, permitting the avoidance of silicon carbide formation. It should be noted that the higher the boiling point of the compound being studied the easier it is to burn it correctly and to obtain good results for carbon and silicon. The results for hydrogen are always completely satisfactory.

The Simultaneous Determination of Carbon, Hydrogen, and Silicon in Compounds That Decompose Without the Formation of Silicon Carbide

The simultaneous determination of carbon, hydrogen, and silicon in organic silicon compounds that contained alkoxy or naphthyl radicals and in all the compounds available to us in which the silicon was bound to a benzene ring was effected by our previously developed rapid method for determining carbon and hydrogen in an empty combustion tube in a rapid stream of oxygen [1, 2]. A weighed sample of the substance was decomposed by the roaring flame of a gas burner under conditions of complete pyrolysis. The carbon and hydrogen were collected and weighed as usual, while the silicon was determined in the form of silicon dioxide by the gain in weight of the weighed-sample capsule. The analytical figures obtained for all three elements were within the limits of permissible accuracy (see Tables 1 and 2).

The Simultaneous Determination of Carbon, Hydrogen, and Silicon in Tetraalkylsilanes

Silicon carbide forms readily when organic silicon compounds in which the silicon is bound with aliphatic radicals are subjected to thermal decomposition (under conditions of rapid combustion). Consequently, a determination of the carbon contained therein cannot be made in the usual

TABLE 2
Compounds With Naphthyl Radicals

Substance	C %			H %			Si %		
	Found	Calc.	Δ	Found	Calc.	Δ	Found	Calc.	Δ
Naphthyltriphenylsilane	86.90	86.99	0.09	5.80	5.74	0.06	7.34	7.27	0.07
	86.78	86.99	0.21	5.82	5.74	0.08	7.49	7.57	0.22
Naphthyltribenzylsilane	86.68	86.89	0.21	6.73	6.58	0.15	6.18	6.54	0.36
	86.86	86.89	0.03	6.51	6.58	0.07	6.36	6.54	0.18
Naphthylditolylbutylsilane	84.95	85.22	0.27	7.45	7.66	0.21	7.31	7.11	0.20
	85.20	85.22	0.02	7.67	7.66	0.01	7.11	7.11	0.00
Dinaphthyldiphenylsilane	87.90	88.03	0.13	5.71	5.54	0.17	6.16	6.43	0.27
	87.71	88.03	0.32	5.82	5.54	0.28	6.53	6.43	0.10
Naphthyltritolylsilane	86.60	86.89	0.29	6.54	6.58	0.04	6.68	6.53	0.15
	86.57	86.89	0.32	6.73	6.58	0.15	6.67	6.53	0.14

TABLE 3
Compounds With Saturated Radicals

Substance	C %			H %			Si %		
	Found	Calc.	Δ	Found	Calc.	Δ	Found	Calc.	Δ
Triethyl(2-methylbutyl)silane	70.82	70.97	0.15	13.77	13.98	0.21	14.18	15.05	0.87
	70.66	70.97	0.31	14.02	13.98	0.04	14.68	15.05	0.37
Trimethyl(2-methylbutyl)silane	66.41	66.67	0.26	14.04	13.90	0.14	-	-	-
	66.40	66.67	0.27	13.96	13.90	0.06	19.20	19.43	0.23
Triethylcyclohexylsilane	72.70	72.64	0.06	13.25	13.21	0.04	13.99	14.14	0.15
	72.76	72.64	0.12	13.08	13.21	0.13	13.80	14.14	0.34
Tripropyl(2-methylpentyl)silane	74.50	74.58	0.08	14.14	14.05	0.09	11.19	11.57	0.38
	74.58	74.58	0.00	14.14	14.05	0.09	11.21	11.57	0.36

TABLE 4

Compounds With Unsaturated Radicals

Substance	C %			H %			Si %		
	Found	Calc.	Δ	Found	Calc.	Δ	Found	Calc.	Δ
Triethyl(2-methyl-4-pentenyl)silane	72.70	72.64	0.06	13.25	13.21	0.04	13.99	14.15	0.16
	72.66	72.64	0.02	13.08	13.21	0.13	13.80	14.15	0.35
Triethyl(2-methyl-2-propenyl)silane	70.34	70.51	0.17	13.05	13.02	0.03	16.82	16.47	0.35
	70.67	70.51	0.16	12.96	13.02	0.06	16.50	16.47	0.03
Dimethylphenylallylsilane	75.24	75.00	0.24	9.16	9.09	0.07	15.97	15.91	0.06
	74.97	75.00	0.03	9.17	9.09	0.08	16.29	15.91	0.38
Triallyl-n-butylsilane	75.31	74.92	0.39	11.67	11.62	0.05	12.91	13.46	0.55
	75.16	74.92	0.24	11.77	11.62	0.15	12.78	13.46	0.68
Allyltributylsilane	74.73	74.92	0.19	13.48	13.41	0.07	10.78	11.67	0.89
	75.15	74.92	0.23	13.64	13.41	0.23	11.18	11.67	0.49
Allyltriethylsilane	69.30	69.15	0.15	12.82	12.90	0.08	17.52	17.95	0.43
	69.17	69.15	0.02	13.05	12.90	0.15	17.31	17.95	0.64
Triallylmethylsilane	72.39	72.29	0.10	11.07	10.85	0.22	16.79	16.86	0.07

manner. The combustion of such compounds can be effected with success, however, at a slow tempo (20-40 min.) with the colorless flame of a gas burner. In this case the capsule for the weighed sample must be of definite dimensions: 60 mm. in length and 7 mm. in inside diameter. Care must be taken that the decomposition proceeds gradually and that it be completed in the capsule. Thus, in all the compounds at our disposal, we were able to determine carbon and hydrogen with satisfactory results and silicon with an accuracy of 0.8% (see Table 3).

The Decomposition of Organic Silicon Compounds in which the Silicon is Bound to Radicals Having Double Bonds

Silicon carbide forms with particular ease upon the thermal decomposition of organic silicon compounds containing radicals with double bonds. Furthermore, the products of the incomplete pyrolysis of these compounds very frequently seem to contain a certain number of volatile silicon compounds. Hence, the determination of carbon presents serious difficulties. These can be overcome, however, by applying the above-mentioned slow combustion technique. When this is done, the silicon can be determined with an accuracy of up to 1% (see Table 4).

CONCLUSIONS

The thermal decomposition of organic silicon componds under conditions of rapid combustion does not result in the formation of silicon carbide among all classes of these compounds. When the decomposition is carried out at a rapid tempo, compounds that contain naphthalene rings or alkoxy groups do not form silicon carbide. On the other hand, silicon carbide usually is obtained under these conditions from tetraalkylsilanes and compounds with unsaturated radicals. Nonetheless, they can be burned completely and without the formation of carbide if the decomposition is effected slowly, under conditions of partial pyrolysis. Under these conditions it is possible to determine not only carbon and hydrogen but also silicon—with an accuracy of up to 1%—at the same time from the same weighed sample.

More than 100 organic silicon compounds of various structures submitted to us by A. D. Petrov, corresponding member of the Academy of Science, U.S.S.R., and his co-workers have been analyzed by this method in the microanalytical laboratories of the Institute of Organic Chemistry, Academy of Science, U.S.S.R. In no case did the deviation from the theoretical data exceed the limits of accuracy that we have been mentioning.

LITERATURE CITED

[1] M. O. Korshun and V. A. Klimova, *Zhur. Anal. Khim.* **2**, 274 (1947); **3**, 176 (1948); **4**, 292 (1949).

[2] V. A. Klimova and M. O. Korshun, *Zhur. Anal. Khim.* **6**, 231 (1951).

CHAPTER 10

Rapid Method for the Determination of Carbon and Hydrogen*

M. O. Korshun

In our previous reports [1] we described a new rapid method for the micro and semimicro determination of carbon and hydrogen and showed that it is possible to achieve quantitative combustion of a substance in a large excess of oxygen without the use of oxidizing agents and catalysts, provided that the weighed sample has been pyrolyzed beforehand with almost no entry of oxygen. When this rapid method was used without pyrolysis, however, quantitative oxidation of organic compounds proved impossible.

Thus, it became clear to us that the procedure used to decompose the substance could play an essential role in the method of determining carbon and hydrogen.

The combustion of the organic substance in an empty tube, at the expense of gaseous oxygen alone, made it possible for us to investigate the role of the decomposition procedure in the course of the oxidation of the substance, i.e., to look more deeply into the decomposition process taking place during the analysis. It was, of course, impossible to determine this role experimentally in work with a tube containing a "universal" filling.

Indeed, the basic concern of analytical chemists working on the micro determination of the elementary composition of organic substances was the quantitative oxidation of the substance by combustion at the expense chiefly of solid

---

**Zhur. Anal. Khim.* 7, 96-100 (1952).

oxidants and catalysts in the presence of oxygen or air, and the question of the role of the decomposition procedure was never thoroughly treated.

Thus, for example, Weygand [2] comes to completely incorrect conclusions in his discussion of the methods for the micro determination of carbon and hydrogen.

In 1948, he wrote that the possibility of burning organic substances in the presence of catalysts "... appears very tempting ...". But he encountered difficulties in this catalytic combustion and hence concludes that "oxidation with the aid of unfailing active chemical agents is always more reliable". Here he proposes the use of the Pregl "universal" combustion tube filling, which consists of CuO, $PbCrO_4$, PbO_2, Ag, and asbestos. Weygand warns, however, that only the silver is completely satisfactory with respect to purity, while the asbestos, cupric oxide, and lead chromate have grave shortcomings, and, further, that completely irreproachable cupric oxide and lead chromate do not exist. Nonetheless, he prefers to effect combustion by the Liebig-Pregl method and describes this method in detail in his book. Meanwhile, it had been shown by our earlier articles [3, 4, 5], that the method of catalytic combustion is more reliable than that of Liebig and Pregl.

The simplest and most reliable method of all, however, is rapid combustion in an empty tube. As to the decomposition procedure, it has, as we were convinced, great importance for any of the existing methods of determining carbon and hydrogen, but this can be demonstrated experimentally only when the substance is burned in an empty tube.

Observation [1] of the course of the decomposition of a substance when rapid pyrolysis was employed provided us with a basis for assuming that the bulk of the volatile thermal decomposition products is oxidized almost completely

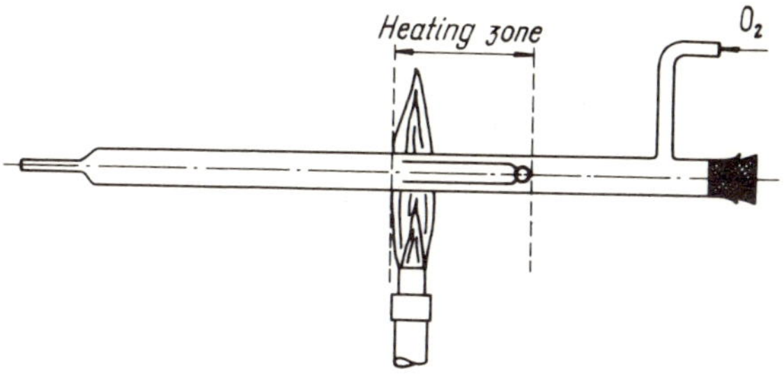

Fig. 1. Combustion Diagram

immediately upon leaving the capsule, in the short, red-hot zone of the combustion tube heated by the movable burner.

Further study of the role of the decomposition procedure under the conditions of our rapid method for the determination of carbon and hydrogen led us to extremely interesting results.

Using the previously described rapid method and setup[1], we conducted experiments to determine carbon and hydrogen in several organic compounds. The sole difference was that in these tests the combustion of the weighed sample was accomplished with only one gas burner, *without electric furnaces.* The burner was not moved outside the limits of the zone designated in the figure by two broken lines and, at all times during the combustion, the heated zone was 2-3 cm. wide, i.e., the width of the gas burner flame. The oxygen was passed through at the rate of 35-40ml/min. and consequently the combustion time did not exceed 10-12 minutes, including sweeping. Such an oxygen rate proved completely adequate for the quantitative removal of water condensing on the cold parts of the combustion tube.

The results obtained surpassed all expectation. It is evident from the examples given in Table 1 (combustion with pyrolysis) that 95 and sometimes 100% of the carbon in the substance being analyzed is oxidized to carbon diox-

TABLE 1

Combustion With Pyrolysis

Name of the substance	Weighed sample in mg.	Found %		Calculated %		ΔC% absolute	ΔH% absolute	ΔC% relative	ΔH% relative
		C	H	C	H				
Naphthalene ($C_{10}H_8$)	5.080	92.05	6.27	93.70	6.30	−1.65	−0.03	−1.77	−0.48
	4.890	92.33	6.17			−1.37	−0.13	−0.47	−2.17
	4.345	93.67	6.23			−0.03	−0.07	−0.04	+1.02
Levulose ($C_6H_{12}O_6$)	5.180	38.11	6.75	39.98	6.67	−1.87	−0.08	−4.68	+1.19
	4.500	35.93	6.77			-4.05	-0.10	-10.14	+11.49
9,10-(2,5-Dimethyl-2-cyclopenten-3-ol-4-one)-1,2,3,4,5,6,7,8,9,10-decahydrophenanthrene ($C_{19}H_{26}O_2$)	6.580	75.96	9.07	79.67	9.15	-3.71	-0.08	-4.66	-0.88
Methylhexahydrophthalic acid ($C_9H_{14}O_4$)	4.922	58.10	7.71	58.06	7.52	+0.04	+0.19	+0.06	+2.52
	4.022	57.76	7.70			-0.50	-0.18	-0.87	+2.39
Picric acid[a] ($C_6H_3O_7N_3$)	4.950	31.12	1.29	31.44	1.32	-0.32	-0.03	-1.02	-2.68
	4.520	30.44	1.39			-1.00	+0.07	-3.19	+5.30

[a] With collection of the nitrogen oxides by means of a solid absorbent prepared according to the specifications of V. A. Klimova (6).

ide, and in all cases the hydrogen is quantitatively converted to water. Meanwhile, when the substance is burned in a boat under the conditions described above, even the hydrogen is not completely oxidized (Table 2).

TABLE 2
Combustion in a Boat

	Naphthalene ($C_{10}H_8$)	Methylhexahydrophthalic acid ($C_9H_{14}O_4$)
Weighed sample in mg.....	4.215	4.330
C:—Found %.............	5.30	7.39
Calculated %.........	93.70	58.06
Δ (absolute) %.	–88.40	–50.66
Δ (relative) %	–94.35	–87.28
H:—Found %.............	0.67	2.62
Calculated %.........	6.30	7.52
Δ (absolute) %	–6.63	–4.90
Δ (relative) %	–89.37	–65.16

Thus, it may be considered that, in decomposition in a boat, oxidation of the substance takes place chiefly in the combustion tube zone heated by the furnace or long burner.

The fact that, in all cases of combustion with pyrolysis, hydrogen is quantitatively transformed to water shows that hydrogen is a primary decomposition product of the substance during pyrolysis of the weighed sample.

The small amount of carbon that is not completely oxidized in combustion with a single burner is detected both qualitatively and quantitatively in the form of carbon monoxide.

The carbon and hydrogen data obtained by burning substance 3 (Table 1)—a condensed ring with an angular

TABLE 3

Name and empirical formula of the substance	% Carbon			% Hydrogen		
	Found	Calculated	Difference	Found	Calculated	Difference
10,14,17-Trimethyl-Δ^9-dodecahydrocyclopentano-	79.82	79.96	−0.14	9.30	9.39	−0.09
phenanthrene-3,15-dione ($C_{20}H_{28}O_2$)	79.71	79.96	−0.25	9.28	9.39	−0.11
Diacetate of 3,5,7,9-tetramethylbicyclopentane-	61.85	61.70	+0.15	6.26	6.33	−0.07
dioxine-$\Delta^{6,11}$-diene-1,2-dione ($C_{18}H_{22}O_7$)	61.63	61.70	−0.07	6.23	6.33	−0.10
Semicarbazone of 9-methyl-6-methoxy-Δ^6-1-octalone	62.10	62.10	0.00	8.51	8.40	+0.11
($C_{13}H_{21}O_2N_3$)	62.38	62.10	+0.18	8.22	8.40	−0.18
10,14,17-Trimethyl-Δ^9-dodecahydrocyclopen-	83.93	83.85	+0.08	10.78	10.80	−0.02
tanophenanthrene-15-one ($C_{20}H_{30}O$)	83.75	83.85	−0.10	10.68	10.80	−0.12
9-Methyl-1-ethynyl-1-decalol-6-one ($C_{13}H_{18}O_2$)	75.91	75.70	+0.21	8.83	8.80	+0.03
	75.88	75.70	+0.18	8.92	8.80	+0.12
2,3-Dimethylbicyclo-(2,2,2)octane ($C_{10}H_{18}$)	87.05	86.88	+0.17	13.28	13.12	+0.16
	86.98	86.88	+0.10	13.08	13.12	−0.04
17-Methyl-$\Delta^{2,16}$-1,4,8-triketotetradecahydrochrysene	76.20	76.48	−0.28	7.53	7.43	+0.10
($C_9H_{22}O_3$)	76.49	76.48	+0.01	7.60	7.43	+0.17
Tri-(α-isopropoxy)ethyl ester of triethanolamine	61.73	61.92	−0.19	11.17	11.06	+0.11
($C_{21}H_{45}O_6N$)	61.88	61.92	−0.04	11.15	11.06	+0.09

TABLE 3—(*continued*)

Name and empirical formula of the substance	% Carbon Found	% Carbon Calculated	% Carbon Difference	% Hydrogen Found	% Hydrogen Calculated	% Hydrogen Difference
Di-(α-isopropoxy)ethyl ester of diethanolamine	60.47	60.61	-0.14	11.26	11.26	0.00
($C_{14}H_{31}O_4N$)	60.41	60.61	-0.20	11.37	11.26	+0.11
1-Methyl-3-ethylcyclobutane (C_7H_{14}), *trans* form, b.p.	85.61	85.63	-0.02	14.39	14.37	+0.02
87.9-88.1°	85.64	85.63	+0.01	14.40	14.37	+0.03
cis form, b.p. 91-91.1°	85.56	85.63	-0.07	14.27	14.37	-0.10
	85.53	85.63	-0.10	14.43	14.37	+0.06
2-Methylpentane (C_6H_{14})	83.78	83.63	+0.15	16.33	16.37	-0.04
b.p. 58.5-59°	83.80	83.63	+0.17	16.44	16.37	+0.07
Pentane (C_5H_{12})	83.12	83.24	-0.12	16.68	16.76	-0.08
b.p. 35-36°	83.61	83.24	+0.37	16.95	16.76	+0.29
Cyclopentane (C_5H_{12})	85.33	85.63	-0.30	14.65	14.37	+0.28
b.p. 40-45°	85.20	85.63	-0.43	14.72	14.37	+0.35
1,3-Dimethylcyclobutane (C_6H_{12})	85.45	85.63	-0.18	14.53	14.37	+0.16
trans form, b.p. 57.4-57.6°	85.42	85.63	-0.21	14.39	14.37	+0.02
3-Methylpentane (C_6H_{14})	83.88	83.63	+0.25	16.36	16.37	-0.01
b.p. 60-64°	83.53	83.63	-0.10	16.35	16.37	-0.02

methyl group—with a single burner are particularly interesting. Even in this difficult case, where the formation of methane might be expected during pyrolysis, we obtained the theoretical values for hydrogen and only 3.7% (absolute) less for carbon.

It is interesting to note that, for the combustion of similar substances, Niederl and Niederl [7] in their recent textbook propose a six-layer (!!) combustion tube filling composed of cupric oxide and platinized asbestos. If combustion of the sample were conducted with pyrolysis, then quantitative oxidation of such substances would undoubtedly be achieved either in the presence of a catalyst alone, or, in the presence of a "universal" Pregl filling alone, even under the conditions of the "classical" methods of analysis.

Many tests in which the combustion of the substance was effected with a *single burner* showed that when an organic substance is pyrolytically decomposed almost complete oxidation is achieved with gaseous oxygen alone during the period that the decomposition products are in the very short zone heated by the gas burner, a residence time of only 2-3 seconds. (When the oxygen rate is 35-40ml./min., the diameter of the combustion tube 9 mm., and the length of the heated zone 20-30 mm., the time of contact with the oxygen is 2-3 seconds.)

All the hydrogen present in the molecule is oxidized to water and 95-100% of the carbon to carbon dioxide. Only up to about 5% of the carbon remains in the form of carbon monoxide.

It is quite obvious that the subsequent oxidation in the zone heated by the electric furnace actually proceeds far more easily than hitherto assumed.

It should be noted that no such method of investigating the decomposition of a substance has been described in the literature. This study was not only of theoretical interest, but also permitted drawing a practical conclusion, namely, that in the rapid determination of carbon and hydrogen large heating zones, i.e., long furnaces and long combustion tubes are unnecessary. Use of a tube 33-38 cm. long instead of 62 cm. and a furnace 8-10 cm. long instead of 25-30 cm. proved satisfactory.

The analytical results of the determination of carbon and hydrogen in substances of different composition and structure are given in Table 3. The combustion of these substances was carried out in a 36 cm. tube with a 10 cm. electric furnace heated to 950°.

In the analysis of the low boiling substances shown in Table 3 we used our previously described [1] combustion technique for volatile substances. The method was changed slightly for the combustion of pentane and cyclopentane. In these cases the weighed sample is taken in a sealed capillary, which is opened in the tube. As usual, the capillary is placed in a capsule sealed end first. The place in the tube where the wide part of the capillary is located is cooled with ice while the sealed end is heated.

After the capillary has been opened, the open end is cooled while the burner is placed at the open end of the capsule in which the capillary is located.

Using this procedure, we were able to obtain satisfactory results for pentane and cyclopentane, which boil within the limits of 34-45°.

CONCLUSIONS

1. It has been shown that the procedure for decomposing a substance during combustion exerts an influence on the ease with which the substance is oxidized. When pyrolytic decomposition is employed, 95-100% oxidation of the substance to CO_2 and H_2O is achieved in the short zone heated by the gas burner, where the substance remains for only a very short time (2-3 sec.).

2. In the rapid method of determining C and H in a considerable excess of oxygen, large heating zones, i.e., long furnaces and long combustion tubes, are unnecessary. Even in the analysis of highly volatile substances, a tube length of 33-38 cm. and a furnace length of 10 cm. are sufficient.

LITERATURE CITED

[1] M. O. Korshun and V. A. Klimova, *Zhur. Anal. Khim.* **2**, 274 (1947); **3**, 176 (1948); **4**, 292 (1949).

[2] C. Weygand, *Metody eksperimenta v organicheskoi khimii* [Experimental Methods in Organic Chemistry], Moscow, 1950, pp. 20, 49, 50.

[3] M. O. Korshun and N. E. Gel'man, *Zavodskaya Lab.* **4**, 114 (1935).

[4] N. E. Gel'man, *Zavodskaya Lab.* **9**, 381 (1940).

[5] M. O. Korshun, *Zavodskaya Lab.* **9**, 615 (1940).

[6] V. A. Klimova and M. O. Korshun, *Zhur. Anal. Khim.* **6**, 230 (1951).

[7] J. Niederl and V. Niederl, *Mikrometody kolichestvennogo organicheskogo analiza* [Micro Methods of Quantitative Organic Analysis]. State Chemical Publishers, Moscow, 1949, p. 101.

CHAPTER 11

Mechanism of the Formation of Silver Sulfate During the Simultaneous Determination of Carbon, Hydrogen, and Sulfur in Organic Compounds*

M. O. Korshun

In 1948 we demonstrated the possibility of applying the principle of rapid combustion in the determination of C and H in an empty tube to the simultaneous determination of C, H, and S from one 4-6 mg. weighed sample[1].

The absorption of all the combustion products outside the tube made it possible to determine the sulfur in the substance simultaneously with the carbon and hydrogen. In this operation the sulfur was collected by metallic silver in a specially constructed tube. It is clearly evident that the quantitative determination of sulfur by this means assumes the conversion of all the "organic" sulfur to sulfate on the silver. The collection of sulfur oxides by means of silver was first undertaken by Dennstedt in his catalytic method of determining carbon and hydrogen.

Not until almost 40 years afterward did the question of the possibility of using the reaction between metallic silver and sulfur oxides for the quantitative determination of sulfur in organic substances start to be discussed in the literature[3]. Later on, several authors used silver in creating a method for the determination of sulfur by the catalytic combustion of the substance in a stream of oxy-

*Zhur. Anal. Khim. 7, 101-103 (1952).

gen. According to this method, when the combustion is over, the silver is removed from the combustion tube and the silver salts obtained are dissolved. Finally, the silver is determined electrolytically from the weight of the silver deposited on the cathode [4], or the percentage content of sulfur is calculated by the loss in weight of the silver gauze after dissolving off the silver salts [5].

Up to now the reaction mechanism by which the quantitative formation of silver sulfate proceeds in the reaction between metallic silver and sulfur oxides has been unknown.

Dennstedt [2] believed that in the catalytic combustion of sulfur-containing organic substances the bulk of the sulfur reacts with the silver to form silver sulfate. This hypothesis has some validity, since it is justified in cases where the substance being tested contains a sufficient amount of hydrogen. Meanwhile, however, silver sulfate is obtained quantitatively even in the analysis of elementary sulfur in a stream of dry oxygen [6].

Other authors [3] studying the mechanism of this reaction consider it more probable that sulfur trioxide reacts directly with the hot silver, forming silver sulfite, which is rapidly oxidized to sulfate in the presence of oxygen.

This hypothesis, too, seems unfounded, since at the temperatures at which silver collects sulfur, the silver sulfate not only is not oxidized to sulfate, but on the contrary, decomposes, liberating SO_2 and losing 30% of its sulfur content [9].

Thus, all the existing hypotheses concerning the mechanism of the reaction between sulfur oxides and silver seem unlikely, especially since they stem from the assumption that the reaction of the formation of Ag_2SO_4 takes place between SO_3 and silver. The formation of Ag_2SO_4 in the reaction between silver and SO_2 is generally not mentioned.

Meanwhile, in the combustion of sulfur-containing organic substances at atmospheric pressure in a stream of oxygen, the predominant formation of SO_2 or SO_3 depends on the temperature, the amount of excess oxygen, and the presence or absence of a catalyst.

According to Knietsch's [7] data on the chemical equilibrium between SO_2 and SO_3 in the presence of oxygen and platinum, 400° is the optimum temperature for a 100% formation of SO_3 according to the equation:

$$2SO_2 + O_2 \rightleftharpoons 2SO_3$$

Not one of the authors cited above [2-6] operated at temperatures below 600°, and at this temperature there is only an 80% formation of SO_3. We work at still higher temperatures and hence the conditions of our method are not the best for the formation of SO_3. At a temperature of 950°, even in the presence of platinum contacts, the equilibrium of the above reaction is strongly displaced to the left, i.e., to the side of the formation of SO_2. Consequently, under our conditions, it is highly probable that sulfur is oxidized to SO_2, which then, as our experimental data have shown, reacts with silver with the quantitative formation of silver sulfate.

Such a treatment of the question refutes the above hypotheses that silver sulfate forms only in the reaction between silver and sulfur trioxide. Meanwhile, we have succeeded in demonstrating just what sulfur oxides enter the silver tube.

In order to solve this problem we undertook a series of experiments with a mercaptan (1-methoxy-5-methylhexane-5-thiol-3-one), assuming that under the conditions of our method its combustion must proceed with a preponderant liberation of SO_2.

These experiments were conducted with collection of the combustion products by 2 ml. of a 0.02 N alkali solution and subsequent iodometric determination of the sulfur dioxide[8].

TABLE 1

The Combustion of the Mercaptan of $C_8H_{11}O_2$ With SO_2 Collection in 0.02 Alkali. Calculated: %S = 18.19

Weighed sample in mg.	Consumption of 0.02 N I_2 in ml.	S in the form of SO_2 in %	Experimental conditions
4.700	2.62	97.85	Combustion with one burner
7.130	3.80	93.43	Combustion with an electric furnace 10 cm. long and heated to 950° C.
6.902	1.55	38.21	Same conditions, but in the presence of a platinum contact.

As is evident from the data presented in Table 1, when this mercaptan is burned with one burner [9], with the electric furnace switched off, 98% of the sulfur contained in the molecule is obtained in the form of SO_2. With a temperature of 950° and a heated zone 10 cm. long, but without a platinum contact, 93.4% of the sulfur is found in the form of SO_2. In the presence of a platinum catalyst under analogous experimental conditions only 38% of the sulfur is found as SO_2.

Thus, we established that the combustion of a mercaptan with a single burner results in the almost quantitative formation of sulfur in the form of SO_2.

Other tests were made with this same mercaptan. Here one burner was used; the electric furnace was switched off. The sulfur dioxide was collected with silver at 700° and the water and carbon dioxide were removed in the usual way. Equally satisfactory results were then obtained for carbon, hydrogen and sulfur (Table 2).

TABLE 2

The Combustion of the Mercaptan of $C_8H_{11}O_2$ With SO_2 Collection on Silver. Calculated: % C=54.50, H=9.16, S=18.19.

Weighed sample in mg.	4.133	4.695
Weight of CO_2 in mg.	8.234	9.333
Weight of H_2O in mg.	3.462	3.880
Weight of SO_4 in mg.	2.218	2.512
Found %: C	54.37	54.25
H	9.37	9.25
S	17.91	17.86
Difference in absolute %: C	−0.13	−0.25
H	+0.22	+0.10
S	−0.28	−0.33

Thus, the following facts have been proved experimentally.

1. There is no necessity of the presence of water for heated silver to react with sulfur oxides to form Ag_2SO_4. Silver sulfate forms quantitatively even when elementary sulfur is burned in a stream of dry oxygen.

2. Ag_2SO_3 is not oxidized to Ag_2SO_4 by a stream of oxygen in the absence of silver; on the contrary, at high temperature (650°) Ag_2SO_3 decomposes with the evolution of about 30% SO_2.

3. Silver heated to 650-800° reacts with the mixture of sulfur oxides and oxygen passing over it with the quantitative formation of Ag_2SO_4.

4. The last reaction proceeds independent of whether it is SO_2, SO_3 or a mixture thereof that reaches the silver.

These facts refute the opinion that the quantitative formation of silver sulfate can take place only in the reaction of silver with sulfur trioxide in a stream of oxygen.

On the basis of these experimental data it seems possible to us to set forth the following hypothesis concerning the mechanism of the absorption of sulfur oxides by silver:

during combustion, metallic silver completely converts any sulfur oxides reaching it into silver sulfate. Thus, it is a unique catalyst, participating in the formation of the end products of the reaction.

CONCLUSIONS

1. It has been established that in the combustion of substances containing C, H, and S, under our experimental conditions, the sulfur separates out as a mixture of SO_2 and SO_3; the ratio thereof is determined by the equilibrium of the reaction

$$2SO_3 \rightleftharpoons 2SO_2 + O_2$$

and is a function of the temperature, amount of oxygen in the mixture, and other experimental conditions.

2. It has been established that, under the conditions of this method, hot silver reacts with SO_2 and SO_3 with the quantitative formation of Ag_2SO_4. The formation of Ag_2SO_4 proceeds independent of whether SO_2, SO_3 or a mixture thereof reaches the silver.

LITERATURE CITED

[1] M. O. Korshun and N. S. Sheveleva, *Doklady Akad. Nauk S.S.S.R.* **60**, 63 (1948).

[2] M. Dennstedt, *Ber.* **30**, 1950 (1897).

[3] W. R. Kirner, *Ind. Eng. Chem., Anal Ed.* **7**, 363 (1935).

[4] E. W. Huffman, *Ind. Eng. Chem., Anal. Ed.* **12**, 53 (1940).

[5] R. Belcher and C. Spooner, *J. Chem. Soc.* **313**, 16 (1943).

[6] G. Stragand and H. Safford, *Anal. Chem.* **21**, 625 (1949).

[7] R. Knietsch, *Ber.* **34**, 4069 (1901).

[8] I. Kolthoff, *Ob'emnyi analiz* [Volumetric Analysis] II, 1932, p. 398.

[9] M. O. Korshun, *Vestnik Akad. Nauk S.S.S.R.* **1949**, No. 6, 111; *Zhur. Anal. Khim.* **7**, 96 (1952).

CHAPTER 12

A New Method for the Simultaneous Determination of Carbon, Hydrogen, and Sulfur in Organic Compounds Composed of Carbon, Hydrogen, Oxygen, Nitrogen, and Sulfur*

M. O. Korshun and N. S. Sheveleva

We have already established [1, 2] that at a temperature of 750° metallic silver quantitatively retains sulfur oxides. Then, as has been proved, the increase in weight of the silver tube is exactly equivalent to the weight of the SO_4, and hence the conversion factor is $S/SO_4 = 0.3338$.

In our previously developed method the accuracy of the results for sulfur determined simultaneously with carbon and hydrogen lay within the limits of –0.5 to –1.5%. As we later explained, the reason that the results obtained for sulfur were low was that small amounts of sulfuric acid condensed at the place in the combustion tube tip where the rubber connection was located. When we changed to a sulfur-collecting tube ground to fit the combustion tube we began to get entirely satisfactory results.

TEMPERATURE CONTROL OF THE SILVER

Literature data concerning the temperature to which the silver is heated in the quantitative collection of sulfur oxides are somewhat contradictory. Some authors believe the optimum temperature is 665°. Huffmann [3] considers it to be 350°, but at the same time says that

Zhur. Anal Khim. 7, 104-111 (1952).

silver becomes more active at 550°. Others collect sulfur oxides with silver at 800° [4].

In our experiments on the determination of sulfur in toluenesulfonamide we collected the sulfur oxides at a silver temperature of 420 to 700°. These tests showed that greatly reduced values for the percentage content of sulfur are obtained in the interval from 420 to 600°. This reduction decreases with an increase in the silver temperature. The data are as follows:

Temperature of the silver in °C	420	500	600	700
Found % of sulfur	6.0	13.0	17.54	18.73
Calculated % of sulfur		18.73		

Further observations showed that 650-700° may be considered thé optimum temperature for the quantitative formation of Ag_2SO_4 under the conditions of rapid combustion in a large excess of oxygen.

We also observed the quantitative formation of Ag_2SO_4 at 800°. It may be necessary to resort to this high temperature when determining C, H, and S without the use of an electric furnace to heat the combustion tube. The possibility of so conducting the simultaneous determination of C, H, and S with heated metallic silver as the only oxidizing zone for the decomposition products of the substance has been demonstrated by experimental data, which have been presented in one of our previously published reports [1]. Nonetheless, it should be noted that at such a high temperature the tube containing the silver corrodes badly and often fails. Consequently, in our work described here we heated the oxidizing zone of the combustion tube with an electric furnace and operated at a silver temperature no higher than 700°.

Temperature Control of the Oxidizing Zone With and Without a Platinum Catalyst Present

We were not interested in making any fundamental changes in our previously developed method of determining carbon and hydrogen in which the quantitative oxidation of organic substances composed of C, H, O, and N proceeds without platinum contacts and exclusively at the expense of a large excess of gaseous oxygen heated to 950°. Instead, we wished to conduct the simultaneous determination of carbon, hydrogen, and sulfur under the same conditions.

TABLE 1

The Determination of Sulfur in Toluenesulfonamide at Various Silver Temperatures. Calculated: % S = 18.73

Silver temperature in °C.	Weighed sample in mg.	Weight of SO_4 in mg.	Found S absolute %	Experimental conditions
420	4.730	0.465	3.28	The combustion was effected in a stream of O_2 passing at the rate of 35-40 ml./min. Length of the heated zone of the combustion tube, 10 cm. Temperature, 950°C.
420	3.640	0.635	5.82	
420	8.860	2.110	11.96	
520	6.070	1.695	9.32	
540	4.905	2.430	16.54	
600	4.102	2.155	17.54	
620	4.580	2.550	18.58	
640	5.450	3.082	18.88	
700	3.636	2.060	18.91	
700	4.522	2.538	18.73	
740	5.359	2.995	18.65	

Substances that contained sulfur in reduced form — sulfides and mercaptans — interested us particularly. It seemed logical that under our analytical conditions the oxidation of divalent sulfur should proceed with more difficulty than that of sulfur in the oxidized form. Tests carried out with ditolyl disulfide, thianthrene, a mercaptan (1-methoxy-5-methylhexane-5-thiol-3-one), and dithiane

TABLE 2

The Simultaneous Determination of C, H, and S With and Without Platinum Contacts

Name and empirical formula of the substance	Platinum contact	Weighed sample in mg.	Weight in mg. CO_2	Weight in mg. H_2O	Weight in mg. SO_4	Found % C	Found % H	Found % S	Difference % C	Difference % H	Difference % S
Thianthrene ($C_{12}H_8S_2$)	No	5.170	12.620	1.760	4.610	66.62	3.81	29.76	0.00	+0.08	+0.11
Calculated %:	No	6.105	14.875	2.030	5.425	66.49	3.72	29.66	-0.13	-0.01	-0.01
C 66.62; H 3.73; S 29.65;	Yes	4.940	12.110	1.700	4.390	66.90	3.35	29.66	+0.28	+0.12	+0.01
The mercaptan, 1-methoxy-	Yes	4.900	9.800	3.990	2.652	54.58	9.11	18.07	+0.08	-0.05	-0.12
5-methylhexane-5-thiol-	No	4.366	8.690	3.580	2.370	54.32	9.18	18.12	-0.18	+0.02	-0.07
3-one ($C_8H_{16}O_2S$) Calc. %:											
C 54.50; H 9.16; S 18.19;											
2,5-Dimethyltetrahydro-1-	Yes	4.348	9.208	3.028	2.205	57.80	7.79	16.93	-0.23	+0.21	-0.25
thio-4-pyranacetic acid	No	4.957	10.550	3.345	2.530	58.08	7.55	17.04	+0.05	-0.03	-0.14
($C_9H_{14}O_2S$) Calculated %:											
C 58.03; H 7.58; S 17.18;											
Dithiane ($C_4H_8S_2$)	Yes	4.452	6.545	2.605	7.122	40.12	6.55	53.40	+0.06	-0.15	+0.26
Calculated %:	Yes	5.290	7.788	3.253	8.410	40.18	6.88	53.06	-0.18	+0.18	-0.28
C 37.97; H 6.70; S 53.34;	No	5.460	8.030	3.320	8.720	40.13	6.80	53.31	-0.03	+0.10	-0.03
	No	4.605	6.790	2.800	7.385	40.24	6.80	53.53	+0.19	+0.10	+0.15
Ditolyl disulfide ($C_{14}H_{14}S_2$)	Yes	4.860	12.135	2.585	3.780	68.14	5.95	25.96	-0.10	+0.23	-0.07
Calculated %:	No	5.470	13.645	2.775	4.270	68.07	5.68	26.05	-0.17	+0.06	+0.02
C 68.24; H 5.73; S 26.03											

showed (Table 2) equally good results for carbon, hydrogen, and sulfur in the presence and in the absence of a platinum catalyst.

In this way, we established that with a temperature of 950° in the oxidizing zone the analysis may be conducted without platinum contacts. Table 3 presents the results that we obtained in a large number of simultaneous determinations of sulfur, carbon, and hydrogen in different organic substances.

It is evident from these examples that equally accurate results are obtained for sulfides, disulfides, mercaptans, sulfones, and other substances, independent of the percentage content of sulfur.

APPARATUS

As has already been mentioned, the determination of sulfur could be performed with an accuracy of -0.5 to -1.5% by our earlier method. These low results for sulfur were due to the condensation of small amounts of sulfuric acid at the place in the combustion tube tip where the rubber connection was located [2].

Therefore, it was necessary to construct a quartz combustion tube connected to the silver tube by means of a ground joint.

Fig. 1 shows our redesigned combustion tube with the ground-joint absorption tube for the collection of sulfur oxides by silver.

Both the combustion and the absorption tube are made of transparent quartz.

This type of ground part is sufficiently rugged for the work and provides complete gas-tightness under heating. Furthermore, such a combustion tube is universal, since, when there is no sulfur in the substance being tested, the water absorption tube may be connected directly by rub-

TABLE 3

Name and empirical formula of the substance	Weighed sample in mg	Weight in mg.			Found %			Difference in %		
		CO_2	H_2O	SO_4	C	H	S	C	H	S
7,13,15-Trimethyl-6-thiaster-9(11)-ene-	5.290	14.520	4.470	1.670	74.91	9.45	10.54	-0.04	+0.18	+0.02
17-one ($C_{19}H_{28}OS$)	5.130	14.060	4.210	1.585	74.80	9.18	10.31	-0.15	-0.09	-0.21
Calc. %: C 74.95, H 9.27, S 10.52										
α-Naphthyldecyl sulfide ($C_{20}H_{28}S$)	4.800	14.060	4.090	1.535	79.94	9.54	10.68	-0.06	+0.20	+0.02
Calc. %: C 80.00, H 9.34, S 10.66	4.980	14.585	4.240	1.595	79.92	9.53	10.69	-0.08	+0.19	+0.03
Cyclohexyldecyl sulfide ($C_{16}H_{32}S$)	6.420	17.090	7.340	2.450	75.20	12.70	12.74	+0.18	+0.22	+0.24
Calc. %: C 74.92, H 12.57, S 12.50	5.990	-	-	2.270	-	-	12.65	-	-	+0.15
2-Methylhexahydro-1-thiochroman-4-	3.915	7.856	2.846	1.542	54.76	8.13	13.15	+0.02	+0.20	-0.13
one semicarbazone ($C_{11}H_{19}ON_3S$)	3.820	7.648	2.710	1.520	54.64	7.94	13.28	-0.10	+0.01	0.00
Calc. %: C 54.74, H 7.93 S 13.28										
2,2′-Diisobutoxydiethyl sulfide	6.833	15.392	6.830	2.790	61.48	11.18	13.63	-0.01	0.00	-0.05
($C_{12}H_{26}O_2S$)	5.070	11.424	5.074	2.059	61.49	11.20	13.55	0.00	+0.02	-0.13
Calc. %: C 61.49, H 11.18, S 13.68										
2,5-Dimethyltetrahydro-1-thiopyran-4-	5.438	11.362	3.070	2.460	57.02	6.32	15.10	+0.17	+0.12	-0.04
cyanoacetic acid ($C_{10}H_{13}O_2NS$)	5.139	10.747	2.938	2.373	57.07	6.40	15.42	+0.24	+0.20	+0.28
Calc. %: C 56.85, H 6.20, S 15.14										

2,5-Dimethyltetrahydrothiopyran-4-	4.310	8.400	3.327	2.020	53.19	8.64	15.65	+0.03	+0.21	−0.12
hydroxy-4-acetamide ($C_9H_{17}O_2NS$)	3.960	7.740	3.028	1.850	53·34	8.56	15.60	+0.18	+0.12	−0.17
Calc. %: C 53.16, H 8.43, S 15.77										
Phenylcyclopentyl mercaptan ($C_{11}H_{14}S$)	6.090	16.577	4.332	3.328	74.28	7.96	18.24	+0.18	+0.04	+0.26
Calc. %: C 74.10, H 7.92, S 17.98	5.297	14.440	3.740	2.840	74.40	7.90	17.90	+0.30	−0.02	−0.08
α-Thionaphthol ($C_{10}H_8S$)	4.200	11.540	1.920	2.525	74.98	5.12	20.07	+0.02	+0.09	+0.06
Calc. %: C 74.96, H 5.03, S 20.01	4.320	11.860	1.980	2.600	74.92	5.13	20.08	−0.04	+0.10	+0.08
2-Methyl-2,3-dihydrothiopyran dioxide	6.700	11.031	3.035	3.995	44.93	5.07	19.90	−0.05	+0.04	−0.12
($C_6H_8O_3S$) Calc. %: C 44.98, H 5.03, S 20.02	5.598	9.185	2.506	3.345	44.78	5.01	19.95	−0.20	−0.02	−0.07
Bis-β-butoxyethylmercapto-1,1-ethane	4.575	9.550	4.217	3.000	56.84	10.30	21.84	−0.25	+0.03	+0.07
($C_{14}H_{30}O_2S$)	7.911	15.440	6.827	4.773	57.01	10.34	21.55	−0.08	+0.07	−0.22
Calc. %: C 57.09, H 10.27, S 21.77										
Thioxane (C_4H_8OS)	4.643	7.850	3.250	4.250	46.14	7.83	30.55	+0.02	+0.09	−0.23
Calc. %: C 46.12, H 7.74, S 30.78	4.986	8.442	3.460	4.562	46.20	7.76	30.54	+0.08	+0.02	−0.24
Vinylethyl sulfide (C_4H_8S)	4.447	8.845	3.624	4.830	54.30	9.14	36.25	−0.19	0.00	−0.12
Calc. %: C 54.49, H 9.14, S 36.37	4.000	8.770	3.550	4.775	54.39	9.03	36.30	−0.10	−0.11	−0.07
Butane-1,3-dithiol ($C_4H_{10}S_2$)	7.190	10.360	5.295	11.300	39.32	8.24	52.46	+0.02	0.00	0.00
Calc. %: C 39.30, H 8.24, S 52.46	4.420	6.380	3.260	6.920	39.39	8.25	52.30	+0.09	+0.01	−0.16
1,2-Dithialan ($C_3H_6S_2$)	3.918	4.005	1.950	7.064	34.16	5.57	60.18	+0.23	−0.12	−0.20
Calc. %: C 33.93, H 5.69, S 60.38	7.655	9.600	3.960	13.500	34.22	5.79	60.23	+0.29	+0.10	−0.15

ber tubing to the exit tube projecting from the ground joint of the combustion tube.

This apparatus has been used successfully in our laboratory for over two years and all the present analyses were carried out on it. As we established earlier, the length of the combustion tube may be reduced to 365 mm. and the length of the heated zone to 100 mm. [5].

The set-up for the simultaneous determination of carbon, hydrogen, and sulfur is diagrammed in Fig. 2.

Sulfur absorption tube. This tube (Fig. 1) is so constructed that it can be weighed accurately on a microbalance. At one end it is equipped with an exit tube 1 mm. in inside diameter, at the other, with a ground joint by which it is connected to the combustion tube. During weighing, the tube is closed with a hollow, ground stopper. Such a tube construction decreases the diffusion of outside air during this period. Like the carbon dioxide and water absorption tubes, this tube is weighed full of oxygen.

The tube is filled with metallic silver through the wide opening. It is desirable to use silver in the form of thin

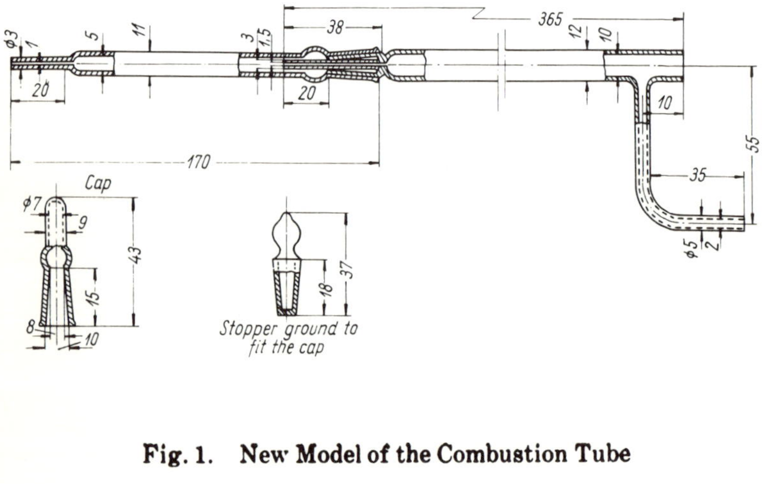

Fig. 1. New Model of the Combustion Tube

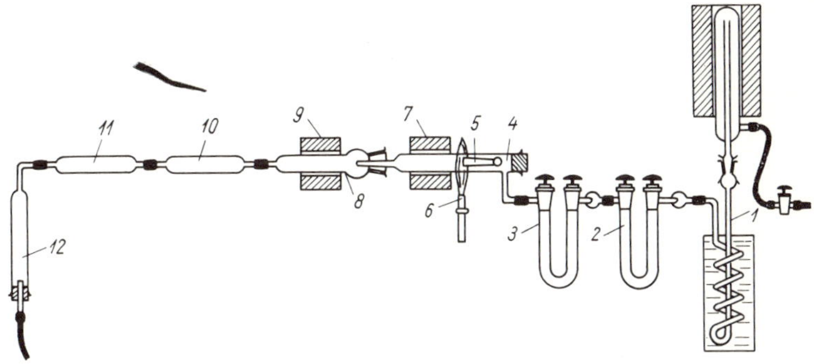

Fig. 2. Set-up for the Simultaneous Determination of Carbon, Hydrogen, and Sulfur. 1—gas scrubber with electric furnace; 2—U-tube with Ascarite; 3—U-tube with Anhydrone; 4—quartz combustion tube; 5—quartz capsule for the weighed sample; 6—burner; 7 and 9—electric furnaces; 8—sulfur oxides absorption tube; 10—water absorption tube; 11—carbon dioxide absorption tube; 12—guard tube with Anhydrone.

ribbon, turnings, or wool, since in this form it has a large surface for a relatively small weight.†

The weight of ribbon silver used to fill the tube is generally from 2.5 to 3.5 g. This amount is enough for 25-30 sulfur determinations. In this number of determinations less than half the layer of silver is exhausted and has to be replaced.

Carrying Out the Analysis

Before work is begun but after the electric furnaces (1, 7 and 9) (Fig. 2) have heated to the requisite temperature, the silver tube is connected by its ground joint to the com-

† We have found that it is expedient to reduce the silver with methyl alcohol before use in order to increase its activity. The catalytic activity of silver reduced with hydrogen is considerably lower; furthermore, when reduced in this manner, silver loses mechanical strength.

bustion tube, and the short 8-10 cm. electric furnace, heated to 700°, is drawn over it. The water and carbon dioxide absorption tubes are connected behind it and a blank test is run. After the blank test, all three tubes are weighed, the silver tube last. Then the tubes are again connected to the combustion tube, the capsule with the weighed sample is introduced into the combustion tube, and the combustion is effected as previously described in the account of the method for the determination of carbon and hydrogen [6].

If halogen and sulfur are both present in the substance, it is not possible to determine each of these elements separately by the above-described method. The presence of nitrogen does not hinder the simultaneous determination of carbon, hydrogen, and sulfur. [Ed. Note: Presumably the reader should infer the insertion of the nitrogen oxides absorber in the train shown in Fig. 2 if nitrogen-containing compounds are going to be analyzed.]

I consider it a pleasant duty to express my profound thanks to I. N. Tits-Skvortsova, lecturer of the chair of petroleum, faculty of chemistry, Moscow State University, for generously providing me with a number of very pure substances for the confirmation of this method.

CONCLUSIONS

1. A new gravimetric method has been developed for the simultaneous determination of carbon, hydrogen, and sulfur from one 4-7 mg. weighed sample. The accuracy of the determination is ±0.2-0.3% for all three elements.

2. The method consists in decomposing a weighed sample as usual in an empty combustion tube in a rapid stream of oxygen (35-40 ml./min.), and in subsequently collecting the oxidation products by means of suitable absorbents. Sulfur oxides are absorbed by metallic silver

in a special quartz tube while carbon dioxide and water are collected, as usual, by Ascarite and Anhydrone.

3. The optimum temperatures of the silver layer for the quantitative absorption of sulfur oxides were found.

These temperatures lie within the limits of 650-800°.

LITERATURE CITED

[1] M. O. Korshun, *Zhur. Anal. Khim.* **7**, 101 (1952).

[2] M. O. Korshun and N. S. Sheveleva, *Doklady Akad. Nauk S.S.S.R.* **60**, 63 (1948).

[3] E. W. Huffman, *Ind. Eng. Chem., Anal. Ed.* **12**, 53 (1940).

[4] R. Belcher and C. Spooner, *J. Chem. Soc.* **313**, 16 (1943).

[5] M. O. Korshun, *Zhur. Anal. Khim.* **7**, 96 (1952).

[6] M. O. Korshun and N. E. Gel'man, *Novye Metody Elementarnogo Mikroanaliza* [New Methods of Elementary Microanalysis], Moscow-Leningrad, p. 49.

CHAPTER 13

New Method for the Simultaneous Microdetermination of Fluorine, Hydrogen, and Carbon in Organic Compounds*

N. E. Gel'man and M. O. Korshun

The elementary analysis of organic substances containing fluorine presents difficulties both because of the high stability of such compounds and because of the ability of fluorine to react with quartz and glass, the materials from which most ordinary chemical equipment is made.

Although there have been a large number of studies devoted to the determination of fluorine itself, only three articles have been published on the simultaneous macro [1] and semimicro determination [2, 3] of several elements (C, H, and F; C, Cl, and F; C and F) in organic fluorides. Authors who worked on this problem burned the substance in a stream of oxygen, thus trying to remove fluorine from the sphere of combustion in the form of a gaseous compound — hydrogen fluoride or silicon tetrafluoride — and to retain it by suitable absorbents in a special tube outside the combustion tube.

The aim of our work was to create a method permitting the determination of three elements — fluorine, hydrogen, and carbon — from one weighed sample.

We considered the separation of fluorine as hydrogen fluoride unacceptable because it necessitated the use of platinum equipment. We also abandoned the conversion of fluorine to silicon tetrafluoride since, after a large num-

Doklady Akad. Nauk S.S.S.R.* **86, 685-7 (1953).

ber of tests, we came to the conclusion that the quantitative separation of fluorine in this form is impracticable in the presence of hydrogen.

Thus we were unable to use either of the two previously proposed methods for determining fluorine simultaneously with other elements.

Continuing our investigation, we reached a fundamentally new solution to the problem, one that permits determining not only carbon but also hydrogen simultaneously with fluorine.

The principle of the micro method that we have developed for the simultaneous determination of fluorine, hydrogen, and carbon consists in the following: the substance is burned in a stream of oxygen or air in a quartz tube, the carbon and hydrogen are determined as usual [4], and, at the moment the weighed sample decomposes, the fluorine is combined with a metal oxide to form a high-melting fluoride. This takes place in the combustion tube at the point where the substance being analyzed burns, i.e., in the quartz weighing capsule. The amount of fluorine in the weighed sample is determined by the change in weight of the capsule.

This principle of determining fluorine simultaneously with hydrogen and carbon was first applied to the analysis of fluorides. The possibility of determining hydrogen simultaneously with fluorine and carbon is also new. This is of interest not only for the analysis of substances containing constitutional hydrogen but also in those cases where it is necessary to determine residual hydrogen in perfluorides. It should be emphasized that, in this analytical method, when the fluorine is converted to a nonvolatile metal fluoride at the moment it separates from the decomposing substance, all errors produced by the reaction of fluorine with quartz and glass and, consequently, all errors associated with the formation and hydrolysis of silicon tetrafluoride are eliminated.

In selecting a substance to absorb fluorine we took into account that the absorbent should be located in the quartz capsule together with the weighed sample where, during combustion, it will undergo heating to temperatures of approximately 1000°. Consequently, the absorbent must maintain a constant weight at this temperature in the presence of carbon and water vapor in an oxygen atmosphere and, at the same time, must not retain carbon dioxide or water. In addition, the fluoride formed during combustion must be high-melting and nonvolatile. An examination of the physical properties of inorganic fluorides and metal oxides showed that magnesium oxide, which melts at 2800° and boils at 3600°, best meets these requirements. The melting and boiling points of magnesium fluoride are also extremely high, being 1225° and 2260° respectively, whereas magnesium carbonate begins to decompose as low as 300°. Magnesium oxide is readily obtained in pure form, quickly reaches constant weight upon heating, and manifests good weight constancy when substances that do not contain fluorine are burned in it presence (see Table 1).

TABLE 1

Combustion of Saccharose in the Presence of Magnesium Oxide

C - 42.12 %		H - 6.48 %		Change in weight of the MgO in mg.
Found	Difference	Found	Difference	
42.01	-0.11	6.42	-0.06	+0.016
42.13	+0.01	6.76	+0.28	-0.003
42.34	+0.22	6.66	+0.18	-0.004
42.33	+0.21	6.73	+0.25	-0.026
42.18	+0.06	6.51	+0.03	-0.006

TABLE 2
Determination of F, H, and C from One Sample

	% F Found	% F Diff.	% H Found	% H Diff.	% C Found	% C Diff.	Total of C, H, and F %
Polyfluorostyrene $(C_8H_7F)_n$	15.06	-0.50	6.06	+0.28	78.75	+0.09	99.87
Calculated %: F=15.56	14.90	-0.66	5.87	+0.09	78.36	-0.30	99.13
H = 5.78	15.76	+0.20	5.91	+0.13	78.75	+0.09	100.42
C=78.66	15.70	+0.14	5.74	-0.04	78.64	-0.02	100.08
	15.85	+0.29	5.76	-0.02	78.79	+0.13	100.40
	15.58	+0.02	5.68	-0.10	78.80	+0.14	100.06
	15.78	+0.22	5.84	+0.06	78.36	-0.30	99.98
	15.68	+0.12	5.82	+0.04	78.80	+0.14	100.30
p-Fluorobenzophenone	9.64	+0.15	4.73	+0.20	78.08	+0.10	-
$C_{13}H_9OF$	9.20	-0.29	4.61	+0.08	77.85	-0.13	-
Calculated %: F = 9.49	9.19	-0.30	4.49	-0.04	77.62	-0.36	-
H = 4.53	8.76	-0.73	4.59	+0.06	77.79	-0.19	-
C=77.98	9.94	+0.45	4.68	+0.15	78.15	+0.14	-
2-Fluoro-5-nitrotoluene	11.87	-0.38	3.83	-0.07	53.96	-0.23	-
$C_7H_6O_2NF$ Calc. %:	12.37	+0.12	3.92	+0.02	54.12	-0.07	-
F=12.25 H=3.90	12.68	+0.43	4.12	+0.22	54.15	-0.04	-
C=54.19	11.77	-0.48	3.90	0.00	54.33	+0.14	-

The absorption of fluorine proceeds according to the equation:

$$MgO + F_2 \rightarrow MgF_2 + O$$

The analysis differs from the usual determination of carbon and hydrogen only in that, before the combustion, 150-200 mg. of freshly dried magnesium oxide is poured into the quartz capsule on top of the weighed sample, and the capsule, closed with a ground glass stopper, is weighed on a microbalance before and after the combustion.

This method is suitable for the analysis of both solids and liquids. It has been tested on substances containing C, H, O, N, and F with weighed samples of 4.5 to 9 mg. The figures obtained in determinations of the three elements from one weighed sample are given in Table 2.

LITERATURE CITED

[1] N. S. Nikolaev, *Izvest. Akad. Nauk S.S.S.R., Otdel. Khim. Nauk*, **1945**, 309; *Zavodskaya Lab.* **1947**, 807.

[2] R. Teston and F. E. McKenna, *Anal. Chem.* **19**, 193 (1947).

[3] O. J. Milner, *Anal. Chem.* **22**, 315 (1950).

[4] M. O. Korshun and V. A. Klimova, *Zhur. Anal. Khim.* **2**, 274 (1947); **4**, 292 (1949); M. O. Korshun and N. E. Gel'man, *Novye Metody Elementarnogo Mikroanaliza* [New Methods of Elementary Microanalysis] Moscow, 1949.

CHAPTER 14

The Simultaneous Determination of Carbon, Hydrogen, and Silicon by Combustion in the Presence of a Catalyst*

V. A. Klimova, M. O. Korshun, and E. G. Bereznitskaya

We have shown [1] that when organic silicon compounds of the aliphatic series are thermally decomposed a certain amount of silicon carbide forms during the rapid combustion. The combustion of such compounds can be effected only if preliminary pyrolysis is omitted. When this is done, however, the analytical figures obtained for silicon are low and the combustion time is prolonged to 45-50 minutes.

The search for conditions under which the principles of rapid pyrolysis of a substance could be preserved while the formation of silicon carbide could be avoided led us to an attempt to add a catalyst to the weighed sample (i.e. in the capsule). First of all we tested a series of catalysts—platinum, cupric oxide, vanadium pentoxide, and chromium sesquioxide. We found that platinum placed in the capsule in the form of net or foil did not prevent the formation of silicon carbide. Work with cupric oxide also gave no positive results. With vanadium pentoxide and chromium sesquioxide†, on the other hand, we obtained good

---

Doklady Akad. Nauk S.S.S.R.* **96, 81-83 (1954).

† P. N. Fedoseev and his colleague used chromium sesquioxide successfully as a catalyst for surface burning in elementary microanalysis [2].

results for carbon and hence concluded that silicon carbide did not form in their presence. Even so, we had to fill at least two-thirds of the capsule with catalyst. With such a large amount of catalyst it is impossible to avoid some dispersion during combustion and consequently it cannot be considered that the silicon is determinable simultaneously with the carbon and hydrogen.

Furthermore, in selecting a catalyst we had to take into account that silicon dioxide forms in finely dispersed state during the decomposition of organic silicon compounds. Hence it was expedient to use a catalyst having a fibrous structure since such a catalyst would easily retain fumes or dust. Proceeding on this theory we used asbestos as the carrier for the catalyst. Inasmuch as vanadium pentoxide and chromium sesquioxide had already given a somewhat positive result we made up two catalysts — vanadium pentoxide on asbestos and chromium sesquioxide on asbestos. From 50-200 mg. of these catalysts proved sufficient to give a fluffy bulk that fills most of the capsule. This makes it possible to heat the catalyst before the substance being analyzed begins to vaporize. Thus, during combustion, vapors of the substance will strike the already hot layer of catalyst and decompose on its surface.

No silicon carbide forms in this decomposition procedure and combustion can be effected with complete preliminary pyrolysis by means of the roaring flame of a gas burner. Meanwhile the asbestos present in the catalyst mechanically retains the finely dispersed silicon dioxide formed during combustion and the analytical results for silicon are thereby greatly improved.

Each catalyst can be considered entirely suitable. Chromium sesquioxide on asbestos has a certain advantage over vanadium pentoxide on asbestos because chromium sesquioxide is completely nonvolatile at high temperatures.

Using these catalysts, we obtained better results in the combustion of compounds with aliphatic radicals and succeeded in performing several analyses on compounds having conjugate double and triple bonds. We also obtained satisfactory results in the analysis of substances boiling at 130, 100, and even 60°. The determination of silicon in such compounds is completely impossible without a catalyst. It should also be noted that it is preferable to use small weighed samples (3-4 mg.) in the analysis of such compounds.

The attempt to use a fibrous catalyst for the analysis of organic silicon compounds thus proved completely justified.

APPARATUS

The apparatus is the same as that for the rapid determination of carbon and hydrogen [3]. The capsule for the weighed sample must have a ground stopper and must be large; viz. 75-80 mm. long and 5.5-6.5 mm. in inside diameter.

CARRYING OUT THE ANALYSIS

A 4-12 mg. sample of the substance being analyzed is weighed out in a quartz capsule. 50-200 mg. of catalyst is added, depending on the size of the sample. The catalyst must fill 4/5ths of the capsule and must lie in loose, porous layers. The capsule is placed in the combustion tube and combustion is carried out as is customary in the rapid C and H determination [3].

When the substance is volatile, the sample is weighed out into a capillary. The capsule containing the catalyst is weighed and the capillary is placed in the capsule and broken there. Combustion is then carried out as usual.

During combustion the vapors of the substance strike the hot layer of catalyst and decompose thereon without forming silicon carbide.

After the combustion is over, the tubes for the absorption of water and carbon dioxide and the capsule containing the residual silicon dioxide are weighed. The silicon content is calculated from the increase in weight due to the silicon dioxide.

TABLE 1

	% C		% H		% Si	
	Found	Calc.	Found	Calc.	Found	Calc.
Dimethylphenylvinyl-ethynylsilane	77.22 77.50	77.42	7.23 7.38	7.53	14.78 14.80	15.05
Dimethylbenzylvinyl-ethynylsilane	78.15 78.25	78.00	7.95 8.04	8.00	14 01 14.03	14.00
Dipropylbis(vinylethy-nyl)silane	78.03 77.70	77.77	9.13 9.11	9.26	12.77 13.10	12.97
Dimethyldiethylsilane b.p. 95°	61.97 61.91	62.07	14.09 13.98	13.79	23.85 24.44	24.19
Trimethylethylsilane b.p. 60°	58.92 58.86	58.82	14.08 13.82	13.73	26.60 26.54	27.51
1,2-Bis(triethylsilyl) acetylene	66.22 66.23	66.15	11.87 12.09	11.81	22.24 21.89	22.04
Naphthyldiphenyl-benzylsilane	87.04 87.15	86.95	5.92 5,96	6.04	7.42 7.22	7.01

Table 1 shows the results that we obtained when we analyzed organic silicon compounds in the presence of chromium sesquioxide on asbestos. We determined all three elements in these compounds — carbon, hydrogen, and silicon. As is evident from the data, the results for carbon and hydrogen lie within the limits of ordinary accuracy, while for silicon the accuracy is 0.2-0.4%.

LITERATURE CITED

[1] V. A. Klimova, M. O. Korshun, and E. G. Bereznitskaya, *Doklady Akad. Nauk S.S.S.R.* 84, 1175 (1952).

[2] P. N. Fedoseev and M. M. Pavlenko, *Zhur. Anal. Khim.* 5, 296 (1950); 6, 317 (1951).

[3] M. O. Korshun and V. A. Klimova, *Zhur. Anal. Khim.* 2, 274 (1947).

CHAPTER 15

The Simultaneous Determination of Carbon, Hydrogen, and Phosphorous in Organic Phosphorus Compounds*

V. A. Klimova, M. O. Korshun, and E. G. Bereznitskaya

The analysis of organic phosphorus compounds is extremely difficult with respect to the determination not only of phosphorus but also of carbon and hydrogen.

The length of the analysis (2-3 days), difficulties arising in the oxidation of many organic phosphorus compounds, especially those in which the phosphorus is bound to a carbon, and complications accompanying the precipitation of phosphoric acid are characteristic of the existing methods of determining phosphorus. The determination of carbon and hydrogen in organic phosphorus compounds, even by the rapid method [1, 2], does not always lead to the desired results.

The reason for these difficulties is that, during combustion in a stream of oxygen, carbonization of the substance is accompanied by the formation of phosphoric anhydride, which coats the carbon with a dense film and prevents its complete combustion. In many cases it is impossible to burn the substance entirely even when it is heated for a long time. This phenomenon was also observed by Dennstedt [3], who proposed interrupting combustion after the bulk of the substance had carbonized, dissolving the phosphoric anhydride, and then burning up the freed carbon.

Doklady Akad. Nauk S.S.S.R.* **96, 287-288 (1954).

This procedure cannot guarantee great accuracy and furthermore it is extremely time-consuming and difficult.

It should be noted that the phosphoric anhydride formed during combustion impedes the determination not only of carbon but also of hydrogen. At high temperatures phosphoric anhydride is rather volatile. Consequently part of it is carried away by the oxygen stream. The P_2O_5 is then precipitated either on the cold parts of the combustion tube or in the water-absorption tube. In the first case low results are obtained for hydrogen; in the second, high. All this leads to entirely indefinite analytical figures and necessitates the search for a better solution of this analytical problem.

For this purpose we made use of a certain analogy in the character of the combustion of organic silicon and phosphorus compounds. It should be emphasized that this analogy is of a formal nature since phosphorus, in distinction to silicon, does not form carbide but only mechanically hinders combustion of the carbon. It is necessary in both cases, however, to isolate the anhydride formed from the carbon and to prevent its vaporization from the combustion capsule.

As mentioned in a previous work [4], we succeeded in achieving good analytical results for organic silicon compounds by using chromium sesquioxide or vanadium pentoxide on asbestos as the catalyst. We place the catalyst directly in the capsule containing the weighed sample. In this case the combustion tube remains uncontaminated.

Just as we expected, the catalysts mentioned proved usable in the analysis of organic phosphorus compounds. We found that when organic phosphorus compounds are burned in the presence of an asbestos-supported catalyst the formation of a phosphoric anhydride coating on the carbon is completely eliminated, since the phosphoric anhydride is absorbed by the catalyst and conditions favorable to the determination of carbon and hydrogen are thereby created. Furthermore, phosphoric anhydride is retained by the catalyst and may be determined gravimetrically in the form of P_2O_5.

For the time being it remains unclear whether the asbestos and phosphoric anhydride combine chemically or whether the collection of P_2O_5 fumes is a mechanical process. Since phosphorus pentoxide is retained in the capsule despite the high combustion temperature (900°), it is more likely that a chemical reaction takes place between the phosphoric anhydride and the silicon oxide in the asbestos and results in the formation of a thermally stable compound. The use of vanadium pentoxide and chromium sesquioxide may prove unnecessary.

TABLE 1

The Simultaneous Determination of C, H, and P.

Substance	% C		% H		% P	
	Found	Calc.	Found	Calc	Found	Calc.
Ethyl α-amino-sec.-butylphosphonate	46.20	45.93	9.68	9.57	14.88	14.83
	46.23	45.93	9.66	9.57	14.50	14.83
Ethyl α-hydroxybenzyl-phosphonate	53.94	54.10	7.19	7.00	12.36	12.70
	54.33	54.10	7.23	7.00	12.50	12.70
Ethyl α-aminoisopropyl-phosphonate	43.10	43.10	9.46	9.20	15.87	15.90
	43.00	43.10	9.38	9.20	16.15	15.90
Diethyamyl α-amino-sec.-amylphosphonate	48.43	48.43	10.19	9.87	13.99	13.90
	48.19	48.43	10.19	9.87	13.93	13.90
α-Amino-sec.-amyl-phosphonate	32.40	32.40	9.14	8.70	16.80	16.75
	32.75	32.40	9.13	8.70	16.95	16.75
Triphenylphosphine	82.16	82.44	5.72	5.72	11.61	11.83
	82.71	82.44	5.98	5.72	11.90	11.83
	82.24	82.44	5.74	5.72	11.60	11.83

In our current work we are pursuing a study on the use of other chemical agents for the quantitative retention of phosphoric anhydride.

Table 1 presents the results obtained in the analysis of organic phosphorus compounds with a catalyst (chromium sesquioxide on asbestos). An attempt was made to determine all three elements in these compounds: carbon, hydrogen, and phosphorus.

It can be seen from the data given that the analytical results for carbon and hydrogen lie within the limits of ordinary error; but for phosphorus the accuracy is somewhat lower. It may be supposed that in addition to the method for the simultaneous determination of C, H, and P, the future will also see the development of a simple method that will replace today's troublesome methods for determining phosphorus in organic phosphorus compounds.

LITERATURE CITED

[1] M. O. Korshun and V. A. Klimova, Zhur. *Anal. Khim.* **2**, 274 (1947).

[2] M. O. Korshun and N. E. Gel'man, *Novye Metody Elementarnogo Mikroanaliza* [New Methods of Elementary Microanalysis] Moscow-Leningrad (1949), p. 27.

[3] M. Dennstedt, *Z. physiol. Chem.* **52**, 181 (1907).

[4] V. A. Klimova, M. O. Korshun, and E. G. Bereznitskaya, *Doklady Akad. Nauk S.S.S.R.* **96**, 81-3 (1954).

CHAPTER 16

The Simultaneous Determination of Elements Through the Decomposition of Organic Substances by Heating With Metallic Potassium*

M. O. Korshun and M. N. Chumachenko

The concept of "rapid methods" embraces not only the idea of a quick procedure for determining one element but also the possibility of the simultaneous determination of several elements by the decomposition of a single weighed sample.

Over the past eight years the idea of the simultaneous determination of several elements has been reflected in a series of gravimetric methods worked out by M. O. Korshun and her colleagues on the basis of the previously developed rapid method for the determination of carbon and hydrogen [1]. The simultaneous determination of C, H, and halogen; C, H, and S; C, H, and Si; and C, H, and F [2] should be noted here.

For the simultaneous determination of other elements besides C and H we used a different procedure for decomposing the organic substance — heating with metallic potassium. In 1941, for the microdetermination of halogen and S, Bürger decomposed organic substances by heating with metallic potassium in a glass ampoule. Other scientists followed the same course [4, 5]. In 1946 M. O. Korshun and V. A. Klimova studied this decomposition

Doklady Akad. Nauk S.S.S.R.* **99, 769-71 (1954).

procedure for semimicro samples and then M. O. Korshun and M. N. Chumachenko developed the semimicro determination of Cl, Br, and I [6]. Here the decomposition was carried out not in glass ampoules but in a metal microbomb.

In the course of our extensive work on this method [6] we became convinced that metallic potassium is an active and reliable decomposing agent. At 80-85° it breaks down polyelemental organic compounds completely within a few minutes. Halogens are bound in the form of potassium salts of hydrohalic acid; sulfur, in the form of potassium sulfide. Most of the carbon separates as thick particles of charcoal; part is converted into cyanogen compounds. If mercury is present, it is reduced to the metal. Since the decomposition takes place in an enclosed space (in a hermetically sealed bomb) there is no possibility of loss even in the decomposition of highly volatile liquids or of obtaining the elements sought in any form other than those mentioned above. Apart from the unusual intensity of the destructive action of potassium, the decomposition described has the advantage that the melt obtained in the bomb can be analyzed in the form of aliquot samples after being dissolved and brought to a definite volume. Hence it is possible to titrate separate aliquots for the simultaneous determination of several elements in one weighed sample.

Over the course of all these years we perfected the original variation of the method 6 and also introduced certain improvements. After having tested many different types of bombs, we regard the construction illustrated in Fig. 1 as the most successful.† In addition, in the determination of Cl, Br, and I, argentometric titration was replaced by mercurometric for Cl and Br, and by

† A. D. Nikiforov helped to make the bomb.

iodometric for I. The advantages of iodometric titration are well known. As to mercurometric titration, the directions available [7, 8] had to be made more specific for the conditions of our methods (See Fig. 1). It was found that the sharpest transition was observed when the acidity of the medium was equivalent to 0.02N HNO_3. In the determination of bromine and chlorine the use of mercurometric titration in the presence of diphenylcarbazone as the indicator made it possible to titrate more accurately and also eliminated the necessity of using $AgNO_3$, which is far from satisfactory. In the case of the analysis of substances containing N and S, potassium cyanide and sulfide form along with the halide salts and, of course, disrupt both the argentometric and the mercurometric titration of the halogens. Consequently, before the titration, the aliquot sample is oxidized with permanganate to

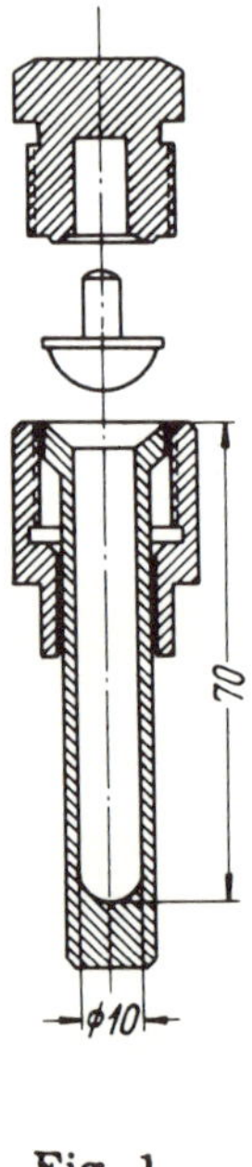

Fig. 1

remove these interfering substances. The excess permanganate is then destroyed with hydrogen peroxide. We found that such a treatment is always necessary in the determination of Cl and Br, for in an alkaline fusion cyanides produced upon the interaction of reaction mixture and atmospheric nitrogen are always present in the bomb. In the iodometric determination of I there is no need of so destroying the interfering substances since when the iodide is oxidized to iodate by bromine in acetic acid medium, the cyanides, sulfides, and phosphides are oxidized at the same time. Meanwhile, because of the differences in oxidizing potential, the chlorides and bromides are not oxidized under these conditions and hence Cl or Br can be determined simultaneously with I in an alkaline solution of the melt.

When determining halogen in organic mercury compounds we noted that the Hg in the alkali fusion was obtained in the form of compact drops, which proved easy to collect quantitatively. This observation served as a basis for the development of the method for the simultaneous

TABLE 1

The Determination of Halogen in Organic Compounds

Substance		Found %	Calc.%	Diff.%
Binary compound n-butoxytitanium trichloride and piperidine $C_9H_{13}ONTiCl_3$	Cl	34.19	34.07	+0.12
		34.25	34.07	+0.18
Chlorobenzylphosphinic acid $C_7H_8O_3PCl$	Cl	17.21	17.15	+0.02
		17.13	17.15	+0.06
Di-(1-methyl-2-ethoxy-1-propen-1-yl) thallium chloride $C_{12}H_{18}O_4TlCl$	Cl	7.71	7.62	+0.09
		7.69	7.62	+0.07
Di-β-chlorovinyllead dichloride $C_4H_4Cl_4Pb$	Cl	35.08	35.36	-0.28
		35.27	35.36	-0.09
α-Iodoethyldiethylsilane $C_6H_{15}SiI$	I	52.42	25.41	+0.01
		52.50	52.41	+0.09

TABLE 2

The Determination of Several Elements from One Weighed Sample

Substance	Found %	Diff. %	Found %	Diff. %	Found %	Diff. %
3,3-Diethyl-6,6-dichlorothiatricarbocyanine	11.50	-0.70	20.71	+0.01	-	-
iodide $C_{25}H_{23}N_2S_2Cl_2I$	11.48	-0.90	20.67	-0.03	-	-
Calculated, % Cl = 11.57; I = 20.70						
Iodoglutaconaldehyde dianil	17.31	-0.21	27.69	-0.19	-	-
hydrobromide $C_{17}H_{16}N_2BrI$	17.40	-0.12	27.77	-0.11	-	-
Calculated % Br = 17.52; I = 27.88						
Ethylmercury chloride C_2H_5HgCl	13.38	+0.01	75.49	-0.17	-	-
Calculated % Cl 13.37; Hg 75.66	13.37	0.00	75.46	-0.20	-	-
β-Hydroxyethylmercaptomercury	11.17	-0.15	10.20	-0.10	64.00	-0.07
chloride $C_2H_5OSHgCl$	11.13	-0.19	10.35	+0.05	64.32	+0.25
Calc. % Cl = 11.32; S = 10.30; Hg = 64.07						

determination of halogen and Hg. In this method, the fusion solution is filtered and the halogens are determined in the filtrate while the Hg which remains on the glass filter is dissolved in strong nitric acid. After the addition of excess sodium chloride, the $Hg(NO_3)_2$ is determined by back titration. When S was present in the substance, we tried to determine it directly by iodometric titration in an aliquot of the fusion solution but we obtained low and widely varying results. The S losses can be explained by the oxidation of the sulfide by atmospheric oxygen that occurs in weakly alkaline, solution. When the melt was dissolved in a strong alkali we observed no oxidation of the sulfide. Tests showed that a 15% potassium hydroxide solution was entirely suitable for our purposes. N and halogen do not disturb the determination of S, and S does not disturb the determination of halogen. Thus, the analytical conditions that we developed supplied the prerequisites for the simultaneous determination of halogen and sulfur from one weighed sample. When Hg is present in the molecule, the method makes it possible to determine halogen, S, and Hg simultaneously by the decomposition of a single sample.

The methods described are widely used in our laboratory, and the results obtained agree with the calculated figures within the limits of 0.1-0.3%. Tables 1 and 2 present examples of our microanalysis of organic substances.

The method will be described in detail in the Zhurnal Analiticheskoi Khimii.

LITERATURE CITED

[1] M. O. Korshun and V. A. Klimova, *Zhur. Anal. Khim.* **2**, 274 (1947); **3**, 176 (1948); **4**, 292 (1949); V. A. Klimova and M. O. Korshun, *ibid* **6**, 231 (1950).

[2] M. O. Korshun and N. S. Sheveleva, *Doklady Akad. Nauk S.S.S.R.* **60**, **63** (1948); *Zhur. Anal. Khim.* **7**, 111 (1952); M. O. Korshun, *ibid*

7, 98 (1952); V. A. Klimova, M. O. Korshun and E. G. Bereznitskaya, *Doklady Akad. Nauk S.S.S.R.* **84**, 1175 (1952); N. E. Gel'man and M. O. Korshun, *ibid.* **89,** 685 (1953).

[3] K. Bürger, Angew. *Chem.* **54,** 479 (1941); *Chemie* **55,** 245 (1942); **57,** 25 (1944).

[4] P. J. Elving and W. B. Ligett, *Ind. Eng. Chem. Anal. Ed.* **14,** 449 (1942).

[5] W. Zimmermann, *Mikrochem.* **31,** 15 (1943); **33,** 122 (1946); **35,** 80 (1950).

[6] M. O. Korshun and N. E. Gel'man, *Novye Methody Elementarnogo Mikroanaliza* [New Methods of Elementary Microanalysis] 1949, p. 77

[7] J. Dubsky and J. Trtilek, *Mikrochem.* **12,** 315 (1933).

[8] L. N. Lapin and V. P. Moroz, *Zavodskaya Lab.* **9,** 1247 (1940).

CHAPTER 17

The Simultaneous Determination of Carbon, Hydrogen, and Phosphorus in Organic Phosphorus Compounds Composed of Carbon, Hydrogen, Oxygen, Phosphorus, and Nitrogen*

M. O. Korshun, E. A. Terent'eva, and V. A. Klimova

INTRODUCTION

The chemistry of organic phosphorus compounds has developed rapidly in recent years. The study of natural phosphorus-containing products, the necessity of checking the phosphorus content of production samples, the many experimental syntheses, all call for rapid, reliable, and universally applicable methods of analyzing for phosphorus — methods that do not require large amounts of material.

The principles of the analysis of organic phosphorus compounds were developed more than 50 years ago. Each of the methods used today for the analysis of these substances can be broken down into the following three steps: 1) mineralization, 2) quantitative isolation of the phosphorus in a definite chemical form, and 3) determination of the amount of this isolated compound.

The characteristic feature of organic phosphorus compounds—their stability—is clearly revealed during mineralization. These compounds possess very strong resistance to the action of oxidants and high temperatures, especially when there is a C-P bond in the molecule.

Zhur. Anal. Khim.* **9, 275-281 (1954).

The most widely accepted method of decomposing organic phosphorus compounds is mineralization with a liquid oxidizing mixture. Mixtures of the following compositions are those most frequently used: H_2SO_4 + $K_2Cr_2O_7$; $KMnO_4$ + HNO_3; and H_2SO_4 + HNO_3; with K_2SO_4, $K_2S_2O_8$, $CuSO_4$, $Al_2(SO_4)_3$ or Se added. Even though extremely rigid oxidation conditions are maintained — high temperature, long heating time (up to 30 or 40 hours), and frequent repetition of the treatment—cases in which complete quantitative mineralization does not occur are still frequently noted [1].

Dry mineralization by fusion of the substance with certain alkaline additives is often resorted to. But even this procedure quite often fails to result in quantitative destruction of the subtance to be analyzed [2]. When the treatment is conducted in the open, there is the possibility that some of the phosphorus may be lost [3].

After the organic phosphorus molecule has been mineralized, the purely inorganic part of the analysis is begun — the quantitative separation of the ionic phosphorus obtained (generally the PO_4^{-3} ion) in a chemical form lending itself to determination. Even if the selected analytical method, e.g., volumetric or colorimetric, does not require actual isolation of the compound under study, the final solution after mineralization contains so many interfering ions in high concentrations that quantitative separation of the phosphorus-containing ion from the other ions is preferable. At present the best specific precipitant of the PO_4^{-3} ion is ammonium molybdate, which combines with it to form the insoluble complex $(NH_4)_3PO_4 \cdot 12MoO_3$. This complex has always appealed to analysts because of its extremely favorable conversion factor for phosphorus. The ratio of its molecular weight to its phosphorus content is 68:1. Micro samples of precipitates having this composition can be weighed on ordinary analytical balances [4]. The good factor compensates somewhat for the negative qualities of the complex: change in composition as a function of the precipitation conditions, the composition of the solution, and the treatment and drying of the precipitate [5]; and the time and tediousness of the operations in the isolation of the precipitate. To make the molecule still heavier, use is made of its ability to add certain organic bases: strychnine [6], hydroxyquinoline [7], quinine [8], and antipyrine [9]. Centrifugation of the precipitate and comparison of its volume with the volume of the precipitate from a PO_4^{-3} ion solution of standard concentration obtained in the same way [10] are resorted to for rough determinations of the order of $+5\%$.

The amount of ammonium phosphomolybdate precipitate is usually determined gravimetrically but can also be determined by titrating the precipitate directly with alkali or back-titrating the excess alkali after dissolving of the precipitate [11]. In addition, the phosphorus content can also be determined by the amount of nitrogen present (Kjeldahl) [12] or by the amount of molybdenum, determined polarographically [13]. Treatment of the precipitate with any reducing agent ($SnCl_2$ + HCl, freshly prepared $FeSO_4$, 1-amino-2-naphthol-4-sulfonic acid, hydrazine sulfate, hydroquinone, or citric acid) [14] puts it into solution in the form of a colored complex (molybdenum blue), the amount of which is determined colorimetrically [15]. Methods have been developed that, on the one hand, make use of the specificity of ammonium molybdate and, on the other, eliminate the difficulties associated with the isolation of phosphomolybdate; the precipitate is dissolved in ammonia and reprecipitated in the form of $MgNH_4PO_4 \cdot 6H_2O$, which is heated to give magnesium pyrophosphate, $Mg_2P_2O_7$. (If the ammonium-magnesium salt is precipitated directly after mineralization, it is very difficult to achieve a quantitative separation of the precipitate [16]. Colored complexes of the PO_4^{-3} ion with cobalt [17] and vanadium [18] salts are also used for colorimetric determinations. Here the concentration of the complex can be determined by amperometric titration with the trivalent iron ion [19] or by titration with uranium [20] or cerium [21] salts.

The use of physico-chemical methods of analysis is usually limited by the unsatisfactory accuracy of the results. The majority of the colorimetric methods can be used only for samples with a small phosphorus content or for the determination of traces. Not one of the above mentioned methods makes use of the quantitative formation of phosphorus pentoxide in an oxygen atmosphere for a direct gravimetric determination of the inorganic phosphorous compounds. Furthermore, no method has been developed for the simultaneous determination, in one weighed sample, of both the phosphorus and the other elements in the molecule.

Our objective was to create a fundamentally new, rapid, and accurate method of determining phosphorus along with the other elements of the molecule. To do this we had to find more favorable conditions for the quantitative breakdown of the organophosphorus molecule and for the isolation of the phosphorus in a form having a definite composition. Success, on the one hand, would result in a distinct reduction in the time consumed by the analysis and, on the other, would open up the possibility of determining the other elements at the same time.

DEVELOPMENT OF THE METHOD

First of all, we set ourselves the task of analyzing compounds composed of C, H, O, P, and N. To accomplish this work we tried to extend to organic phosphorus compounds the principles of combustion applicable in the analysis for carbon and hydrogen by the methods developed in the Microanalytical Laboratory of the Institute of Organic Chemistry, Academy of Sciences, U.S.S.R. [22]. The substance is submitted to pyrolysis in an open quartz capsule placed in a quartz tube and the pyrolytic products are burned in a rapid stream of oxygen. Under these conditions organic phosphorus is converted to its highest oxide — phosphorus pentoxide [23]. Experiments showed, however, that phosphorus pentoxide is difficult to determine in the form of the sol that remains in the capsule since, on the one hand, it is very hygroscopic and does not maintain a constant weight and, on the other, owing to its volatility, it penetrates the water and carbon dioxide absorption tubes, destroying the balance of all three elements — C, H, and P. Furthermore, phosphorus pentoxide can form a thin film on the carbon residue of the sample, which is obtained as a result of pyrolysis in a capsule. This film keeps the carbon from burning. The ability of phosphorus pentoxide to corrode quartz is another of its unpleasant features.

In order to prevent the escape of phosphorus pentoxide from the capsule and to avoid the possibility of its forming a film on the carbon residue, we decided to use an internal absorbent — to cover the weighed sample in the capsule with a layer of a preparation meeting definite requirements. Such a preparation must be thermally stable under the combustion conditions, must react quantitatively with the phosphorus pentoxide to form a stable product, and must at the same time be inert to the other pyrolytic and combustion products.

We tested purified, fibrous asbestos and granulated magnesium oxide as absorbents of this type. These preparations had proved quite satisfactory in the analysis of such troublesome substances as organic fluorine and silicon compounds [24, 25]. We found that asbestos impregnated with vanadium pentoxide [26] has no advantage over pure asbestos. Both asbestos and magnesium oxide enabled us to determine phosphorus in the form of phosphorus pentoxide by the increase in weight, but the results were not accurate enough. The two materials are difficult to bring to constant weight upon heating, they corrode the quartz of the capsule, and, because of their hygroscopic nature [27], it is hard to establish the correction for the blank when they are used. In view of the fact that only 1.0-1.5 mg. of phosphorus pentoxide is formed from the average weighed sample of organic phosphorus material, the fluctuation of the correction from 100 to 200 μg. produces a large error in the phosphorus determination. However, in the analysis of organic fluorine and silicon compounds, because of their specific peculiarities, it has not yet been possible to find more satisfactory absorbents than magnesium oxide and asbestos.

In the case of organic phosphorus compounds we tried to find a material that would have no blank, i.e., that first of all would not be hygroscopic. Pumice is a material analogous to asbestos in composition but different in physical structure. Thanks to its porosity, it has a large surface, but at the same time it is not hygroscopic like fluffy fibrous asbestos. When finely ground, it pours easily from capsule to capsule; it quickly attains constant weight upon heating; and it does not corrode quartz.

Combustion tests of organic phosphorus compounds with pumice showed that the phosphorus pentoxide quantitatively formed under the combustion conditions is all kept in the capsule by the pumice. Since this is the case, a possibility arises of determining carbon and hydrogen simultaneously.

We found, however, that to achieve complete combustion of the compounds it is necessary to have an excess of oxygen in the capsule.

PREPARATION OF THE PUMICE

Pieces of pumice were ground and the fraction having a grain size of about 1 mm. was screened off. The granules were treated with 30% hydrochloric acid for 3-4 hours in a round-bottomed flask on a boiling water bath. They were then washed in a funnel with boiling water until the chloride ion disappeared from the wash water. The air-dried pumice was first heated in a muffle, afterwards it was placed in a quartz tube and heated under dry oxygen in a furnace at 900° for a working day. The prepared pumice is stored in a rubber-stoppered jar.

CARRYING OUT THE ANALYSIS

Enough pumice is scooped up with an ordinary quartz combustion capsule so that, after the side of the capsule is tapped, the level is 10 mm. below the rim. The capsule is then wiped with a chamois, passed through the oxidizing flame of a burner, and placed in the combustion tube of the set-up for the microdetermination of carbon and hydrogen. A second, empty capsule is also placed there. The tube is closed with a stopper, an electric furnace at 900° is moved up to the capsules, and the absorption tubes for carbon dioxide, nitrogen oxides (if the substance contains nitrogen), and water are connected to the set-up. 500 ml. of oxygen is then passed through the set-up. After the furnace has been moved away from the capsules, the absorption tubes are disconnected, wiped, and transferred to the balance. The capsules are removed, placed on stands and blocks, and also transferred to the balance. The absorption tubes are weighed at the 15th and 20th minutes and again connected to the set-up. A 6-8 mg. sample of the substance is weighed out in the empty capsule (or a capillary, afterwards placed in the capsule) and covered with pumice from the other capsule. A reading of the total weight of the capsule containing the sample and pumice (with account taken of the zero point of the balance) is taken at the 25th minute after the removal of the pumice-containing capsule from the tube. The capsule with the weighed sample is placed in the combustion tube. Oxygen is passed through the tube at the rate of 30-35 ml./min. and combustion is effected with a burner.

Upon the completion of the combustion, i.e., when a black zone of carbon no longer forms on the pumice when the burner flame approaches the weighed sample, the end of the capsule is thoroughly heated with the burner, the burner is then taken away, and the capsule is let cool for 5 min. (oxygen is "sucked through"). After this, the furnace is moved up to the capsule for 5 min. The operation is repeated. The furnace is moved away, the capsule is let cool a little, the combustion tube plug is removed, the oxidizing flame of a burner is brushed along the end of the tube 2 or 3 times, and the capsule is then extracted, care being taken that no pumice falls out of it. After this, the procedure is the same as at the start of the analysis; the absorption tubes are weighed at the 15th and 20th minutes and the capsule at the 25th min. The increase in weight of the phosphorus pentoxide is calculated with account taken of the change in the zero point of the balance during the experiment.

CALCULATION OF THE RESULTS

The carbon and hydrogen content of the weighed sample is calculated as usual; the phosphorus is calculated with the formula:

$$\%P = a.f/S \cdot 100,$$

where *a* is the increase in weight of the phosphorus pentoxide. This is determined as the difference between the weight of the capsule before and after combustion (subtraction of the weighed sample); *f* is the conversion factor $= 2P/P_2O_5 = 0.4364$; *S* is the weight of sample.

DISCUSSION OF THE RESULTS

The proposed analytical method saves a great deal of time because it combines the first two steps of ordinary phosphorus determinations mentioned in the introduction, i.e., mineralization, which takes place under highly favorable conditions, and isolation of a definite chemical form. The third step — determination of the isolated compound — amounts to no more than a simple weighing of the capsule immediately after combustion. The combustion conditions used permit the simultaneous determination of carbon and hydrogen with the usual acceptable accuracy.

Under these combustion conditions the fundamental action of the pumice granules is the cracking effect of the large, porous, incandescent surface, which is full of sharp edges and angles [28, 29]. The phosphorus pentoxide formed reacts chemically with the pumice in the immediate vicinity of the weighed sample. We cannot speak of the mechanical retention of phosphorus pentoxide fumes here since the compound sublimes at about 300-400°. SiO_2, M_2O (alkalies), Al_2O_3, and iron oxides make up the composition of pumice. All these components are capable of reacting with phosphorus pentoxide to give thermally

The Simultaneous Deter

No.	Substances	Weighed sample mg.	CO_2 mg.
1	$(C_6H_5)_3P$	4.815	14.555
	C—82.42; H—5.76; P—11.82	4.567	13.780
2	$(C_6H_5)_3CPO(OC_4H_9i)_2$	3.399	9.193
	C—74.29; H—7.62; P—7.09	6.301	17.167
3	$C(_6H_5)_3CPO(OCH_3)_2$	5.493	14.366
	C—71.57; H—6.01; P—8.79	5.253	13.760
4	$C_6H_5NPC_2H_5(OC_2H_5)_2$	6.300	13.721
	C—59.73; H—8.36; P—12.84	5.490	11.932
5	$(C_6H_5)_3CPO(OC_3H_7i)_2$	4.288	11.579
	C—73.51; H—7.16; P—7.58	5.702	15.314
6	$(CH_3)_2(CH_3CONH)CPO(OC_2H_5)_2$	7.470	12.501
	C—45.56; H—8.50; P—13.06	8.109	13.596
7	$CH_3COHNCH(C_6H_5)PO(OC_2H_5)_2$	5.112	10.224
	C—54.73; H—7.07; P—10.86	9.099	18.300
8	$(C_4H_9O)_2POCON(C_2H_5)_2$	3.951	7.749
	C—53.22; H—9.63; P—10.56	4.601	8.992
9	$(C_6H_5)NHCONHCH(C_6H_5)PO(OC_2H_5)_2$	7.455	16.256
	C—59.66; H—6.41; P—8.54	6.140	13.452
10	$(C_2H_5)_2NCOPO(C_2H_5)OC_4H_9$	4.947	9.592
	C—52.99; H—9.70; P—12.44	3.450	6.698

stable compounds; $SiO_2 \cdot P_2O_5$, as is known, is one of the characteristic compounds of phosphorus. The modification of the silicyl (SiO_2^{+2}) metaphosphate that forms at the temperature prevailing where the substance is burning (about 1000°) is extremely stable. The reaction proceeds equimolarly and thus the calculation may be made as if for an increase in weight in the form of P_2O_5.

It is best to pyrolyze the weighed sample quickly as

mination of C, H, and P

H_2O mg.	P_2O_5 mg.	%C	Δ%C	%H	Δ%H	%P	Δ%P
2.470	1.280	82.49	+0.07	5.76	0.00	11.62	-0.20
2.320	1.258	82.34	-0.08	5.67	-0.08	12.02	+0.20
2.268	0.568	74.33	+0.04	7.47	-0.15	7.29	+0.20
4.157	0.995	74.35	+0.06	7.38	-0.24	6.89	-0.20
2.835	1.101	71.37	-0.20	5.76	-0.23	8.77	-0.02
2.970	1.037	71.48	-0.09	6.33	+0.32	8.62	-0.17
4.660	1.852	59.43	-0.30	8.28	-0.08	12.83	-0.01
4.130	1.622	59.51	+0.22	8.42	+0.06	12.89	-0.05
2.650	0.757	73.70	+0.19	6.92	-0.24	7.71	+0.13
3·390	0.991	73.31	-0.20	6.65	-0.51	7.56	-0.02
5.687	2.240	45.67	+0.11	8.52	+0.02	13.09	+0.03
6.144	2.386	45.75	+0.19	8.48	-0.02	12.84	-0.22
3.306	1.263	54.58	-0.15	7.24	+0.17	10.78	-0.08
5.752	2.259	54.89	+0.16	7.07	0.00	10.48	-0.02
3.427	0.971	53.52	-0.30	9.71	+0.08	10.73	+0.17
4.043	1.118	53.33	-0.11	9.83	+0.20	10.60	+0.04
4.239	1.457	59.51	-0.15	6.36	-0.04	8.53	-0.01
3.580	1.219	59.79	+0.13	6.52	+0.11	8.67	+0.13
4.350	1.389	52.91	-0.08	9.84	+0.14	12.26	-0.18
3.008	1.001	52.98	-0.01	9.76	+0.06	12.65	+0.21

in the rapid combustion of the Korshun and Klimova method [22]. When this is done, the film of liberated carbon is distributed over a large surface, i.e., it is thinner and can thus be easily burned up. There is then no need to fear that the pyrolytic products will burn near the capsule opening. As experiments have shown, burning outside the absorption layer does not result in low figures for phosphorus. Apparently the volatile pyrolytic products do not contain phosphorus; instead, the phosphorus is converted exclusively into pentoxide, which reacts quantitatively with pumice, as is evidenced by the change in color of the latter from rose to gray in the closed end of the capsule.

For the overwhelming majority of compounds that we analyzed two final heatings with the furnace proved enough. For certain highly heat-resistant compounds, however (derivatives of triphenylmethylphosphinic acid; see Nos. 3, 5, and 2 in the table), it was necessary to resort to stepwise combustion by repeated movement of the burner up to and away from the sample. When this is done the area around the sample is enriched with oxygen and complete combustion of the charred part of the sample is thereby promoted.

Certain precautions must be taken during the analysis. When the sample is being weighed out, especially if it is a liquid, grease, or easily electrified solid, care must be taken to see that the substance does not strike the sides of the capsule far from the bottom, for if it does, low figures will be obtained for phosphorus. Furthermore, it must be remembered that when the humidity is low, the light pieces of pumice are strongly electrified and may fall out of the capsule when it is moved and extracted from the combustion tube. Consequently the electrostatic charge must be removed with the oxidizing flame of a gas burner as mentioned above.

If necessary, phosphorus can be determined alone. In this case one gasometer with oxygen and a purifying system comprising 2 or 3 combustion tubes are all that is needed. Under these conditions the determinations may be made in series.

All the weighings were made on a Russian balance, the VM-20, which came out in 1953. Despite the fact that this model is officially considered a semimicrobalance, it may be used as a microbalance when properly adjusted.

At present we are successfully developing methods for the simultaneous microdetermination, in one weighed sample, of C, H, P, and S in organic compounds.

CONCLUSION

A new method has been developed for the simultaneous determination of carbon, hydrogen, and phosphorus in compounds composed of C, H, O, P, and N. One determination takes about an hour and a half. The deviation from the theoretical is no more than 0.3%.

The advantages of the method are its rapidity, the possibility of analyzing substances that cannot be decomposed by other methods, the determination of carbon and hydrogen along with phosphorus in one weighed sample, and the preservation of the quartz tubes from corrosion.

LITERATURE CITED

[1] C. Marie, *Compt. rend.* **129**, 766; L. Jolly, *Compt. rend* **126**, 531 (1898); C. Marie, *Bull. soc. chim. France* [3] **23**, 44.

[2] M. Jacobson and S. A. Hall, *Anal. Chem.* **20**, 736 (1948); *Chem. Abstr.* **42**, 7657.

[3] Garola, [Miss], *Ann. chim. anal. appl.* [2] **3**, 326 (1921); *Chem. Abstr.* **18**, 955.

[4] R. Kuhn, *Z. Physiol. Chem., Hoppe-Seyler's* **129**, 64 (1923); F. Pregl, Kolichestvennyi Organicheskii Mikroanaliz [Quantitative Organic Microanalysis], 1934, p. 130.

[5] W. Saschek, *Ind. Eng. Chem. Anal. Ed.* **9**, 491 (1937); *Chem Abstr.* **31**, 8441.

[6] G. Bergold and L. Pister, Z. Naturforsch. **3**b, 332 (1948); *Chem. Abstr.* **43**, 5699.

[7] G. I. Deshalit, N. M. Prosvirnina, and A. Gurevich, *Brennstoff-Chem.* **17**, 130 (1936); *Chem. Abstr.* **30**, 7305.

[8] J. A. Street, *Australian Chem. Inst. J. & Proc.* **6**, 395 (1939); *Chem. Abstr.* **34**, 1588.

[9] S. I. Gusev, *Trudy Komissii Anal. Khim. Akad. Nauk S.S.S.R.* **3**, 215 (1951); *Chem. Abstr.* **47**, 2641.

[10] A. Vila, *Compt. rend.* **198**, 657 (1934); *Chem. Abstr.* **28**, 6085.

[11] R. Strebinger, *Mikrochemie* [N.S.] **1**, 129 (1929).

[12] W. Diemair and R. Baier, *Z. anal. Chem.* **133**, 7 (1951); *Chem. Zentr.* **1952**, 1380.

[13] J. V. A. Novak, Sbornik Mezinarod Polarog. Sjezdu Praze, 1st Congr. **1951** Pt. III, Proc. 439; *Chem. Abstr.* **47**, 10408.

[14] B. Vasarhelyi, Mikrochem. Pregl Festschr., 329 (1929); *Chem. Abstr.* **24**, 2398.

[15] F. L. Schaffer, J. Fong, and P. L. Kirk, Anal. Chem. **25**, 343 (1953); *Chem. Abstr.* **47**, 5307.

[16] K. Schuhecker, *Z. anal. Chem.* **116**, 14 (1939); *Chem. Abstr.* **33**, 3718.

[17] R. Belcher and A. L. Godbert, *Analyst* **66**. 194 (1941); *Chem. Abstr.* **35**, 5064.

[18] T. S. Ma and J. D. McKinley, Jr., *Mikrochim. Acta* **1953**, 4-13.

[19] D. H. Templeton and L. G. Bassett, Natl. Nuclear Energy Ser. Division VIII, 1, Anal. Chem. Manhattan Project 321 (1950); I. Bay, Compt. rend. **146**, 814 (1908); *Chem. Abstr.* **2**, 2057.

[20] M. K. Saikina and V. F. Toropova, *Trudy Komissii Anal. Khim. Akad. Nauk S.S.S.R. Otdel Khim. Nauk* **4**, (7), 141 (1952); *Chem. Abstr.* **48**, 1889.

[21] J. Thiry, *J. pharm. Belg.* [N.S.] **1**, No. 5, 81; No. 6 144; No. 7.8, 161 (1946); *Chem. Abstr.* **41**. 5409.

[22] M. O. Korshun and N. E. Gel'man, *Novye Metody Elementarnogo Mikroanaliza*, Moscow, 1949.

[23] D. I. Mendeleev, *Osnovy Khimii* [Fundamentals of Chemistry] Vol. II, 1947, p. 170.

[24] N. E. Gel'man and M. O. Korshun, *Doklady Akad. Nauk S.S.S.R.* **89**, 685 (1953).

[25] V. A. Klimova, M. O. Korshun, and E. G. Bereznitskaya *Doklady Akad. Nauk S.S.S.R.* **84**, 1175 (1952).

[26] V. A. Klimova, M. O. Korshun, and E. G. Bereznitskaya, *Doklady Akad. Nauk S.S.S.R.* **96**, 287 (1954).

[27] J. Lindner, *Ber.* **76B**, 701 (1934).

[28] D. Colaitis and M. Lesbre, *Bull. soc. chim. France* **1952**, 1069; *Chem. Abstr.* **47**, 3754.

[29] J. Gauthier, *Bull. soc. chim.* [5] **2**, 322 (1935); *Chem. Abstr.* **29**, 3627.

CHAPTER 18

The Simultaneous Quantitative Microdetermination of Phosphorus, Sulfur, Carbon, and Hydrogen in Organic Phosphorus Compounds Composed of C, H, O, N, S, and P.*

M. O. Korshun and E. A. Terent'eva

The chemistry of organic phosphorus compounds has developed rapidly in the last few years, and an urgent need has arisen for methods of analyzing organic phosphorus compounds of complex structure and specific properties. Judging from the literature, however, quantitative organic analysis has, up to now, lacked rapid, accurate, and universally applicable methods of determining phosphorus; even more, it has lacked methods for the simultaneous determination of phosphorus along with the other elements making up the molecule.

The methods now used for determining phosphorus alone (their principles were developed more than 50 years ago) suffer from numerous shortcomings. Despite the creation of extremely rigid conditions, the procedures of wet or dry mineralization of the organic phosphorus molecule often fail to give a quantitative breakdown of the substance. Compounds with a C-P bond and heavy condensed and branched aliphatic radicals are distinguished by particular stability. A favorite analytical form in which to isolate mineralized phosphorus is that of a heavy insoluble complex of PO_4^{-3} ion with oxides of certain metals

Doklady Akad. Nauk. S.S.S.R.* **100, 707-9 (1955).

and ammonia molecules. The basic merit of such forms is the favorable conversion factor for phosphorus in either a gravimetric or a volumetric finish to the analysis. At the same time, however, these forms possess a number of undesirable qualities: variability of structure as a function of the isolation conditions; necessity of protracted ripening of the precipitates and close observation of their drying conditions; and instability of the reagents used. With such procedures, the determination of phosphorus takes 2 or 3 days.

Recently, physico-chemical instrumental methods of finishing the analysis — colorimetry, spectrophotometry, polarography, etc. — have achieved great popularity. They save a certain amount of time in the final stage of the procedure, but they are not very accurate and can be used only for substances having a low phosphorus content or for trace analysis.

Up to the present the basic difficulties involved in the analysis of organic phosphorus compounds — incomplete mineralization, and length and tediousness of the operation — had not been overcome. No methods had been developed for the simultaneous determination of phosphorus along with the other elements of the molecule.

Under the supervision of M. O. Korshun, we at the Microanalytical laboratory of the Institute of Elemento-Organic Compounds of the Academy of Sciences of the USSR have conducted a whole cycle of investigations aimed at creating rapid quantitative micromethods for the simultaneous determination of several elements in an organic molecule [1-3]. On the basis of this work we very recently developed a micromethod for the simultaneous determination of carbon, hydrogen, and phosphorus in substances composed of C, H, O, N, and P, the phosphorus being determined as a sol [4, 5]. By applying the principle of the pyrolytic decomposition of an organic substance in a quartz capsule and combustion of the pyrolytic products in a large excess of oxygen passing through an empty combustion tube at high speed, we were able to create very favorable conditions for the mineralization of such difficultly combustible substances as organic phosphorus compounds. Under these conditions phosphorus is completely converted to the pentoxide, which is collected quantitatively in the combustion capsule by means of purified ground pumice. Carbon and hydrogen are determined as usual [1].

As expected, however, when we attempted to burn sulfur-containing organic phosphorus compounds with pumice, we obtained high figures for the percent of phosphorus. The oxides of iron and alkali and alkaline earth metals in the pumice combine with the sulfur oxides evolved upon combustion of the compound to form thermally stable sulfates. Preliminary treatment of the pumice with sulfuric acid produces no improvement.

For the absorption of phosphorus pentoxide in a capsule we selected a preparation that reacted quantitatively with this compound and was inert to oxides of sulfur — viz. crushed quartz treated with a strong alkali. The use of this preparation opened up the possibility of simultaneously determining not only phosphorus, carbon, and hydrogen but also a fourth element — sulfur — along with them in a single 3-5 mg. sample.

The procedure for the simultaneous determination of phosphorus, sulfur, carbon, and hydrogen is a combination of two methods that we had developed earlier: the determination of phosphorus [5] (but with quartz, not pumice) and the simultaneous determination of carbon, hydrogen, and sulfur [2]. One analysis takes about 2.5 hours.

TABLE 1

Formula No.	Sample mg.	C%	ΔC%	H%	ΔH%	S%	ΔS%	P%	ΔP%
I	4.428	33.50	-0.29	6.52	-0.08	30.01	-0.06	14.68	+0.15
	6.409	33.88	+0.09	6.48	-0.12	30.03	-0.04	14.48	-0.05
II	6.845	24.02	+0.15	5.05	+0.04	40.03	+0.16	15.49	+0.10
	3.139	23.83	-0.04	4.94	-0.07	39.84	-0.03	15.48	+0.09
III	7.590	41.79	-0.15	7.94	-0.15	22.51	+0.12	10.70	-0.12
	5.401	41.95	+0.01	8.13	+0.04	22.13	-0.26	10.74	-0.08
IV	5.680	48.47	-0.06	6.56	-0.13	18.59	+0.07	8.74	-0.20
	7.763	48.56	+0.03	6.59	-0.10	18.41	-0.10	8.61	-0.33
V	5.280	57.07	-0.06	6.15	+0.02	8.39	-0.08	8.19	0.00
	6.685	56.91	-0.22	6.15	+0.02	8.80	+0.33	8.32	+0.13
VI	6.596	49.29	-0.06	6.66	-0.03	9.97	-0.16	10.03	+0.24
	8.049	49.22	-0.13	6.57	-0.12	10.08	-0.05	9.96	+0.17
VII	5.279	42.89	+0.21	8.10	+0.14	12.34	-0.32	12.20	-0.08
	5.431	42.52	-0.16	7.90	-0.06	12.92	+0.26	12.48	+0.25

As can be seen from the examples presented in Table 1, the substances analyzed have a complex structure. Most of them have a C-P bond, many of them contain benzyl groups and saturated radicals of iso structure; i.e., these substances are among those difficult to decompose. Nonetheless we achieved quantitative destruction of the compounds tested in every case. Despite the small weight gains of the forms analyzed in the case of all four elements determined, — phosphorus pentoxide, in particular (1-3 mg.) — the deviation from the theoretical figure does not exceed 0.3%. All the weighings were made on a Russian VM-20 balance.

It is also possible that an absorbent for phosphorus pentoxide more suitable than quartz will be found and that an analytical method based on the same principles will be

FORMULAE AND PERCENTAGE COMPOSITION

I

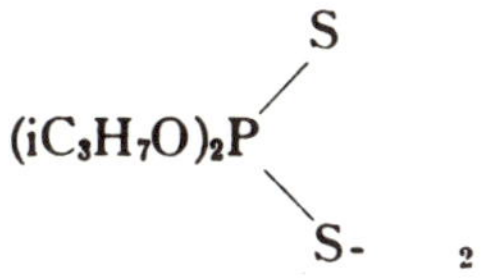

C-33.79; H-6.61; S-30.07; P-14.53

II

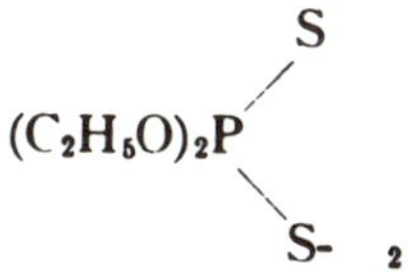

C-23.87; H-5.01; S-39.87; P-15.39

III

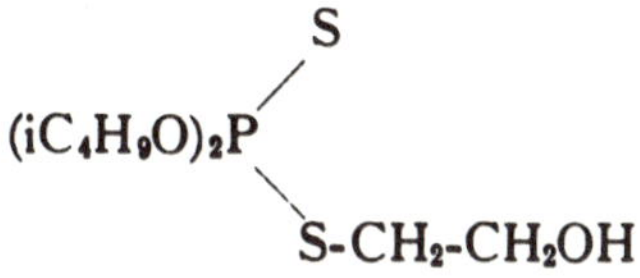

C-41.94; H-8.09; S-22.39; P-10.82

IV

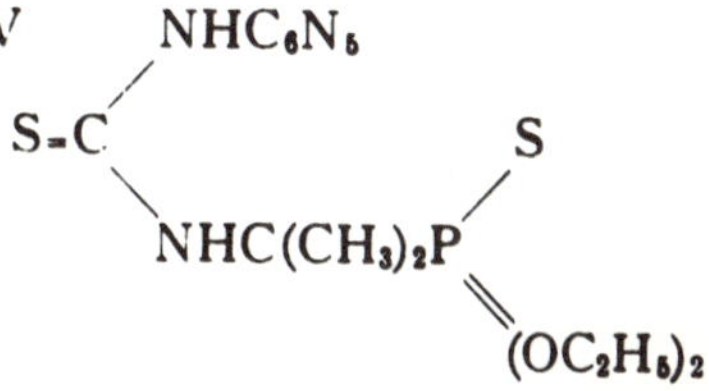

C-48.53; H-6.69; S-18.51; P-8.94

V

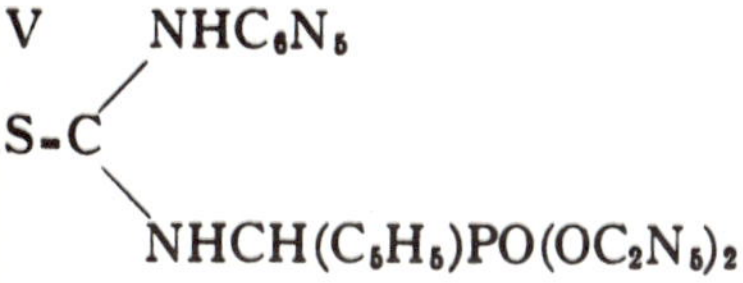

C-57.13; H-6.13; S-8.47; P-8.19

VI

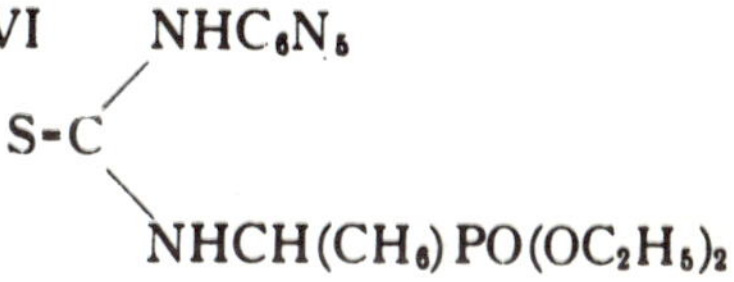

C-49.35; H-6.69; S-10.13; P-9.79

VII

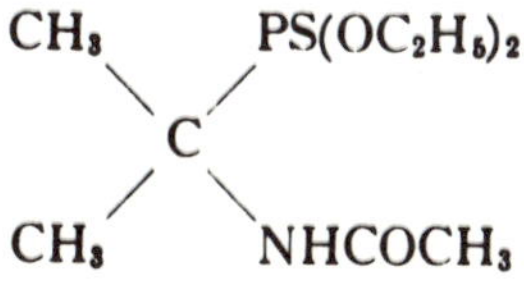

C-42.68; H-7.96; S-12.66; P-12.23

developed for organic phosphorus compounds of even more complex elementary composition.

In brief, we have developed a fundamentally new micromethod for the simultaneous determination of four elements — C, H, S and P — in such difficult-to-analyze substances as organic phosphorus compounds. The method saves a great deal of time even in comparison with the accepted analytical processes for phosphorus alone, and requires a very small amount of substance (of the order of 20 mg.).

The method is especially important in the analysis of difficult-to-decompose, unstable, and liquid substances, including highly volatile ones.

LITERATURE CITED

[1] M. O. Korshun and N. E. Gel'man. *Novye metody elementarnogo mikroanaliza* [New Methods of Elementary Microanalysis]. Moscow, 1949.

[2] M. O. Korshun and N. S. Sheveleva, *Zhur. Anal. Khim.* **7**, 104 (1952).

[3] V. A. Klimova, M. O. Korshun, and E. G. Bereznitskaya, *Doklady Akad. Nauk SSSR* **84**, 1175 (1952).

[4] V. A. Klimova, M. O. Korshun, and E. G. Bereznitskaya, *Doklady Akad. Nauk SSSR* **96**, 287 (1954).

[5] M. O. Korshun, E. A. Terent'eva, and V. A. Klimova, *Zhur. Anal. Khim.* **9**, 275 (1954).

CHAPTER 19

The Determination of Fluorine in Organic Compounds*

M. O. Korshun, V. A. Klimova, and M. N. Chumachenko

The analysis of fluorine-containing organic compounds demands that we solve the problem not only of determining the fluoride ion but also of how to decompose the organic fluorine compound. Particular difficulty is encountered in the decomposition of those substances that contain more than one fluorine atom attached to a single carbon. These compounds often are highly heat stable and therefore rupture of the C-F bond requires especially severe decomposition conditions. In addition, the formation of the extremely stable gaseous compound CF_4 must be expected during the thermal decomposition of carbon-fluorine compounds.

It has been noted in the literature [1] that CF_4 and carbon fluorides of higher molecular weight react with alkali metals at about 400° to form alkali fluoride and carbon. This fact has formed the basis of various methods developed for the determination of halogens, including fluorine [2-6].

In 1946 we developed a semimicro method for the quantitative determination of fluorine based on the reaction of organic fluorine-containing compounds with metallic potassium in a closed system but at higher temperatures (800-850°). The subsequent determination of the fluoride ion was performed by the volumetric method. Quantitative decomposition is achieved in a few minutes irrespective

Zhur. Anal. Khim.* **10, 358-63 (1955).

of the elementary composition and structure of the molecule. The metal reacts with the thermal decomposition products of the substance in the vapor phase, which ensures a quite rapid reaction rate and high effectiveness of the reaction for the analysis of "hetero-element" organic compounds.

Decomposition. An 8- to 25-mg sample, depending on the percentage content of fluorine in the substance under analysis, is weighed out and heated with a 4-5-fold amount of potassium metal in a steel microbomb at 800-850° for 10-15 minutes. The heating is performed in a special electric furnace holding 6 bombs at a time (see Fig. 1). The design of the bomb and the subsequent treatment of the melt were described earlier when we presented the method for the determination of halogens [3] and the simultaneous determination of other elements [6]. The bombs are made of the following types of stainless steel: EYal-T for the test tube, EZh-4 for the hemispherical sur-

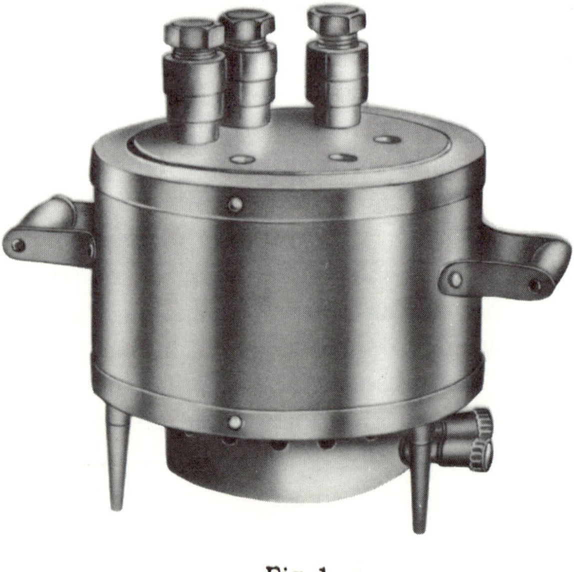

Fig. 1

face, and 40-Kh for the plug and union. A bomb of good gas-tightness will last for a year or more of daily use.

DETERMINATION OF THE FLUORIDE ION

The most widely accepted method for determining fluorine is the gravimetric method in which fluorine is precipitated in the form of lead chloride fluoride [2, 5, 7, 8]; it may also be regarded as the best, owing to the favorable chemical conversion factor. This method, however, has serious shortcomings, viz., the length of time required for filtration of the precipitate and the appreciable difficulties arising when other ions are present, these being adsorbed on the surface of the precipitate or reacting with the lead, thus reducing the accuracy of the analysis. Of the volumetric methods, titration of the fluoride ion with thorium nitrate [9-14] seems the most attractive to us.

The conditions under which the titration is performed should be exactly the same as those of the blank, which also serves as verification during the fluoride titration. 0.4 ml of a 0.1% aqueous solution of sodium alizarinsulfonate is added to each 10-25-ml aliquot as the indicator, the solution is neutralized by the dropwise addition of 1:50 nitric acid to a lemon yellow color, 2.5 ml of buffer is added, the volume is made up to 50 ml with water, and the mixture is titrated with 0.025 N $Th(NO_3)_4$ solution to the appearance of a pale pink color. The titer of the thorium nitrate is established on the basis of dry ThO_2 residue. A 10-ml microburette with 0.02-ml graduations is used for the titration. The reading is made to an accuracy of 0.005 ml with the aid of a magnifying glass.

The hydrogen ion concentration of the solution being titrated is of essential importance to a correct end point determination. In our study of this problem we found that the most accurate results are obtained at a pH of 2.9-3. The results of these experiments are given below.

Effect of pH on the Titration of Fluoride with 0.025 N $Th(NO_3)_4$
50 ml of solution containing 1.203 mg of fluorine was titrated

pH	2.4	2.6	2.8	2.9	3.0	3.2	3.4	3.6
Found F, mg	1.311	1.286	1.240	1.212	1.206	1.143	1.107	1.025

In order to create the necessary pH a buffer is made up as follows: 50 mg of 1 N monochloroacetic acid solution is made neutral to phenolphthalein with a 1 N sodium hydroxide solution. 50 ml more of acid is added and the volume is made up to 250 ml. If measurement of the pH of the buffer solution shows it to be lower or higher than the requisite value, then it is necessary to add the proper amount of acid or alkali in order to obtain a pH of precisely 2.9-3.

It is well known that when the titration is performed with thorium nitrate in the presence of sodium alizarinsulfonate the color change of the indicator appears somewhat after the equivalence point, and the volume of the solution consumed to form the lake must be subtracted from the total consumption of $Th(NO_3)_4$. In addition, the color change of the indicator is related to fluoride ion concentration in the solution being titrated. Consequently, in order to avoid obtaining high results, a correction must be introduced. We found these corrections experimentally by titrating solutions of pure sodium fluoride containing various amounts of fluorine (see Table 1).

1 ml of 0.25 N $Th(NO_3)_4$ solution is equivalent to 0.475 mg of fluorine.

Despite the fact that the corrections are quite large in relation to the total consumption of $Th(NO_3)_4$ solution, their use under the titration conditions that we developed made it possible for us to determine fluorine with sufficient accuracy in the concentration range from 0.1 to 3 mg per 50 ml solution being titrated (see Table 2).

TABLE 1

Volume of Solution Being Titrated 50 ml.

ml of 0.025 N Th $(NO_3)_4$ consumed	Correction for 0.025 N Th $(NO_3)_4$, ml	ml of 0.025 Th $(NO_3)_4$ consumed	Correction for 0.025 N Th $(NO_3)_4$, ml
0.200	0.075	3.000	0.130
0.500	0.130	3.500	0.100
1.000	0.200	4.000	0.100
1.500	0.190	5.000	0.100
2.000	0.180	8.000	0.100
2.500	0.160		

TABLE 2

Titration of Aliquot Samples After the Decomposition of Teflon Sample, 23.919 mg; volume of the solution after decomposition of the melt, 250 ml; volume of the solution being titrated, 50 ml.

Aliquot, ml	Fluorine content. mg	Found fluorine, mg
25	1.819	1.817
15	1.091	1.090
10	0.727	0.730
5	0.363	0.367
2	0.146	0.149
1	0.073	0.073

EFFECT OF OTHER IONS DURING THE TITRATION OF FLUORIDE WITH $Th(NO_3)_4$

All ions capable of reacting with fluoride or thorium [15-17] interfere with the titration with thorium nitrate. The presence of sulfate and phosphate ions [17-19] is particularly disturbing. As to chlorides and nitrates, their interference is to a large degree a function of the concentration of these ions in the solution being titrated. According to Geyer [20], for example, chlorides, nitrates, and perchlorates present in concentrations equivalent to

TABLE 3

Determination of Fluorine in the Presence of Halogen, Sulfur or Nitrogen

Substance	% Fluorine Calc.	Sample, mg	Ml of 0.025 N Th $(NO_3)_4$ solution consumed per 0.1 g portion of sample	% Fluorine Found	% Fluorine Difference
$C_4H_4N_3F_3$	46.34	14.218	1.389	46.43	+0.09
		21.562	2.096	46.18	-0.16
		25.775	2.513	46.33	-0.01
$CF_3CFBrCOOCH_3$	31.80	18.178	1.207	31.64	-0.16
		24.020	1.593	31.59	-0.21
Mixture of $(C_2F_4)_n$ and $C_{10}H_6(SO_3H)_2$	75.98	22.095 + 23.273	3.520	75.87	-0.12
		20.789 + 25.977	3.308	75.80	-0.18

TABLE 4

Simultaneous Determination of Fluorine and Chlorine

Substance	Sample, mg	Ml of 0.025 N Th $(NO_3)_4$ solution consumed per ¼ portion of the sample	% Fluorine Found	% Fluorine Difference	Ml of 0.025 N Th $(NO_3)_4$ solution consumed per ¼ portion of sample	% Chlorine Found	% Chlorine Difference
$C_4H_3Cl_3FNO$ Calculated:	28.265	1.511	10.09	+0.11	11.300	56.10	+0.23
%F=9.98 %Cl=55.87	25.587	1.343	9.98	0.00	10.110	56.01	+0.14
$C_{10}H_9ClFN$ Calculated:	25.200	0.458	8.63	-0.27	1.187	16.70	+0.09
%F=8.90 %Cl=16.61	26.048	0.488	8.91	+0.01	1.229	16.73	+0.12
	28.589	0.514	8.55	-0.35	1.335	16.55	-0.06

0.1-0.2 N solutions do not interfere with the titration. In the solutions that we obtained after washing of the melt, the ion concentration was equivalent to a normality of 0.01 to 0.005. Chlorides, and also bromides and iodides, not only do not interfere with the fluoride determination, but may be quantitatively determined along with fluoride by the titration of separate aliquots [6]. If sulfur is present in the substances under analysis, it is quantitatively transformed to sulfide ion during the decomposition and does not disturb the fluorine analysis. Most of the nitrogen of the organic substance is removed in the form of the cyano ion, which also has no effect on the fluoride titration. The analytical figures for fluorine in the presence of halogens, sulfur, and nitrogen, and also the results for the simultaneous determination of fluorine and chlorine are given in Tables 3 and 4. Since no substances containing both fluorine and sulfur were available to us, we performed the analysis on a mixture of Teflon and naphthalenedisulfonic acid.

EFFECT OF PHOSPHOROUS

The determination of fluorine in the presence of phosphorus has been studied by many authors [17, 19] in their gravimetric determinations of fluorine by precipitation in the form of lead chloride fluoride.

During the course of research it was found that the end point in the titration for the fluoride ion is quite sharp in the presence of phosphorus, but it appears after the equivalence point for fluoride, the magnitude of lag being a function of the amount of phosphorus present. It is most likely that when phosphorus-containing substances are decomposed with metallic potassium, the phosphorus changes to the trivalent state. However, strong evolution of heat and sometimes even explosion, which is observed when the melt is dissolved in water, may cause partial oxidation of the phosphorus and some loss of it in the form of phosphine. Previously, all this kept us from being able to draw any final conclusion as to whether any reproducible quantity of oxidized phosphorus was present in each sample under analysis. In fact, in the decomposition of three samples of one and the same organic phosphorus compound, different amounts of thorium nitrate were consumed (Table 5).

TABLE 5

Sample, mg	24.710	23.268	21.233
ml of 0.025 N TH $(NO_3)_4$ consumed per 0.1 portion of the sample	0.365	0.315	0.400

As can be seen from the data of Table 5, the amount of thorium nitrate consumed for the titration of phosphorus compounds must be determined separately, i.e., for each solution of the melt obtained as a result of decomposing each individual weighed sample. Only when this is done is it possible to make an accurate determination of the amount of thorium nitrate consumed for the fluoride titration. The solution of this problem is facilitated by the possibility of taking aliquots for titration in our method. The total thorium nitrate consumption and the amount required to titrate phosphorus compounds are determined in parallel aliquots. Two samples are titrated as usual, and the fluorine is completely removed from two others as follows: the samples are placed in a porcelain dish and evaporated to dryness on a water bath. This treatment is repeated 2 or 3 times and then the residue is dissolved in water, transferred into the titration flask, and titrated exactly like the first samples. The difference between the results obtained in the titrations corresponds to the thorium nitrate consumption for the fluoride titration. In

TABLE 6
Determination of Florine in the Presence of Phosphorus

Mixture of substances	mg sample or control (+)	ml of 0.025 N $Th(NO_3)_4$ consumed per 0.1 g. sample or control		Found F %	Difference %
Teflon and triphenylphosphine	15.121	2.627		75.85	−0.13
Calculated, %F = 75.98	+14.142		0.210		
	12.141	2.168		76.10	+0.12
	+15.145		0.215		
α-Bromo-p-trifluoromethylacetophenone	12.421	0.738		21.15	−0.16
and triphenylphosphine	+12.425		0.158		
Calculated, %F = 21.31	17.220	0.971		21.40	−0.09
	+13.140		0.195		
	18.142	0.989		20.95	−0.36
	+13.045		0.190		
α-Acetamino-β-hydroxy-p-trifluoro-	20.217	1.100		20.80	+0.10
methylpropiophenone	+15.140		0.215		
Calculated, %F = 20.70	12.145	0.734		20.90	+0.20
	+14.610		0.200		
p-Fluorobezophenone and	15.145	0.513		9.35	−0.14
triphenylphosphine	+15.190		0.215		
Calculated, %F = 9.49	12.415	0.441		9.40	−0.09
	+13.150		0.195		

this case it is unnecessary to introduce any other corrections, and the results obtained are more accurate than usual. The method was checked on mixtures of organic fluorine-containing compounds with triphenylphosphine. The experiments, which are summarized in Table 6, show that fluorine may be determined in the presence of phosphorus.

SUMMARY

We have described a semicro method of determining fluorine in organic compounds by heating the substance with metallic potassium in a steel microbomb at 800-850° and titrating the fluoride obtained with thorium nitrate. The presence of nitrogen, sulfur, and halogens does not interfere with the titration.

We have demonstrated the possibility of determining fluorine and chlorine simultaneously and have also described a method of determining fluorine in the presence of phosphorus.

LITERATURE CITED

[1] J. H. Simons and L. P. Block, *J. Am. Chem. Soc.*, **61**, 2962 (1939).

[2] P. J. Elving and W. B. Ligett, *Ind. Eng. Chem. Anal. Ed.*, **14**, 449 (1942).

[3] M. O. Korshun and N. E. Gel'man, *Novye metody mikroelemenarnogo analiza* [New Methods of Elementary Microanalysis], pp. 77-87, Goskhimizdat [State Publishing House of Chemical Literature], 1949.

[4] R. Kimball and L. Tufts, *Anal. Chem.*, **19**, 150 (1947).

[5] R. Belcher and J. C. Tatlow, *Analyst*, **76**, 593 (1951).

[6] M. O. Korshun and M. N. Chumachenko, *Doklady Akad. Nauk S.S.S.R.*, **99**, 769 (1954).

[7] F. G. Hawley, *Ind. Eng. Chem.*, **18**, 573 (1926).

[8] I. V. Tananaev, *Z. anal. Chem.*, **99**, 21 (1926).

[9] H. H. Willard and O. B. Winter, *Ind. Eng. Chem. Anal. Ed.*, **5**, 7 (1933).

[10] W. D. Amstrong, *Ibid.*, **8**, 384 (1936).

[11] W. M. Hoskins and C. A. Ferris, *Ibid.*, **8**, 6 (1936).

[12] L. M. Dubnikov and I. F. Tikhomirov, *Zavodskaya Lab.*, **13**, 787 (1947).

[13] A. A. Vasil'ev, *Ibid.*, **13**, 797 (1947).

[14] R. H. Shell and R. L. Craig, *Anal. Chem.*, **26**, 996 (1954).

[15] J. Huffmann, and G. Lundell, *Chem. Zentr.*, **109**, II, 2798 (1938).

[16] D. Dahle, H. J. Wichmann and R. U. Bonner, *J. Assoc. Offic. Agr. Chemists*, **21**, 468 (1938).

[17] D. S. Raynolds and W. L. Hill, *Ind. Eng. Chem. Anal. Ed.*, **11**, 21 (1939).

[18] M. P. Matszak and O. R. Brown, *Ibid.*, **17**, 148 (1945).

[19] R. Belcher, E. F. Caldas and S. J. Clark, *Analyst*, **77**, 602 (1952).

[20] R. Geyer, *Z. anorg. Chem.*, **252**, 42 (1944).

CHAPTER 20

The Simultaneous Determination of Oxygen and Halogen (Chlorine, Bromine, and Iodine) in Organic Compounds*

M. O. Korshun and E. A. Bondarevskaya

In the course of an investigation of a method for the direct determination of oxygen based on the method of M. O. Korshun [1-4], we found the direct simultaneous determination of oxygen and halogen to be possible.

As is well known, the essence of the direct determination of oxygen is thermal decomposition of the material in a stream of nitrogen, conversion of all the oxygen to CO over incandescent carbon, further oxidation of the CO to CO_2, and gravimetric or volumetric determination of the oxygen. Depending on the elemental composition of the compounds to be analyzed, other compounds, in addition to nitrogen and carbon monoxide, are found in the gaseous effluent from the reaction tube, among them acidic gaseous decomposition products. Since acid gases, for example hydrogen chloride or gaseous chlorine, can interfere with the determination of oxygen, it is necessary to remove them from the mixture of gases prior to oxidation of the CO to CO_2 in the oxidizer. Therefore, directly after the decomposition products, in admixture with nitrogen, leave the reaction tube, they pass through an alkaline absorbent where they are freed from acidic components.

Doklady Akad. Nauk S.S.S.R.* **110, 220-22 (1956).

Proceeding on the assumption that during the decomposition of halogen-containing compounds composed of C, H, O, and X in the presence of a sufficient amount of hydrogen the halogen must be liberated in the form of hydrogen halide, that is, a substance which has an acid character and is absorbed by alkali, we conducted experiments on the simultaneous determination of oxygen and halogen. The determination was carried out in the following manner. A weighed sample was subjected to pyrolytic decomposition. The determination was carried out in a quartz tube, as in the determination of carbon and hydrogen by the rapid method [4], and the active decomposition products, which react more easily and rapidly with carbon than do the vapors of the original material, acted on the catalyst (platinized carbon, heated to 900°). This made it possible to decrease the carbon layer to 3 cm (Oita [5] used a contact layer 9 cm long) and to decrease the length of the heating zone correspondingly. The decomposition products were carried by the stream of nitrogen into the absorption apparatus, which was filled, as in the determination of C and H, with Ascarite and Anhydrone, where the hydrogen halide was absorbed. The carbon monoxide was then passed to a copper oxide tube heated to 100° and oxidized to CO_2, which was absorbed in the same absorption apparatus as was the halide previously. The oxygen was determined by the weight of the CO_2. The results, obtained in this manner, of the simultaneous determination of oxygen and halogen are presented in Table 1.

Since the formation of hydrogen halide is due to the hydrogen formed in the reaction tube by the decomposition of the material being analyzed, as would be expected, in the analysis of material which does not contain hydrogen or which contains it in insufficient amount, difficulty was en-

TABLE 1

Material	Weight mg	Wt. increase, mg CO_2	HCl	O% Found	Diff.	Cl % Found	Diff.
Dimethyl ester of 4-chloro-5-hydroxy-1,2 decalin-dicarboxylic acid $C_{14}H_{15}O_5Cl$ (11.89 % Cl; 26.80 % O)	10.120	7.368	1.217	26.58	-0.22	11.69	-0.20
	10.318	7.506	1.233	26.44	-0.36	11.62	-0.27
γ-Methyl 1-ω'-dichloroethylene-α-pyrone $C_8H_6O_2Cl_2$ (34.64% Cl; 15.61% O)	13.582	6.082	4.818	15.96	+0.35	34.51	-0.13
Trichlorovaleric acid $C_5H_7O_2Cl_3$ (51.96 % Cl; 15.61 % O)	12.883	5.654	6.837	15.88	+0.27	51.83	-0.13
	8.514	3.632	4.631	15.51	-0.10	51.70	-0.26
Chlorotridecanoic acid, $C_{13}H_{25}O_2Cl_2$ (14.29 % Cl: 12.88 % O)	17.916	6.231	2.718	12.65	-0.23	14.64	+0.35
	17.805	6.322	2.603	12.86	-0.02	14.19	-0.10

countered. In these cases, the oxygen determination was good, but the results for halogen were quite a bit too low. As we established later, a considerable portion of the halogen was retained by the carbon, and was slowly and incompletely recovered even on lengthy passage of nitrogen through the reaction tube. This difficulty was overcome by the addition to the substance being analyzed of a material with a high content of hydrogen (for example, paraffin). This proved to be sufficient for quantitative liberation of halogen in the form of hydrogen halide. It was also shown that, during the decomposition of the weighed portion of paraffin, the halide previously retained by the packing was quantitatively recovered from the reaction tube. Thus, for example, during an analysis of hexachloroethane conducted as usual, 57.06% chlorine was found instead of 89.89%. Subsequent passage of nitrogen through the reaction tube four times for 30 minutes gave, respectively, 4.19, 4.65, 5.74, and 0.00% chlorine for a total of 71.58%, which is still 18.31% less than the calculated value. Some paraffin was then decomposed in the reaction tube, and an amount of halogen corresponding to 18.08% of the chlorine, based on the weight of the hexachloroethane previously decomposed, was recovered. Results which were too low by several percentage units were also obtained in the determination of iodine in iodobenzoic acid. In this case, liberation of elemental iodine was observed during decomposition of the sample, the iodine being easily detectable by the violet color of the vapors issuing from the reaction tube. Decomposition of iodobenzoic acid in the presence of paraffin gave correct results for both elements determined. The results of the determinations of halogen and oxygen in the presence of paraffin are presented in Table 2.

TABLE 2

Material	Weight mg	Wt. of parafin	Wt. increase CO2	Wt. increase X	O % Found	O % Diff.	X % Found	X % Diff.
Hexachloroethane C_2Cl_6 (89.89% Cl)	9.916	-	-	5.794	-	-	58.43	-31.46
	6.027	-	-	3.440	-	-	57.06	-32.83
	5.839	-	-	5.346	-	-	89.05	-0.84
	5.236	-	-	4.802	-	-	89.20	-0.60
	6.735	-	-	6.254	-	-	89.09	-0.88
Tetrachloroethylene C_2Cl_4 (85.53 % Cl)	7.746	4.209	-	6.782	-	-	85.15	-0.38
Iodobenzoic acid $C_7H_5O_2I$ (51.20 % I; 12.90%O)	9.484	0.000	3.278	4.294	12.57	-0.33	44.92	-6.28
	10.270	0.000	3.719	4.601	13.17	+0.27	44.45	-6.75
	8.440	7.638	2.991	4.391	12.89	-0.01	51.62	+0.23
	9.081	4.289	3.182	4.701	12.74	-0.16	51.36	+0.16
Bromobenzoic acid, $C_7H_5O_2Br$ (39.76Br15.93O	8.823	-	3.997	1.949	16.28	+0.35	28.82	-10.95
	8.734	4.366	3.800	3.493	15.82	-0.11	39.48	-0.29
	8.420	4.209	3.626	3.462	15.66	-0.27	39.66	-0.11

The possibility of directly determining oxygen and halogen simultaneously was first established by us. The halide can also be determined volumetrically. This extends considerably the application of the method. Moreover, in the presence of a sufficient amount of hydrogen and under the same conditions, it is apparently possible to determine simultaneously oxygen, sulfur, and halogen.

LITERATURE CITED

[1] M. O. Korshun, *Zavodskaya Lab.* **10**, 241 (1941); Dissertation, Direct Microdetermination of Oxygen in Organic Compounds (1943); Reports of the Commission on Analytical Chemistry 1 (IV), 168 (1947).

[2] E. A. Druyan-Rempel and S. A. Babushkin, *Zavodskaya Lab.* **16**, 238 (1950); **17**, 1418 (1951).

[3] N. I. Stognii and F. P. Dorosh, *Zavodskaya Lab.* **19**, *No. 1* (1953).

[4] M. O. Korshun and N. E. Gel'man, *New Methods of Microelemental Analysis* (1949) pp. 56, 27.

[5] J. Oita and H. G. Conway, *Anal. Chem.* **26**, 600 (1954).

CHAPTER 21

Simultaneous Microdetermination of Fluorine, Carbon, and Hydrogen in Elementoorganic Compounds*

M. O. Korshun, N. E. Gel'man, and K. I. Glazova

In a previously published work [1], we showed that quantitative combustion of organic fluorides is possible during their pyrolytic decomposition in a bed of incandescent magnesium oxide in the presence of gaseous oxygen. The remarkable property of magnesium oxide to retain the fluorine completely was proposed as a method for the simultaneous microdetermination of carbon, hydrogen, and fluorine, the carbon and hydrogen being determined as usual [2] and the fluorine being determined from the change in weight of the quartz tube containing the magnesium oxide in which the fluorine is absorbed in the form of magnesium fluoride.

Further work on this method, carried out by N. S. Shevalova on a large number of experimental substances, showed that combustion of the pyrolysis products in the magnesium oxide bed gave exceptionally reliable results for carbon and hydrogen of organic fluorine compounds of varying composition and properties. The gravimetric determination of fluorine in contrast was only approximate in spite of its quantitative absorption by magnesium oxide, owing to the unfavorable weight factor in the calculation and to the insufficient constancy of weight of the

Doklady Akad. S.S.S.R.* **111, 1255-56 (1956)

quartz tube containing the magnesium oxide, in which the combustion was carried out.

Our further investigations were directed toward finding a route for the quantitative determination of the fluorine, which is in the magnesium oxide bed in the form of magnesium fluoride, and thereby to obtain a reliable method for the simultaneous determination of all three elements. For this purpose, we investigated the ability of the fluorides of certain metals to undergo pyrohydrolysis, i.e., to decompose under the action of superheated steam with the liberation of HF and metal oxide [3, 4]. The most favorable conditions for the completion of pyrohydrolysis of several milligrams of MgF_2 in 20-25 minutes were a temperature of 1000° and a steam rate corresponding to 0.5 ml of condensate per minute. At a lower temperature (900°), the time required for the pyrohydrolysis increased sharply and was measured in hours; an increase in the temperature above 1000° during the work with magnesium oxide in a quartz apparatus required caution, owing to rapid destruction of the quartz tubes. An increase in the steam rate led to the formation of a dense plug in the granular magnesium oxide material which prevented rapid pyrohydrolysis; lowering the steam rate to 0.3 ml of condensate per minute considerably retarded the process.

These observations formed a basis for a method for the simultaneous determination of carbon, hydrogen, and fluorine which consists of the following. The material to be analyzed (4-10 mg) is burned in a stream of oxygen in a bed of granular magnesium oxide, 15-18 cm long, heated to 1000°. The magnesium oxide is contained in a quartz capsule with a screen bottom, the capsule being placed in an empty tube for the combustion. At the conclusion of the combustion, the absorption apparatus is weighed as usual, and the capsule with the magnesium

TABLE 1

Name and formula of substance	C % Found	C % Calc.	C % Diff.	H % Found	H % Calc.	H % Diff.	F % Found	F % Calc.	F % Diff.
p-Fluorocinnamic acid, $C_9H_8O_2F$	65.15	65.06	+0.09	4.22	4.25	−0.03	11.30	11.44	−0.14
	65.26	65.06	+0.21	4.25	4.25	-	11.41	11.44	−0.03
Trifluoroacetamide $C_2H_2ONF_3$	21.38	21.25	+0.13	2.01	1.78	+0.23	50.31	50.42	−0.11
	21.31	21.25	+0.06	1.88	1.78	+0.11	50.29	50.42	−0.13
Polytetrafluoroethylene C_2F_4N	23.96	24.02	−0.06	0.10	-	+0.10	76.19	75.98	+0.21
	24.04	24.02	+0.02	-	-	-	76.06	75.98	+0.09
Perfluorotriethylamine $C_6F_{15}N$	19.60	19.42	+0.18	0.11	-	+0.11	77.09	76.80	+0.29
	19.53	19.42	+0.11	0.19	-	+0.19	76.61	76.80	−0.19
Diphenylenephenyloxonium borofluoride $C_{18}H_{13}OBF_4$	64.90	65.11	−0.21	4.09	3.94	+0.15	22.83	22.88	−0.05
	65.17	65.11	+0.05	3.97	3.94	+0.03	22.85	22.88	−0.03
β-Fluoroethylnitrophenyl methylthiophosphonate $C_9H_{11}O_4NPSF$	38.73	38.69	+0.04	4.03	3.96	+0.07	6.88	6.80	+0.08
	38.71	38.69	+4.02	4.01	3.96	+0.05	6.91	6.80	+0.11
α-Hydroperfluoropropionamide + trimethyl(γ-diethylaminopropyl)silane $C_3H_6ONF_4 + C_{10}H_{25}$Sin	44.76	44.96	−0.20	7.93	7.91	+0.02	25.60	25.54	+0.06
	45.92	45.72	+0.20	8.27	8.14	+0.13	24.43	24.52	+0.09

Note: All analyses were carried out on samples of 5-9 mg.

oxide is treated with superheated steam under the conditions indicated above. As the pyrohydrolysis proceeds, the hydrogen fluoride enters the receiver along with the condensate, and the fluorine is determined by titration. A single determination of C, H, and F requires 1.5 hours. The magnesium oxide, after a short drying, is reused.

If the material being analyzed contains other elements besides C, H, O, N, and F, for example phosphorus, boron, or silicon, which are retained during the combustion by the magnesium oxide in the form of compounds not decomposable by steam, the fluorine in the condensate can be titrated either by alkali or thorium nitrate. In the presence of other halogens or sulfur, which form pyrohydrolyzable salts with magnesium oxide, the fluorine can be titrated only with thorium nitrate. According to our observations, sulfur is completely retained by the magnesium oxide, while chlorine and bromine are only partially retained. Metallic silver is used, as usual, for the absorption of halides [2]. The titer of the alkali or thorium nitrate is established by combustion of pure organofluorine compounds.

By means of the method described, it is possible to determine carbon, hydrogen, and fluorine in a single sample of an organic compound containing phosphorus, boron, silicon, halogen, sulfur, and nitrogen. The results of some analyses are presented in Table 1.

LITERATURE CITED

[1] N. E. Gel'man and M. O. Korshun, *Doklady Akad. Nauk S.S.S.R* **89**, 685 (1953).

[2] M. O. Korshun and N. E. Gel'man, *New Methods of Elementa Microanalysis* (Moscow, 1949).

[3] J. C. Warf, W. D. Cline and R. D. Tevebaugh, *Anal. Chem.* **26**, 343 (1954).

[4] C. D. Susano, J. C. White and J. E. Lee Jr., *Anal. Chem.* **27**, 453 (1955).

CHAPTER 22

The Simultaneous Determination of Carbon, Hydrogen, and Silicon in Organic Silicon Compounds*

V. A. Klimova, M. O. Korshun, and E. G. Bereznitskaya

The chemistry of organic silicon compounds has now become a large independent field of organic chemistry, and many organic silicon compounds have acquired not only theoretical but also practical significance. All this has naturally led to an increase in the demands with respect to methods for the quantitative analysis of organic silicon compounds. Nevertheless, despite the rather large number of publications concerned with the analysis of organic silicon compounds, present methods do not always satisfy research workers. This is particularly true in the category of elementary analysis.

The analysis for carbon and hydrogen was generally performed by combustion methods based on the Liebig-Pregl techniques [1]. In the course of the analyses it was discovered that, when submitted to thermal decomposition, organic silicon compounds form silicon carbide and as a result cannot be quantitatively oxidized. It should be noted that the formation of silicon carbide takes place more easily the higher the initial decomposition temperature of the compound [2].

In 1941 Rochow [3] proposed a macromethod for the catalytic combustion of silicon-containing high polymers in a stream of oxygen for the simultaneous determination

Zhur. Anal. Khim.* **11, 223-232 (1956).

of carbon, hydrogen, and silicon, and in 1948 [4] he presented a semimicro modification of his method. Taking account of the peculiar behavior of organic silicon compounds during their thermal decomposition, Rochow combats silicon carbide formation by performing the combustion in two stages. He recommends that the sample be heated to only 300-500° at first. When this is done, the substance vaporizes slowly and then the vapors, mixed with oxygen, pass on to a platinum gauze heated to 850° where they are oxidized to carbon dioxide, water, and silicon dioxide. The carbon and hydrogen are determined in the usual manner and the silicon dioxide formed is weighed together with the combustion tube, in which it is retained by means of a plug of glass wool.

In a number of cases this method of analysis gives positive results, i.e., it does make it possible to avoid the formation of silicon carbide. Nonetheless, it has serious shortcomings, for silicon dioxide, precipitating on the platinum, quickly deactivates it. Furthermore, the necessity of weighing the whole combustion tube in order to determine the silicon forced Rochow to reduce the dimensions of the tube to a minimum. Its length for this method is 18 cm. The combustion of easily volatile compounds in such a short tube is practically impossible, and the weighing of an 18 cm tube with a very small quantity of silicon dioxide can lead to appreciable errors in the silicon determination, even when a microbalance is used.

The drawbacks enumerated were evidently the reason that the Rochow method failed to receive wide acceptance, especially for the determination of silicon. We feel, however, that his research constitutes a tremendous step forward in comparison with previous methods. Indeed, Rochow was one of the first scientists to stop mechanically applying the method of analyzing organic compounds to organic silicon compounds and to take into consideration the specific properties of the latter in developing his method. In addition, Rochow quite rightly followed the course of simultaneously determining three elements in a single sample, mentioning in his paper that in the quantitative combustion of organic silicon compounds in a stream of oxygen nothing interferes with a simultaneous determination of silicon along with carbon and hydrogen, since the silicon is completely converted to silicon dioxide.

Except for Rochow, there has not been a single author who has attempted to develop a method for the simultaneous determination of carbon, hydrogen, and silicon. Inasmuch as the Rochow method could not satisfy us any more than it could other scientists, we made it our goal to develop a reliable method of simultaneously determining carbon, hydrogen, and silicon in organic silicon compounds. Our work was begun in 1949 in the microanalytical laboratory of the Institute of Organic Chemistry of the Academy of Sciences of the USSR, and by 1950 we had already performed several analyses [5].

As a foundation for our work we took our previously developed micro and semimicro method for the rapid determination of carbon and hydrogen in a fast flowing stream of oxygen in an unfilled combustion tube [6]. The basic principle of the rapid method is as follows: the organic substance is subjected to rapid pyrolytic decomposition almost without admission of oxygen at a temperature of 850-900°. The decomposition products are mixed with a large excess of oxygen and are oxidized by it at 900-950° (the flow rate of the oxygen is 35-50 cc per min). Slow vaporization of the substance is not used in analysis by the rapid method. Absorption of the interfering elements (N, S, Cl, Br, I) is accomplished outside the combustion tube. O, and X in the presence of a sufficient amount of h The apparatus used is simpler than the one for the determination of carbon and hydrogen by the Pregl method [7].

Use of an unfilled combustion tube and absorption of the interfering elements outside of it enabled us to undertake the problem of creating methods for the simultaneous determination of several elements with a single sample of substance [8].

Making use of this principle, we investigated first the behavior of various classes of organic silicon compounds during their pyrolytic decomposition and at the same time tried to discover why the above-described Rochow procedure with its two-stage combustion makes it possible to avoid the formation of silicon carbide.

As our earlier research [9] showed, silicon carbide does not form from all classes of organic silicon compounds during combustion as ordinarily performed in the rapid method. Compounds with naphthyl radicals, which contain little hydrogen, as a rule burn without carbide formation, while carbide forms particularly easily in compounds of the aliphatic series. Furthermore, if some water is added to a compound with naphthyl radicals before combustion, silicon carbide then forms during the combustion of the compound so moistened. For example, the following results were obtained during the combustion of methylnaphthylsilanediol. A black unburned residue is left in the capsule after combustion with water.

Substance	Calculated		Found	
			For ordinary combustion	For combustion with water
$CH_3C_{10}H_7Si(OH)_2$	C %	64.90	65.19	64.56
	Si %	13.70	13.78	14.50

In the light of these data it may be presumed that the following process for the oxidation of the substance takes place during combustion by the Rochow method. Heating a sample of the substance under investigation to 300-500° leads not only to its vaporization but also to its partial decomposition. In this case, oxidation of a large part of the hydrogen in the compound takes place first, followed by vaporization of the product water from the combustion zone. This process may be recognized by the formation of drops of water in the tip of the combustion tube, these drops appearing there at the very start of the combustion.

Removal of the water from the decomposition zone apparently promotes further quantitative oxidation. Hence we may conclude that the presence of a large amount of water in the decomposition zone promotes the formation of silicon carbide.

Since, in rapid combustion, pyrolysis of the substance being analyzed takes place in a quartz capsule, there is a possibility, in the analysis of organic silicon compounds that do not form carbide, of simultaneously determining not only carbon and hydrogen but also silicon, by weighing the capsule containing the silicon dioxide formed. In this way it is possible to determine C, H, and silicon simultaneously in a single 3-15-mg sample. It is, of course, essential that all the silicon dioxide should remain in the capsule. This is actually what happened when the pyrolysis was performed rapidly and completely, and we easily succeeded in making such determinations with compounds having naphthyl and alkoxyl radicals.

On the other hand, tetraalkylsilanes required special combustion techniques; the pyrolysis had to be performed at lower temperatures and hence more slowly and not quite completely. In the case of these compounds it was impossible to accomplish the decomposition without silicon carbide formation in any other way. However, this method of decomposition could not entirely satisfy us, for part of the silicon dioxide was carried out of the capsule and the error of its determination reached 0.8%.

In order to avoid the formation of silicon carbide during rapid and complete pyrolysis and to obtain better analytical results for silicon, we went over to an operation in the presence of a catalyst [10]. As the catalyst, we used a special preparation of chromic sesquioxide on asbestos, which was placed dircetly in the decomposition capsule.

Fedoseev and Pavlenko [11] had used chromium sesquioxide as a catalyst for surface combustion in macroelementary analysis, and in our case as well, it proved a sufficiently active oxidant, affording the opportunity of avoiding the formation of silicon carbide in the thermal decomposition of organic silicon compounds. Owing to its fibrous structure, asbestos made an excellent filter, collecting the finely dispersed silicon dioxide. By decomposing the substance by the rapid method in a decomposition capsule filled with Cr_2O_3-on-asbestos, we succeeded in performing the simultaneous determination of C, H, and Si in many organic silicon compounds of very different structure. The presence of oxygen and nitrogen in the substance did not interfere with the analysis.

In 1951-1952, Kreshkov and Guretskii [12] simultaneously but independently developed a method of analyzing for C and H in organic silicon compounds. Unlike us, they did not try to perform the silicon determination jointly with the C and H analysis. The authors, in their own words, started with our rapid method for C and H analysis in organic compounds, but in their method they departed from the fundamental principle of the rapid method, not having understood it clearly enough. Indeed, Kreshkov and Guretskii effect slow vaporization of the substance in the heated zone. Yet, in describing the rapid method, we had repeatedly emphasized that the procedure for decomposing the substance was the decisive factor in this method [13]. In order that the oxidation with a rapid flow of oxygen should take place quantitatively, it is absolutely essential that the substance should not be vaporized but instead should be pyrolyzed beforehand so that pyrolytic decomposition products and not sample vapor should undergo oxidation. Inasmuch as Kreshkov and Guretskii recommend slow vaporization, it is not understandable why in this case they use a capsule instead of a boat, which would unquestionably be more suitable for vaporization, and why they treat the sample with cupric oxide, which cannot promote oxidation of the compound at low temperatures.† In applying the combustion method that they describe, Kreshkov and Guretskii actually resorted to the Rochow method, having complicated it with the extra operation of the cupric oxide addition and having eliminated the silicon determination. The rapid stream of oxygen and the absence of a platinum catalyst obviously do not change the state of affairs, but only force the authors to lengthen the heated zone and raise its temperature to 1050°.‡

Four methods have now been proposed for determining the elements in organic silicon compounds by combustion of these compounds in a stream of oxygen. In Table 1 we have collated the most characteristic features of each of these methods. To date, there has not been one reliable method of analyzing for carbon, hydrogen, and silicon. The method that we developed permits an easy and rapid performance of such a determination. We describe the

† According to our observations [10], cupric oxide, and platinum as well, do not prevent the formation of silicon carbide, even at high temperatures.

‡ We take this opportunity to express our perplexity over the confused and incorrect presentation of our methods in the book by Kreshkov and his coauthors [14].

TABLE 1

Methods	Pregl or Friedrich	Rochow	Kreshkov and Guretskii	Ours
Elements being determined	C and H	C, H, and Si	C and H	C, H, and Si
Oxygen flow, ml/min	4-5	4-5	30-50	30-50
Combustion tube charge	Universal or platinum catalyst	Platinum catalyst	Quartz	None
Length of oxidation zone, mm	200	100-120	335	110
Temperature conditions:				
a) temperature of vaporization or decomposition of the substance	750°	300-500°	300-500°[a]	850-900°
b) temperature of the oxidation zone of the combustion tube	750°	850°	900-1050°	900-950°
Combustion time, min	35-40	35-40	35-40[b]	10-20
Oxidants mixed with the sample of substance	$K_2Cr_2O_7$	None	CuO	Cr_2O_3+asbestos
Decomposition procedure	Slow vaporization (boat)	Slow vaporization (boat)	Slow vaporization (capsule)	Rapid pyrolysis (capsule)

[a] Value assumed from authors' description of their heating.
[b] Estimated time.

analytical method below and present the results that we obtained in the determination of C, H, and Si.

THE ANALYTICAL PROCEDURE

Preparation of the Catalyst. 1-2 g of ammonium chromate or dichromate are heated by a small flame in a 50-ml crucible. The chromium sesquioxide powder obtained is mixed with fibrous asbestos previously calcined at 1200° for 4-6 hours [15]. The catalyst is screened on a sieve with openings 2-3 mm in diameter in order to separate it from the asbestos dust and excess chromic oxide. The catalyst prepared in this manner is green and has a fibrous structure. To remove from it any moisture that might have been adsorbed during screening, it is calcined in the combustion tube in a stream of oxygen at 900-950°. A Pregl tube filled with Anhydrone is connected to the end of the combustion tube, and the former is weighed every hour. When it begins to maintain a constant weight within the limits of 100-300 μg, the drying is discontinued. This process generally lasts 3-4 hours.

Apparatus. The usual apparatus for the rapid method of C and H determination is used (Fig. 1).

Performance of the Analysis. The combustion is performed with the rapid method for C and H analysis at an O_2 flow rate of 35-50 ml/min and a furnace temperature of 900-950°, in an unfilled combustion tube.

In the case of solid compounds or nonvolatile liquids a 3-12-mg sample is weighed out in a quartz capsule 60-80 mm long and 7-8 mm in diameter. Samples of volatile liquids are taken in a capillary. The sample should be carefully placed on the bottom of the capsule. If the material strikes the sides of the capsule, silicon losses may occur during combustion. After the sample has been weighed out and the weight of the empty capsule and the

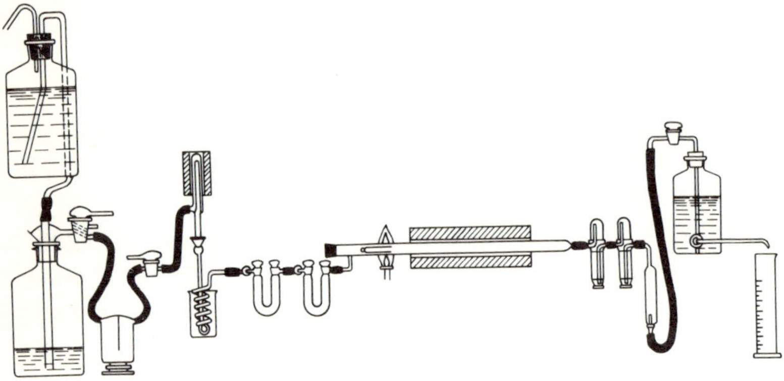

Fig. 1. Diagram of the apparatus for the simultaneous determination of carbon, hydrogen, and silicon.

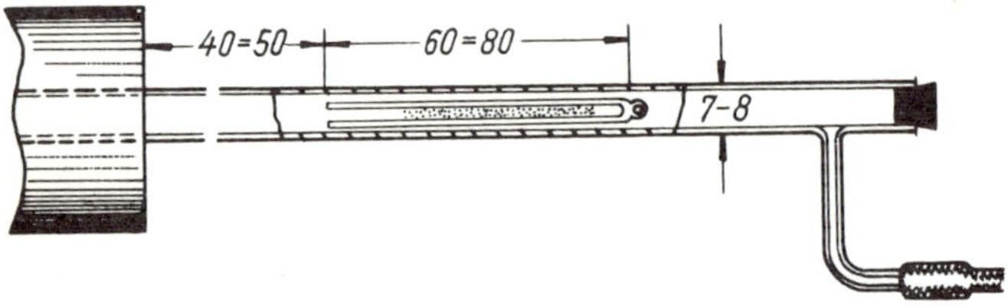

Fig. 2. Location of capsule with sample and catalyst inside the combustion tube.

capsule plus sample has been recorded, the catalyst is added, the capsule being filled to about 3/4 of its length (Fig. 2) and then weighed again. The catalyst should lie in a loose layer and 100-200 mg is usually needed. If the sample is taken in a capillary, the capillary is inserted into the capsule with the catalyst and broken in it. The capillary should be inserted all the way into the capsule; otherwise the glass may fuse to the combustion tube. The capsule is placed in the combustion tube at a distance of 4-5 cm from the furnace. Before the capsule is inserted, the H_2O and CO_2 absorption tubes are connected, and if nitrogen is present in the compound, a tube for absorbing nitrogen oxides is placed between them [16], and the combustion is begun with a roaring burner flame, the burner being moved gradually from the open end of the capsule to the closed end, i.e., counter to the oxygen flow. The combustion of the substance is accomplished somewhat more slowly than in the case of the rapid determination of C and H in organic compounds, so that abrupt explosion of the substance will not occur, for this would eject the catalyst from the capsule. The progress of the combustion is followed by the evolution of water vapor, which condenses in the tip of the combustion tube. 10 to 20 minutes is generally required for complete combustion. A control determination should be made by burning an organic substance, e.g., saccharose, under the same conditions and, if necessary, introducing suitable corrections for C, H, and Si into the calculations.

When the combustion is over, the water and carbon dioxide absorption tubes, and also the capsule with the catalyst and silicon dioxide remaining in it, are weighed.

Calculation: The calculations are made with the formula

$$\%\,C = \frac{K_{CO_2} \cdot a \cdot 100}{N}\ ;\quad \%H = \frac{K_{H_2O} \cdot b \cdot 100}{N}\ ;\quad \%\ Si = \frac{K_{SiO_2} . m . 100}{N}$$

where $K_{CO_2} = 0.2729$; $K_{H_2O} = 0.1119$; $K_{SiO_2} = 0.4672$; a is the CO_2 weight gain in mg; b is the H_2O weight gain in mg; m is the SiO_2 weight gain in mg; and N is the sample weight in mg.

Analytical results. Using this method, we performed C, H, and Si analyses on more than 100 organic silicon compounds of very different structures (Tables 2, 3, and 4).

TABLE 2

Compounds Containing C, H, and Si

No.	Substance	% C Found	% C Calc.	% C Diff.	% H Found	% H Calc.	% H Diff.	% Si Found	% Si Calc.	% Si Diff.
1a	Dibutyldiallylsilane	75.01	74.92	+0.09	12.62	12.58	+0.04	12.37	12.50	-0.13
	$(C_4H_9)_2Si(CH_2CH=CH_2)_2$	74.98	74.92	+0.06	12.66	12.58	+0.08	12.70	12.50	+0.20
2a	Methylnaphthyldiallylsilane	80.83	80.89	-0.06	8.02	7.99	+0.03	11.25	11.12	+0.13
	$C_{10}H_7CH_3Si(CH_2CH=CH_2)_2$	80.84	80.89	-0.05	8.11	7.99	+0.12	11.04	11.12	-0.08
3a	Methylphenyldiallylsilane	77.27	77.17	+0.10	8.82	8.96	-0.14	13.90	13.87	+0.03
	$C_6H_5CH_3Si(CH_2CH=CH_2)_2$	77.11	77.17	-0.06	8.94	8.96	-0.12	13.60	13.87	-0.27
4a	Tetravinylsilane	70.49	70.52	-0.03	8.91	8.88	+0.03	20.82	20.60	+0.22
	$(CH_2=CH)_4Si$	70.71	70.52	+0.19	8.92	8.88	+0.04	20.71	20.60	+0.11
5a	1,2-bis(trimethylsilyl)ethylene	56.06	55.75	+0.31	11.82	11.69	+0.13	32.59	32.56	+0.03
	$(CH_3)_3SiCH=CHSi(CH_3)_3$	55.99	55.75	+0.27	11.69	11.69	+0.00	32.35	32.56	-0.21
6b	Diethyltetramethylenesilane	67.77	67.53	+0.24	12.85	12.75	+0.10	19.70	19.72	-0.02
	CH_2-CH_2 / | $Si(C_2H_5)_2$ / CH_2-CH_2	67.70	67.53	+0.17	12.60	12.75	-0.15	19.55	19.72	-0.17
7c	Butyldiallylsilane	71.55	71.35	+0.20	11.95	11.97	-0.02	16.25	16.68	-0.43
	$C_4H_9SiH(CH_2CH=CH_2)_2$	71.36	71.35	+0.01	11.90	11.97	-0.07	16.60	16.68	-0.08
8c	1-Trimethylsilyl-1-butene	65.64	65.54	+0.10	12.76	12.57	+0.19	22.14	21.88	+0.26
	$(CH_3)_3SiCH=CH-CH_2CH_3$	65.67	65.54	+0.13	12.41	12.57	-0.16	22.01	21.88	+0.13

9c	2-Trimethylsilyl-1-butene $(CH_3)_3SiC(C_2H_5){=}CH_2$	65.67 65.73	65.54 65.54	+0.13 +0.19	12.41 12.45	12.57 12.57	−0.16 −0.12	21.97 22.00	21.88 21.88	+0.09 +0.12
10c	2-Trimethylsilyl-2-butene $(CH_3)_3SiC(CH_3){=}CH{-}CH_3$	65.41 65.51	65.54 65.54	−0.13 −0.03	12.47 12.43	12.57 12.57	−0.10 −0.14	21.76 22.00	21.88 21.88	−0.12 +0.12
11d	Tripropylvinylethynylsilane $(C_3H_7)_3SiC\equiv C{-}CH=CH_2$	74.81	74.92	−0.12	11.70	11.61	+0.09	12.85	13.46	−0.61
12d	Dimethyldivinylethynylsilane $(CH_3)_2Si(C\equiv C{-}CH=CH_2)_2$	74.58	74.96	−0.38	8.08	7.96	+0.12	17.12	17.50	−0.38
13d	Dipropyldivinylethynylsilane $(C_3H_7)_2Si(C\equiv C{-}CH=CH_2)_2$	77.70	77.67	+0.03	9.11	9.24	−0.13	13.28	12.94	+0.32
14d	Dimethylphenylvinylethynylsilane $(CH_3)_2C_6H_5SiC\equiv C{-}CH=CH_2$	77.50	77.35	+0.15	7.38	7.52	−0.14	14.80	15.08	−0.28
15d	Dimethyltolylvinylethynylsilane $(CH_3)_2(CH_3C_6H_4)SiC\equiv C{-}CH=CH_2$	77.89	77.91	−0.02	7.93	7.99	−0.06	14.20	14.00	+0.20
16d	Triethyl-1-octynylsilane $(C_2H_5)_3SiC\equiv C{-}CH_2CH_2C_4H_9$	74.98	74.92	+0.06	12.62	12.57	+0.05	12.37	12.51	−0.14
17d	Triethyl-(γ-triethylsilylbutynyl-δ-butyl) butynylsilane $(C_2H_5)_3SiC\equiv C{-}CH(CH_2{-}C_4H_9)CH_2CH_2C\equiv C{-}Si(C_2H_5)_3$	73.77	73.77	0.00	11.78	11.87	−0.09	14.43	14.36	+0.07

TABLE 2—continued

Compounds Containing C, H, and Si

No.	Substance	% C Found	% C Calc.	% C Diff.	% H Found	% H Calc.	% H Diff.	% Si Found	% Si Calc.	% Si Diff.
18d	Triethyl(2-ethyl-S-butylbutynyl) silane $(C_2H_5)_3SiC\equiv C-CH(C_2H_5)CH_2-C_4H_9$	76.01	76.11	-0.10	12.62	12.77	-0.15	11.27	11.11	+0.16
19d	1,3-bis(triethylsilylethynyl)—l-n-pentyl-3-ethylpropane $(C_2H_5)_3Si-C\equiv C-CH-C_2H_5$ / CH_2 / $(C_2H_5)_3Si-C\equiv C-CH-C_5H_{11}$	74.17	74.50	-0.33	11.96	12.03	-0.07	13.69	13.47	+0.22

Note:

a) these compounds were kindly supplied us by V. F. Mironov, M.Sc., from A.D. Petrov's (corresponding member of the Academy of Sciences of the USSR) Laboratory of the Chemistry of Hydrocarbons, Institute of Organic Chemistry of the Academy of Sciences of the USSR; b) these compounds by N. A. Momma from B. A. Kazanskii's Laboratory of Catalytic Synthesis, Inst. Org. Chem., Acad. Sci. USSR; c) these compounds by G. I. Nikishin from A. D. Petrov's Laboratory; d) cf. *Izvest. Akad. Nauk SSSR, Otdel. Khim. Nauk* 1954, 723. A. D. Petrov, S. I. Sadykhzade and Yu. P. Egorov.

TABLE 3
Compounds Containing C, H, Si, and O

1b	1,2-Dibutyltetraethyldisiloxane	63.81	63.51	+0.30	12.68	12.67	+0.01	18.73	18.55	+0.18
		63.73	63.51	+0.22	12.61	12.67	-0.06	18.59	18.55	+0.04
2c	Hexamethyldisiloxane	44.37	44.39	-0.02	11.12	11.18	-0.06	34.74	34.57	+0.17
	$(CH_3)_3SiOSi(CH_3)_3$	44.14	44.39	-0.25	11.22	11.18	+0.04	34.72	34.57	+0.15
3e	Trimethyl-α-hydroxyethylsilane	50.63	50.80	-0.17	11.79	11.85	-0.06	23.54	23.79	-0.25
	$(CH_3)_3SiCH(OH)CH_3$	50.50	50.80	-0.30	12.01	11.85	+0.16	23.49	23.79	-0.30
4f	Sec.-Amyltriethylsilyl- acetal	63.67	63.36	+0.31	12.20	12.27	-0.07	10.96	11.38	-0.42
	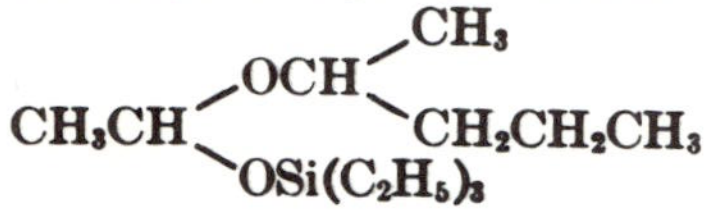	63.67	63.36	+0.31	12.17	12.27	-0.10	11.26	11.38	-0.12
5f	Sec.-Amyldimethylethylsilyl-	60.57	60.49	+0.08	12.10	12.00	+0.10	13.16	12.85	+0.31
	acetal 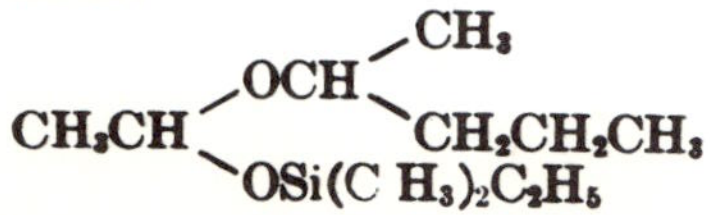	60.48	60.49	-0.01	12.08	12.00	+0.08	12.82	12.85	-0.03

TABLE 3—continued

Compounds Containing C, H, Si, and O

6f	Sec.-Amyldiethylmethylsilyl acetal	62.06 62.21	62.01 62.01	+0.05 +0.20	12.03 12.11	12.14 12.14	-0.11 -0.03	12.04 12.06	12.07 12.07	-0.03 -0.01
	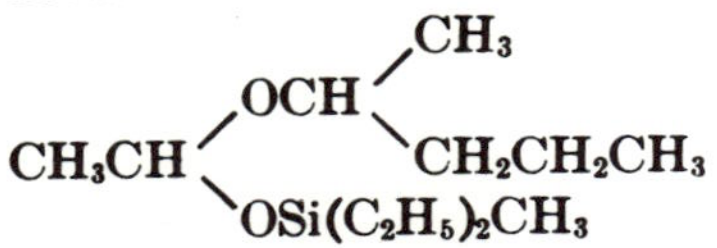									
7f	Methyl(trimethylsilylpropyl)-acetal	56.53 56.50	56.73 56.73	-0.20 -0.23	11.32 11.50	11.65 11.65	-0.33 -0.15	14.74 14.45	14.74 14.74	0.00 -0.29
	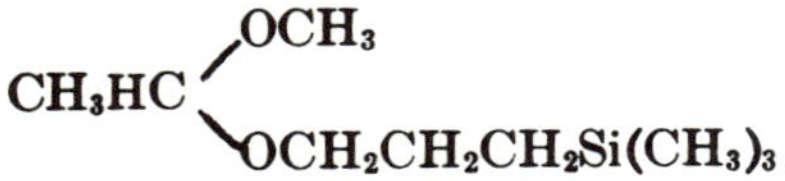									

Note:

b) See the note to Table 2; c) see the note to Table 2; e) these compounds were kindly supplied us by V. A. Ponomarenko, M. Sc., from A. D. Petrov's (Corresponding Member of the Academy of Sciences of the USSR) Laboratory of the Chemistry of Hydrocarbons, Institute of Organic Chemistry of the Academy of Sciences of the USSR; f) these compounds were kindly supplied us by I. A. Shikhiev, M. Sc., from Dr. M. F. Shostakovskii's Laboratory of Vinyl Compounds, Institute of Organic Chemistry of the Academy of Sciences of the USSR.

TABLE 4—Organic Silicon Compounds Containing Nitrogen

1	β-(Dimethylethylsilylmethoxy) propionitrile $(CH_3)_2C_2H_5SiCH_2OCH_2CH_2CN$	56.13 56.12	56.10 56.10	+0.03 +0.02	10.00 10.10	10.00 10.00	0.00 +0.10	16.29 16.26	16.38 16.38	−0.09 −0.12
2	β-(Trimethylsilylpropoxy) propionitrile $(CH_3)_3SiCH_2CH_2CH_2OCH_2CH_2CN$	58.41 58.18	58.33 58.33	+0.08 −0.15	10.54 10.53	10.44 10.44	+0.10 +0.09	14.71 15.03	15.14 15.14	−0.43 −0.01
3	Methyldimethoxysilylmethylethyl-amine $CH_3Si(OCH_3)_2CH_2NHC_2H_5$	44.16 44.37	44.18 44.18	−0.02 +0.19	10.31 10.41	10.49 10.49	+0.18 −0.08	17.11 16.86	17.19 17.19	−0.08 −0.33
4	Trimethylsilylmethylethylamine $(CH_3)_3SiCH_2NHC_2H_5$	54.79 54.91	54.88 54.88	−0.09 +0.03	13.16 13.07	13.07 13.07	+0.09 0.00	21·31 21.52	21.38 21.38	−0.07 +0.14
5	β-(Trimethylsilylmethoxy) propionitrile $(CH_3)_3SiCH_2OCH_2CH_2CN$	53.61 53.46	53.46 53.46	+0.15 0.00	9.77 9.61	9.61 9.61	+0.16 0.00	17.43 17.84	17.84 17.84	−0.41 0.00
6	Siliconeopentylethyl-β-cyanoethyl-amine $(CH_3)_3SiCH_2N\left\langle\begin{matrix}C_2H_5\\CH_2CH_2CN\end{matrix}\right.$	58.66 58.66	58.62 58.62	+0.04 +0.04	10.95 11.04	10.95 10.95	0.00 +0.09	15.45 15.42	15.24 15.24	+0.21 +0.18

Note:
The compounds in this table were generously furnished us by V. F. Mironov, M. S., from A. D. Petrov's (Corresponding Member of the Academy of Sciences of the USSR) Laboratory of the Chemistry of Hydrocarbons, Institute of Organic Chemistry of the Academy of Sciences of the USSR.

SUMMARY

A method was developed for the simultaneous direct determination of C, H, and Si in organic silicon compounds of any structure. The error of the determination does not exceed ±0.3% for C and H, or ±0.4% for Si.

LITERATURE CITED

[1] a) G. Meier [Meyer], *Analiz i opredelenie stroeniya organicheskikh veshchesty* [Analysis and Determination of Structure of Organic Substances], p. 176, 1935; b) E. G. Rochow, *An Introduction to the Chemistry of the Silicones*, p. 162, New York, 1951.

[2] H. Gilman, B. Hofferth, H. W. Melvin and G. E. Dunn, *J. Am. Chem. Soc.*, **72**, 5767 (1950).

[3] E. G. Rochow, W. F. Gilliam, *J. Am. Chem. Soc.*, **63**, 798 (1941).

[4] E. G. Rochow and W. F. Gilliam, *Ibid.*, **70**, 2170 (1948).

[5] a) A. D. Petrov and V. S. Chugunov, *Doklady Akad. Nauk SSSR*, **73**, 323 (1950), b) A. D. Petrov and V. F. Mironov, *Ibid.*, **75**, 707 (1950).

[6] M. O. Korshun and V. A. Klimova, *Zhur. Anal. Khim.*, **2**, 274 (1947).

[7] M. O. Korshun and N. E. Gel'man, *Novye Metody Elementarnogo Mikroanaliza* [New Methods of Elementary Microanalysis], Goskhimizdat [State Publishing House of Chemical Literature], Moscow, (1949).

[8] a) M. O. Korshun and N. S. Sheveleva, *Doklady Akad. Nauk SSSR*, **60**, 63 (1948); b) M. O. Korshun and N. S. Sheveleva, *Zhur. Anal. Khim.*, **7**, 104 (1952); c) N. E. Gel'man and M. O. Korshun, *Doklady Akad. Nauk. SSSR*, **89**, 685 (1953; d) V. A. Klimova, M. O. Korshun and E. G. Bereznitskaya, *Ibid.*, **96**, 287 (1954); e) M. O. Korshun E. A. Terent'eva and V. A. Klimova, *Zhur. Anal. Khim.*. **9**, 275 (1954).

[9] V. A. Klimova, M. O. Korshun and E. G. Bereznitskaya, *Doklady Akad. Nauk SSSR*, **84**, 1175 (1952).

[10] V. A. Klimova, M. O. Korshun and E. G. Bereznitskaya, *Ibid.*, **96**, 81 (1954).

[11] P. N. Fedoseev and M. M. Pavlenko, *Zhur. Anal. Khim.*, **5**, 296 (1950); **6**, 318 (1951); **8**, 158 (1953).

[12] a) I. Ya. Guretskii, M. S. Thesis, The D. I. Mendeleev Moscow Institute of Chemical Engineering, 1951; b) A. P. Kreshkov and I. Ya. Guretskii, *Trudy Moskov. Khim.-Teknol. Inst. im. D. I. Mendeleeva*, **17**, 7 (1952).

[13] M. O. Korshun, *Zhur. Anal. Khim.*, **7**, 96 (1952).

[14] A. P. Kreshkov, V. A. Bork, L. V. Myshleeva and G. D. Nessonova, *Analiz kremniiorganicheskikh soedinenii* [Analysis of Organic Silicon Compounds], Goskhimizdat [State Publishing House of Chemical Literature], 1954.

[15] K. K. Strelkov, *Zhur. Priklad. Khim.*, **26**, 1091 (1953).

[16] V. A. Klimova and M. O. Korshun, *Zhur. Anal. Khim.*, **6**, 232 (1951).

CHAPTER 23

Simultaneous Determination of Carbon, Hydrogen, Silicon, and Halogen*

V. A. Klimova and E. G. Bereznitskaya

Determination of halogens in halogen-containing organosilicon compounds is carried out independently of C, H, and Si, or simultaneously with Si. Methods for halogen determination differ according to whether halogen in the compound is directly connected to silicon or is found in the organic radical [1-4]. In the first instance, the material is usually first hydrolyzed and then the halogen is determined gravimetrically or volumetrically. Nevertheless, such determinations can be troublesome if there are two halogen atoms connected to the silicon, and even more so when there are three halogen atoms involved. The polymerized siloxanes formed in such cases coat the still undecomposed part of the material and prevent quantitative hydrolysis. Dissolving the compound in ether or alcohol beforehand does not always help one to get better results.

In cases where the halogen is not linked to silicon but to carbon, one cannot just resort to simple hydrolysis; more drastic decomposition is required here; usually this is carried out by fusing the material in a nickel bomb with sodium peroxide [5, 6]. When this melt has been dissolved, it is possible to determine halogen and silicon simultaneously; this is the advantage of this particular method. In addition, since fusion is carried out in a

Zhur. Anal. Khim.* **11, 292-298 (1956).

bomb, it is possible to analyze volatile compounds. On the other hand, the large amount of reagents (in comparison with the amount of test sample) which has to be used in the fusion will introduce errors as a result of the large amount of salts in the solution of melt, and the insufficient purity of the reagents used.

With such compounds, if one wants to determine only halogen, good results can be obtained by means of Korshun and Chumachenko's method [7] of fusing the material in a steel bomb with metallic potassium.

Determination of carbon and hydrogen in halogen-containing organosilicon compounds is carried out by classical methods, but the results obtained are often unsatisfactory. Usually therefore, research workers confine themselves to determining only the halogen.

Thus, at present, there is no method available which is fairly simple and reliable for the determination of the elements in halogen-containing organosilicon compounds. Another difficulty the analyst comes up against when dealing with these compounds is the extreme instability of some of them with respect to moisture. In such cases, the possibility of carrying out a simultaneous determination of all the elements, or, at least a few of them, on one aliquot assumes particular importance, since such a determination would allow one to confine oneself to a small amount of test material. The present article is devoted to simultaneous determination of the elements—carbon, hydrogen, silicon, and halogen. In a previous article we described a method for simultaneous determination of carbon, hydrogen, and silicon, in compounds not containing halogens [8, 9]. Combustion of the material is carried out by the rapid method in the presence of chromium sesquioxide on asbestos as catalyst. The catalyst is placed directly in the decomposition capsule. Carbon and hydrogen are determined as usual, while the silicon is calculated from the weight of silicon dioxide which is formed in the capsule. This article is a logical development of that method, and also of the method of Korshun and Sheveleva [10], for the simultaneous determination of carbon, hydrogen, and halogen by rapid combustion. In this method, halogen is fixed by silver at 550° in a special apparatus outside the combustion tube, and is determined from the increase in weight of the silver. A platinum contact is placed in the combustion tube, for without it, results for carbon determination are low.

Halogen-containing organosilicon compounds, as our experiments showed, exhibit their own characteristic thermal decomposition under the conditions of rapid combustion; accordingly, simple combination of the methods described above is out of the question.

Thermal decomposition of halogen-containing organosilicon compounds in the absence of oxidizing agents, and of sufficient oxygen—the state of affairs which holds when combustion is carried out by the rapid method—differs from one class of compounds to the other. Thus, compounds containing C, H, Si, and Cl, and in which the halogen is connected to carbon, and not to silicon, burn without formation of silicon carbide, and results for all four elements are satisfactory, even without addition of an oxidizing agent. The halogen which is split off during thermal decomposition, apparently, itself plays the part of an oxidizing agent, and prevents carbide formation.

Decomposition takes a different path when the organosilicon compound contains halogen connected to silicon. In these cases during decomposition, silicon tetrahalide is formed; in addition, in some cases, carbide formation also occurs. The silicon tetrahalide may be carried out of the combustion capsule and absorbed by the silver; as a result determination of both silicon and halide becomes impossible (Table 1). Nevertheless, if the molecule contains enough hydrogen — as our experiments showed, not less than 3% —then the water which is formed during combustion will decompose the silicon tetrahalide to give SiO_2 and the corresponding hydrogen halide, and its formation will not interfere with the determination. Materials which contain less hydrogen than that indicated above can often be analyzed by adding a small amount of water (3-5 mg) to the test aliquot (Table 2). In such cases, however, decomposition of the compound by water may precede thermal decomposition of the molecule. As a result of interaction with water, the chlorosilane will change over into polymerized siloxanes, and the thermal decomposition of the latter, as Rochow has already pointed out, is usually accompanied by formation of silicon carbide; consequently, combustion of these compounds can only be carried out in the presence of oxidizing agents or catalysts. Accordingly we had to clarify the possibility of carrying out the analysis in the presence of a catalyst which we had adopted, viz. chromium sesquioxide on asbestos‡.

Chromium sesquioxide is known to react with chlorine at high temperatures [11]. It may be postulated therefore, that the catalyst which we have used before—chromium sesquioxide on asbestos—may react with the halogen of an organic compound during the thermal decomposition of the latter. Actually, when combustion of hexachloroethane is carried out in the presence of the catalyst Cr_2O_3 on asbestos, we observed the formation of dark brown drops which settle out on the cold parts of the combustion tube. The weight of capsule plus catalyst decreased by 100-500 μg, while no such brown vapors were formed when halogen-containing organic compounds which also contained hydrogen were decomposed under these same conditions, or when samples of hydrogen chloride were evaporated; in these cases the weight of the capsule changed inappreciably. This can ap-

‡ Since polymerized siloxanes are not volatile, a thin layer of chromium sesquioxide on asbestos is sufficient for their complete oxidation.

TABLE 1

Material	% Halogen		% Silicon		% C		% H	
	Found	Calc.	Found	Calc.	Found	Calc.	Found	Calc.
Bis(trichlorosilyl)dichloromethane	108.80	80.63	0.57	15.96	3.59	3.41	0.00	0.00
$Cl_3Si—Cl_2—SiCl_3$	104.50	80.63	0.00	15.96	3.41	3.41	0.13	0.00
Bis(trichlorosilyl)chloromethane	81.92	78.20	Trace	17.71	3.78	3.77	0.78	0.32
$Cl_3SiCH(Cl)SiCl_3$	102.17	78.20	0.00	17.71	3.87	3.77	0.28	0.32

TABLE 2

Material	% Halogen		% Silicon		% C		% H	
Bis(trichlorosilyl)dichloromethane	80.39	80.65	16.05	15.96	3.68	3.41	0.00	0.00
	80.26	80.65	16.37	15.96	3.47	3.41	0.00	0.00
Bis(trichlorosilyl)chloromethane	78.05	78.20	17.67	17.68	3.77	3.79	0.39	0.32
	78.53	78.20	17.76	17.68	3.56	3.79	0.28	0.32

parently be explained by the hydrolyzing effect of water on the halogen compounds of chromium formed (Table 3).

The results obtained show that analysis of halogen-containing organosilicon compounds should be carried out differently, according to the structure of the test compound, and also permit one to develop a method for simultaneous determination of carbon, hydrogen, silicon, and halogen in the various classes of these compounds.

Analysis of Organosilicon Compounds in Which Halogen is Linked to Carbon

The test sample is weighed out into a quartz capsule 70-80 mm long and 4-5 mm diameter, or into a glass ampoule, if the material is volatile. 100-200 mg of asbestos ignited at 1200° is introduced into this beaker and the whole reweighed. When an ampoule is used, it is placed in a capsule with asbestos, with its open end on the bottom of the capsule, so that all the ampoule is inside the capsule. If the ampoule was weighed sealed, then the tip can be broken on the bottom of the capsule.

Capsule plus sample are placed in the combustion tube which contains a contact made from platinum foil. Three absorption tubes are then attached: the first to be connected up is made of high melting glass or quartz; it is filled with metallic silver and is heated in a special oven to 500-550°; the next in order, is a tube for water adsorption; finally the apparatus for absorbing carbon dioxide. In this case, when nitrogen is present in the test material, an absorption tube for nitrogen oxides is connected between the last two absorption tubes. Combustion is then begun. When combustion is complete, the three absorption tubes and capsule used for combustion are weighed. Silicon is calculated from the increase in weight of the capsule as a result of silicon dioxide formation; halogen from the increase

TABLE 3

Material	Sample weight in mg.	Change in weight of capsule in μg.	% Cl		% H		% C	
			Found	Calculated	Found	Calculated	Found	Calculated
Naphthalene	8.92	- 20			93.80	93.71	6.26	6.29
	10.01	- 30			93.84	93.71	6.33	6.29
Hexachloroethane	10.53	-560	83.95	89.87	10.13	10.12		
	8.27	-200	86.94	89.87	10.23	10.12		
	10.26	-110	91.62	89.87	10.08	10.12		
	8.23	- 90	90.42	89.87	10.12	10.12		
$C_{13}H_{17}ClO_4$	10.52	- 60	12.36	13.00	57.23	57.27	6.23	6.23
	10.68	- 30	12.76	13.00	57.32	57.27		6.34
Hydrochloric acid	17.65	- 10						
	18.73	0						
	20.13	- 10						
	19.24	+ 20						

in weight of the silver; and carbon and hydrogen, as usual (Table 4).

Analysis of Compounds in Which Halogen is Connected to Silicon

The majority of compounds of this type can be analyzed by the method described above (Table 5). Materials which contain less than 3% hydrogen are exceptions to this. When compounds of this type are analyzed, a drop or two of water (3-5 mg) is added to the capsule containing the test material plus asbestos, and the whole reweighed. During final calculations, the weight of water added is subtracted from the weight found. For the rest, the analytical procedure is the same as usual (Table 6).

Formation of silicon carbide was observed during attempts to analyze some compounds by the usual method. Silicon carbide was most often formed during analysis of compounds containing three atoms of halogen on one silicon atom, and during analysis of materials which have both hydrogen and halogen on the silicon atom. Such compounds are the most difficult from an analytical point of view, since their combustion cannot be carried out in the absence of chromium sesquioxide. Nevertheless, as already pointed out, in the presence of water, chromium sesquioxide does not interact or hardly interacts at all with halogen, while the small amount of the halogen compound of chromium formed is completely broken down by the water under the experimental conditions. When such compounds are to be analyzed therefore, the following procedure should be adopted.

To the aliquot of test material is added a small amount of chromium sesquioxide on asbestos, a layer of asbestos is then laid down and the whole weighed. A drop of water is placed on the asbestos and the whole reweighed. Combustion is then carried out as usual. When calculations are made of the water content on the basis of the increase in

TABLE 4

Material	C % Found	C % Calc.	C % Diff.	H % Found	H % Calc.	H % Diff.	Si % Found	Si % Calc.	Si % Diff.	Halogen % Found	Halogen % Calc.	Halogen % Diff.
Dimethylchloro-	44.04	43.93	+0.11	10.04	9.58	+0.46	20.21	20.55	-0.34	25.98	25.94	+0.04
methylethylsilane	44.52	—	+0.59	9.78	—	+0.20	20.16	—	-0.39	26.13	—	+0.19
Trimethyl-γ-	47.76	47.81	-0.05	9.88	10.03	-0.05	18.83	18.63	+0.20	23.50	23.52	-0.02
chloropropylsilane	47.96	—	+0.15	10.02	—	-0.01	18.51	—	-0.12	23.40	—	-0.12
Methyliodomethyl-	35.80	35.56	+0.24	7.01	7.09	-0.08	10.47	10.39	+0.02	47.31	46.96	+0.35
dipropylsilane	35.72	—	+0.16	7.08	—	-0.01	10.41	—	+0.02	46.63	—	-0.33
1, 2-Bis(trimethyl-	28.94	28.91	+0.03	6.14	6.07	+0.07	16.38	16.91	-0.53	48.18	48.11	+0.07
silyl)1,2-dibro-	28.96	—	+0.05	6.08	—	+0.01	16.28	—	-0.63	48.15	—	+0.04
moethane												

TABLE 5

Material	C % Found	C % Calc.	C % Diff.	H % Found	H % Calc.	H % Diff.	Si % Found	Si % Calc.	Si % Diff.	Cl % Found	Cl % Calc.	Cl % Diff.
Dimethylbutyl-	47.88	47.81	+0.07	10.00	10.03	-0.03	18.94	18.64	+0.30	23.27	23.52	-0.25
chlorosilane	47.77	—	-0.04	10.00	—	-0.03	18.84	—	+0.20	23.03	—	-0.49
Methylnaphthyl-	54.51	54.77	-0.16	4.27	4.18	+0.09	11.92	11.65	+0.24	29.10	29.40	-0.30
dichlorosilane	54.96	—	+0.19	4.22	—	+0.04	11.78	—	+0.13	29.70	—	+0.30
Propyldichlorosilane	25.32	25.18	+0.14	5.54	5.63	-0.09	20.07	19.63	+0.44	49.11	49.56	-0.45
	25.08	—	-0.10	5.68	—	+0.05	19.64	—	+0.01	49.61	—	+0.05
n-Butyl(β-chloro-	32.98	32.81	+0.17	6.02	5.97	+0.05	12.33	12.79	-0.46	48.43	48.43	0.00
ethyl)dichlorosilane	33.08	—	+0.22	6.03	—	+0.06	12.91	—	+0.12	48.67	—	+0.24
Phenyl-α-cyεnoiso-	46.41	46.16	+0.25	4.46	4.26	+0.20	10.87	10.80	+0.07	27.08	27.25	-0.17
propoxydichloro-	46.49	—	+0.33	4.49	—	+0.23	10.75	—	-0.05	27.00	—	-0.25
silane												

TABLE 6

Material	C % Found	C % Calc.	C % Diff.	H % Found	H % Calc.	H % Diff.	Si % Found	Si % Calc.	Si % Diff.	Cl % Found	Cl % Calc.	Cl % Diff.
Bis(trichlorosilyl)-	3.68	3.41	+0.27	0.00	0.00	—	16.05	15.97	+0.08	80.39	80.61	-0.22
dichloromethane	3.47	—	+0.06	0.00	0.00	—	16.37	—	+0.40	80.26	—	-0.35
Bis(trichlorosilyl)-	3.77	3.78	-0.01	0.39	0.32	+0.07	17.67	17.70	-0.03	78.05	78.20	-0.15
chloromethane	3.56	—	-0.22	0.28	—	-0.04	17.76	—	+0.06	78.53	—	+0.33
1,2-Bis(trichloro-	7.17	7.25	-0.08	1.01	0.91	+0.10	16.87	16.95	-0.08	75.30	74.89	+0.41
silyl)chloroethane	7.21	—	-0.04	0.96	—	+0.05	16.92	—	-0.03	75.64	—	+0.75
β-chloroethoxytri-	11.37	11.23	+0.14	1.98	1.89	+0.09	13.14	13.11	+0.03	65.67	66.29	-0.72
chlorosilane	11.41	—	+0.18	2.02	—	+0.13	13.10	—	-0.01	65.62	—	-0.67

TABLE 7

Material	C % Found	C % Calc.	C % Diff.	H % Found	H % Calc.	H % Diff.	Si % Found	Si % Calc.	Si % Diff.	Cl % Found	Cl % Calc.	Cl % Diff.
1- Trichlorosilyl-1-	9.34	9.15	+0.19	1.69	1.92	-0.23	21.34	21.40	-0.07	67.94	67.53	+0.41
dichlorosilylethane	9.31	—	+0.16	1.67	—	-0.25	21.19	—	-0.21	68.21	—	+0.68
1 4-bis(trichlorosilyl)-	15.10	14.78	+0.32	2.38	2.48	-0.10	17.46	17.29	+0.17	65.62	65.45	+0.17
butane	14.91	—	+0.13	2.29	—	-0.19	16.79	—	-0.50	65.12	—	-0.33
α-Methylisobutyl-	23.13	23.14	-0.01	4.43	4.37	+0.06	13.50	13.52	-0.02	50.65	51.24	-0.59
chlorosilane	22.99	—	-0.15	4.58	—	+0.21	13.50	—	-0.02	51.02	—	-0.22
Diethylisobutyl-	47.55	47.81	-0.26	10.04	10.03	+0.01	18.39	18.64	-0.25	23.50	23.52	-0.02
chlorosilane	47.52	—	-0.29	9.95	—	-0.08	18.09	—	-0.55	23.32	—	-0.20
Chlorobutyl-	21.14	21.44	-0.30	2.36	2.71	-0.35	12.36	12.54	-0.18	62.97	63.34	-0.34
trichlorosilane	21.44	—	0.00	2.58	—	-0.13	12.43	—	-0.11	63.45	—	+0.04
Propyldichlorosilane	25.32	25.18	+0.14	5.54	5.63	-0.09	20.07	19.63	+0.34	49.11	49.56	-0.45
	25.08	—	-0.10	5.68	—	+0.05	19.64	—	+0.01	49.61	—	+0.05

weight of the apparatus used for absorbing water, the weight of water added at the beginning is subtracted. By means of this method we succeeded in determining the four elements in compounds for which we could not get satisfactory results by means of other methods (Table 7)§.

SUMMARY

1. Decomposition of halogen-containing organosilicon compounds under the conditions used for rapid combustion has been studied.

2. A method has been worked out for the simultaneous determination of carbon, hydrogen, silicon, and halogen in various organosilicon compounds. The accuracy of the method for carbon and hydrogen is ±0.3%, for silicon ±0.4%, and for halogen ±0.5-0.6%.

LITERATURE CITED

[1] J. F. Hyde and R. C. Delong, *J. Am. Chem. Soc.*, **63**, 1194 (1941).

[2] E. G. Rochow, *An Introduction to the Chemistry of the Silicones*, New York, 1946, p. 162.

[3] D. Gibbons and A. Sykes, *Mikrochemie. ver. Mikrochim. Acta* **40**, 76 (1952).

[4] C. Tseng and T. Chao. *J. Am. Chem. Soc.*, **66**, 842 (1944).

[5] W. F. Gilliam and H. A. Liebhafsky, *J. Am. Chem. Soc.*, **63**, 801 (1941).

[6] B. A. Fiekers and E. M. Geronimo, *J. Am. Chem. Soc.*, **70**, 1654 (1948).

[7] M. O. Korshun and M. N. Chumachenko, *Doklady Akad. Nauk S.S.S.R.*, **99**, 769 (1954).

[8] V. A. Klimova, M. O. Korshun, and E. G. Bereznitskaya, *Doklady Akad. Nauk S.S.S.R.*, **96**, 81 (1954).

[9] V. A. Klimova, M. O. Korshun, and E. G. Bereznitskaya, *Zhur. Anal. Khim.*, **11**, 223 (1956).

[10] M. O. Korshun and N. S. Sheveleva, *Doklody Akad. Nauk S.S.S.R.*, **60**, 63 (1948).

[11] A. Lacourt, *Mikrochemie*, **23**, 308 (1938).

§ Most of the compounds which are listed in the article were kindly given to us by A. D. Petrov and his coworkers.

CHAPTER 24

The Simultaneous Determination of Carbon, Hydrogen, and Halogen in Halogen-containing Organic Compounds*

M. O. Korshun and N. S. Sheveleva

In 1948 we demonstrated the possibility of determining carbon, hydrogen, and halogens [1] simultaneously in halogen-containing compounds. It was noted then that in the absence of oxidants and catalysts carbon was not quantitatively oxidized to CO_2 during the pyrolysis of organic substances that contain halogen, even when there was a large excess of oxygen and a high oxidation zone temperature† (900-950°); and that consequently the results obtained for carbon were low and variable. Only the presence of a platinum catalyst assured complete oxidation of the thermal decomposition products of these substances and the attainment of good results [2].

Other authors [3, 4], however, when they analyzed for carbon, hydrogen, and halogens by rapid combustion in an empty tube, observed quantitative oxidation of the organic halogen compounds. Furthermore, it was found that at the temperature previously used by us for heating the silver (400-450°) the silver quantitatively retained halogen, but very quickly became exhausted and had to be frequently replaced.

All these facts forced us to make a more careful study of the temperature conditions employed in heating the silver and the reasons for the incomplete oxidation of the organic

Zhur. Anal. Khim.* **11, 376-382 (1956).

† By the oxidation zone we mean the section of the combustion tube heated by the electric furnace.

halogen compounds under the conditions of rapid pyrolytic combustion in an unfilled, i.e., empty, tube.

Temperature Conditions of the Silver in the Collection of Halogens

The published data on the temperature to which silver should be heated in order to retain halogens completely during the elementary analysis of organic halogen compounds are extremely contradictory. Some authors collect halogen with silver heated to 600-800° [3-6] while others use silver heated to 290-300° [7, 8]. The complete absorption of halogens also depends on the surface area of the heated layer of silver. Colson [5], for example, used a layer of granular metallic silver 150 mm long. This amount of silver weighs 30 g and is good for only 50 microdeterminations. The other authors mentioned above use silver gauze weighing from 6 to 12 g.

On the basis of six years of experiments on the simultaneous microdetermination of carbon, hydrogen, and halogen, it may be considered established that although silver retains halogens at relatively low temperatures, it soon tarnishes (its whole surface becoming covered with a thin film) and loses its activity. Consequently, the amount of silver with which we generally work is inadequate for even one analysis. On subsequent heating of the silver to higher temperatures (800°) the film disappears within a few seconds, the surface acquires its usual shine, and the silver again becomes active; but the weight of the tube remains unchanged. Determinations of chlorine in butylchloral hydrate showed (Table 1) that even at 550-600° silver quantitatively retains halogens and no film formation is observed. It may be presumed that at still lower temperatures (from 200 to 450°) the halogens for the most part react with silver only on the surface and, forming silver

TABLE 1
Determination of Chlorine in Butylchloral Hydrate at Different Heating Temperatures for the Silver. Calculated Cl: 54.98%.

Temperature °C	Sample mg.	Cl. mg.	%Cl Found
200	7.135	1.778	24.92
220	7.290	1.905	26.13
320	5.497	1.260	22.92
320	6.418	1.550	24.15
440	5.290	2.586	48.89
460	5.958	3.130	52.54
480	6.088	3.126	52.83
500	6.807	3.708	54.48
550	5.908	3.270	55.24
600	6.908	3.795	54.94

Note: The combustion was performed in a stream of O_2 at a rate of 35-40 ml/min; the length of the heated zone of the combustion tube was 10 cm, the temperature 950°; the length of the heated zone of the silver tube was 10 cm.

halide as a very thin film, prevent any further reaction of the halogen with metallic silver. At lower temperature the silver reacts in bulk. Consequently, at 550-600° considerably less silver is used to absorb the halogen and 2-3 mg is quite enough for 30 or 40 determinations of chlorine, bromine, and iodine. We generally use silver tape, foil, or gauze bent into a spiral 60-70 mm long, which is placed in a quartz tube ground to fit the combustion tube.‡

The halogen is determined after combustion by weighing the silver tube, just as in the case of sulfur [9]. For better cooling of the tube we place it on a long metal block, which is illustrated in Fig. 1.

‡ Silvered pumice or silica gel may be used instead of metallic silver.

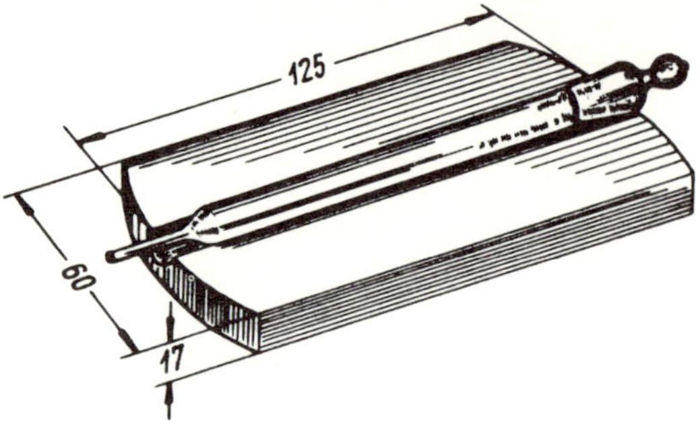

Fig. 1. Metal block

The Problem of Carbon Monoxide Formation during the Pyrolytic Combustion of Halogen-containing Organic Compounds.

As was noted above, we encountered unexpected difficulty in the determination of carbon by the rapid method in substances containing chlorine, bromine, and iodine. Of the many compounds studied, bromoanthracene gave the lowest carbon yield on combustion. Instead of the calculated 65.45%, only 45 to 47% carbon was obtained (Table 2). The length of the heated zone exerts no influence on the quantitative yield of carbon dioxide; equally low results for carbon being obtained when the combustion tube is heated by a short furnace 10 cm long or with a longer one 20-25 cm in length.

Assuming that the carbon losses were the result of the formation of carbon monoxide, the complete oxidation of which was prevented by the halogen present in the substance, we first conducted a qualitative reaction for CO in the gaseous combustion products of bromonaphthalene leaving the combustion tube, and we actually did detect carbon monoxide in them. Later on, we attempted to determine CO quantitatively, collecting all the products formed in order to obtain the total balance of the percentage content of all the elements present. In this operation bromine was

TABLE 2—Determination of C and H under Various Conditions of Pyrolytic Combustion

Substance	Expt. No.	Catalyst	Found % C	Found % H	Difference % C	Difference % H
3,3,3,1-Tetrachloro-1-propylene,	1	None	13.99	1.32	-6.04	+0.20
$C_2H_2Cl_4$. C = 20.03; H = 1.12	2	None	11.82	1.26	-8.29	+0.14
	3	Present	20.32	1.22	+0.29	+0.10
	4	Present	20.18	1.16	+0.15	+0.04
β-Chlorovinyl-*p*-chlorophenyliodonium	5	None	20.83	1.73	-7.81	-0.07
chloride, $C_8H_6Cl_3I$. C = 28.64; H = 1.80	7	None	18.50	1.75	-10.14	-0.05
	7	Present	28.40	1.67	-0.24	-0.13
	8	Present	28.44	1.64	-0.20	-0.16
4-Bromo-4-methyl-2-pentene,	9	None	40.63	6.82	-3.57	+0.01
$C_6H_{11}Br$. C = 44.20; H = 6.81	10	Present	44.22	6.95	+0.02	+0.14
	11	Present	44.33	6.91	+0.12	+0.06
Bromoanthracene,	12	None	47.58	3.72	-17.81	+0.19
$C_{14}H_9Br$. C = 65.39; H = 3.53	13	None	45.27	3.70	-20.12	+0.17
	14	Present	65.45	3.69	+0.06	+0.16
	15[a]	None	65.31	3.61	-0.08	+0.08
Hexachlorocyclohexane,	16	None	23.08	2.33	-1.76	+0.27
$C_6H_6Cl_6$. C = 24.74; H = 2.06	17	Present	25.00	2.32	+0.26	+0.26
	18[a]	None	25.01	2.35	+0.27	+0.29
Bromomonaphthalene,	19	None	54.08	3.51	-3.92	+0.10
$C_{10}H_7Br$. C = 58.00; H = 3.41	20	Present	58.07	3.47	+0.07	+0.06
	21[a]	None	57.87	3.44	-0.13	+0.14
	22[b]	None	57.78	3.52	-0.22	+0.11
	23[b]	None	57.87	3.49	-0.13	+0.08

[a]In experiments 15, 18, and 21 the temperature of the oxidation zone was 1100°.

[b]In experiments 22 and 23 the combustion tube contained metallic silver, as is shown in Fig. 2.

collected by silver at 550-600° in a tube connected to the combustion tube by means of a ground joint [9]. Connected to this tube in series were tubes for the collection of water and carbon dioxide. Then came a tube with cupric oxide heated to 800° and a second tube, filled with Ascarite. The amount of CO in the form of CO_2 was determined from the weight gain of this tube. From the results presented below it can be seen that the sum of the percentage content of carbon obtained as a result of combustion in the form of CO_2 and CO differs from the calculated amount by only 0.2 absolute per cent, while the sum of all the elements satisfies the empirical formula of bromonaphthalene. In this manner it was possible to confirm the presence of carbon monoxide, which governs the carbon losses in the pyrolytic combustion of halogen-containing organic compounds.

Combustion of Bromonaphthalene

Sample, mg	6.735	8.080
CO_2, mg	14.100	15.787
CO_2, mg; from CO	0.230	1.385
H_2O, mg	2.090	2.510
Br, mg	2.630	3.123
Found %		
C as CO_2	57.13	53.32
C as CO	0.93	4.68
C (total)	58.06	58.00
H	3.47	3.47
Br	38.90	38.65
Total	100.43	100.12

We then conducted a series of experiments in which the halogen was collected by silver at 550-600° right in the combustion tube (as shown in Fig. 2) so that the halogen was retained immediately after the decomposition of the substance, i.e., before the thermal decomposition products, in mixture

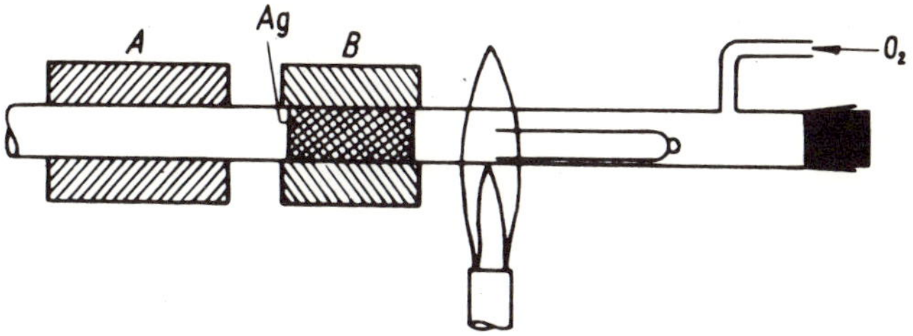

Fig. 2. Electric furnace.
A) Length 110 mm for 900°; B) length 60 mm for 600°

with oxygen, entered the oxidation zone. When this was done correct results were obtained for carbon in the combustion of bromonaphthalene (Table 2).

Good results are also obtained for carbon when these substances are burned in the presence of a platinum catalyst heated to 900-950° (Table 2). We use platinum foil, gauze, or wire 4-5 cm long and bent into a spiral as the catalyst.

In the light of the current research it may be presumed that under the conditions of the pyrolytic decomposition of halogen-containing organic substances, simple halogenated molecules or even radicals will appear in the decomposition products.§ As is well known, such substances as, for example, dichloroethane, ethyl iodide and bromide, dichloro- and trichloroethylene, and also hydrohalic acids and the like, are inhibitors of the oxidation reaction when present in even insignificant concentrations. Depending on the concentration of these substances, the oxidation of carbon may result in carbon monoxide to a greater or lesser extent. It may then be concluded that the presence of carbon monoxide, which we detected in the gaseous combustion products of various organic halogen derivatives, is not unexpected. It is obvious that in this case we are confronted with the phenomenon of inhibition of the oxidation process, which is characteristic of cases

§ A qualitative reaction for phosgene in the gaseous pyrolytic products gave negative results.

of combustion of halogen-containing substances. No such phenomenon is observed during the pyrolysis of the majority of other organic compounds that do not contain halogen.

It seems to us that the point of view expressed is not contradicted by the fact that all the carbon is quantitatively converted to CO_2 in the oxidation zone when the halogen is isolated immediately after the decomposition of the substance (Fig. 2). However, with such an operating procedure halogen cannot be determined simultaneously with carbon and hydrogen. Therefore we believe it most advisable to work in the presence of a platinum catalyst, which will provide complete oxidation of the carbon to CO_2 during the pyrolysis of halogen-containing organic compounds without any other changes in the conditions of the rapid method [9]. The simultaneous determination of halogen, carbon, and hydrogen then becomes possible.

It should be noted that when bromoanthracene, bromonaphthalene, and hexachlorocyclohexane are burned according to the customary procedure but with a temperature of 1100° in the oxidation zone, good results are obtained for carbon (Table 2). This operating procedure is not recommended, however, since the quartz tubes are frequently damaged.

A platinum catalyst is not needed for the determination of halogen alone.

Table 3 presents the values for halogens as determined in the usual manner. The simultaneous determination of carbon, hydrogen, and halogen, except for the use of a platinum catalyst and heating the silver to 550-600°, is in no way different from the method for the simultaneous microdetermination of C, H, and S [9].

It should be mentioned, however, that when carbon, hydrogen, and halogen are determined by the method given the figures for carbon and hydrogen have the usual ac-

TABLE 3
Determination of Halogens

Substance	Sample mg	Halogen mg	Halogen % Found	Halogen % Calc.	Halogen % Diff.
n-Chloroacetylguanine	4.705	0.720	15.30	15.57	-0.27
Methyl-2-(3-chlorobicyclo)-					
2,2,1-(5-heptenyl)ketone	9.585	1.955	20.40	20.78	-0.38
$C_9H_{11}OCl$	5.005	1.025	20.48	20.78	-0.30
o-Chloronitrobenzene	5.608	1.259	22.45	22.50	-0.05
C-H-$ClNO_2$	7.450	1.668	22.39	22.50	-0.11
2-. .-Ethoxy-l-ethoxybromo-					
cyclohexane	4.670	1.495	32.00	31.82	+0.18
C-$H_{19}O_2Br$	6.075	1.930	31.79	31.82	-0.03
Bromonaphthalene	6.735	2.620	38.90	38.59	-0.31
C_{10}H-Br	8.080	3.123	38.65	38.59	+0.06
3,5-Dichloro-1-cyclopentene	3.885	2.020	51.99	51.78	+0.21
C_5H-Cl_2	5.740	2.970	51.75	51.78	-0.05
Isobutyl-. .-iodovinyl ketone	5.590	3.000	53.67	53.31	+0.36
C-$H_{11}OI$	6.635	3.545	53.47	53.31	+0.16
Dibromoethylene	7.790	6.640	85.24	85.07	+0.17
$C_2H_4Br_2$	6.180	5.280	85.44	85.07	+0.06

curacy, but those for halogen are sometimes less accurate, probably because of weighing errors. In this case it is recommended that larger samples be used (of the order of 7-8 mg). When there is metal in the substance as well, four elements can be determined simultaneously, i.e., carbon, hydrogen, halogen, and metal.

Tables 4 and 5 present the results of the simultaneous determination of carbon, hydrogen, and halogen (chlorine, bromine, or iodine) and also carbon, hydrogen, chlorine, and iron in single samples (4-11 mg).

The accuracy of the carbon and hydrogen analysis lies between ±0.1 and 0.3%, that of the halogen analysis between ±0.1 and 0.6%.

TABLE 4

Simultaneous Determination of Carbon, Hydrogen, and Halogen

					Found %			Difference %		
Substance	Sample	CO_2	H_2O	Halogen	C	H	Halogen	C	H	Halogen
1. β-Endo-9,10-dihydroanthrol-chloroamyl vinyl ketone	9.083	25.930	5.660	0.942	77.91	6.85	10.37	−0.06	+0.01	−0.04
$C_{22}H_{23}OCl$ C — 77.97; H — 6.84; Cl — 10.41	11.100	31.769	6.900	1.183	78.11	6.96	10.66	+0.14	+0.12	+0.25
2. Benzodiglycine ester of ethanolamine hydrochloride[a]	6.285	11.375	3.390	0.710	49.37	6.04	11.30	−0.08	+0.30	+0.07
$Cl_{12}H_{18}O_4N_3Cl$ C — 49.45; H — 5.74; Cl — 11.23	7.175	12.960	3.820	0.810	49.30	5.95	11.21	−0.15	+0.21	+0.06
3. o-Chloronitrobenzene[a]	5.870	9.910	1.426	1.330	46.07	2.72	22.66	+0.34	+0.16	+0.16
$C_6H_4ClNO_2$. C — 45.73; H — 2.56; Cl — 22.50	6.180	10.362	1.520	1.390	45.74	2.75	22.49	+0.01	+0.19	+0.02
4. Ethyl ester of γ-bromopropyl-	7.370	12.832	4.155	2.120	47.54	6.31	28.77	−0.12	+0.12	−0.06
oxocyclopentane carboxylic acid	7.030	12.310	3.985	2.020	48.78	6.31	28.59	+0.12	+0.12	−0.24
$C_{11}H_{17}O_3Br$; C — 47.66; H — 6.19; Br — 28.83	4.910	8.570	2.780	1.435	47.63	6.31	29.23	−0.03	+0.12	+0.40

5. Semicarbazone of γ-bromopropylcyclopentanone[a]	6.390	9.650	3.565	1.950	41.21	6.24	30.52	−0.02	+0.09	+0.03
$C_9H_{16}NO_3Br$; C — 41.23; H — 6.15; Br — 30.49	6.650	9.930	3.680	2.000	41.31	6.28	30.49	+0.08	+0.03	0
6. (3-(p-bromophenyl)-1,1-dichloro-1-propene $BrC_6H_4CH_2CH — CCl_2$; C — 40.64; C — 40.64; H — 2.65; Cl + Br — 56.71	11.475	17.075	2.779	6.512	40.61	2.71	56.77	−0.03	−0.06	+0.06
7. β-Chlorovinylphenyliodonium chloride $C_8H_7Cl_2I$. C — 31.92;	5.700	6.700	1.295	3.772	32.08	2.52	66.18	+0.16	+0.28	+0.60
H — 2.34; Cl + I — 65.58	6.018	7.000	1.320	3.930	31.74	2.45	65.31	−0.18	+0.11	−0.27
8. 1,2,3,3-Tetrachloro-1-propene $C_3H_2Cl_4$ C — 20.03; H — 1.12;	8.925	6.449	0.900	7.080	19.77	1.13	78.77	−0.26	+0.01	−0.07
Cl — 78.84	9.603	6.988	1.263	7.593	19.86	1.47	79.07	−0.17	+0.35	+0.23
9. 1 2 3,3-Pentachloro-1-propene [sic] C_3HCl_5. C — 16.79; H — 0.47; Cl — 82.63	8.692	5.347	0.475	7.169	16.79	0.61	82.48	0	+0.14	−0.15
10. 1,1,1,2,2,3-Hexachloropropane $C_3H_2Cl_6$. C — 14.37;	6.600	3.445	0.645	5.595	14.29	0.92	84.77	−0.08	+0.12	−0.06
H — 0.80; Cl — 84.83	8.510	4.392	0.707	7.204	14.09	0.93	84.65	−0.28	+0.13	−0.17

[a]The oxides of nitrogen were collected with a dry absorbent prepared according to the specifications of Klimova and Korshun [10].

TABLE 5

Simultaneous Determination of Carbon, Hydrogen, Chlorine, and Iron

Substance	Sample	CO_2	H_2O	Cl	Fe_2O_3	Found % C	Found % H	Found % Cl	Found % Fe	Difference % C	Difference % H	Difference % Cl	Difference % Fe
2-Isobutylnaphthopy-rylium ferrochloride $C_{17}H_{17}OFeCl_4$	9.390	16.190	3.090	3.050	1.695	47.05	3.68	32.48	12.62	+0.11	-0.26	-0.13	-0.19
	5.540	9.532	1.870	1.825	1.005	46.96	3.74	32.65	12.57	+0.02	-0.20	-0.04	-0.24
C=46.94; H=3.94; Cl=32.61; Fe=12.81													
2-n-Amylnaphthopy-rylium ferrochloride $C_{18}H_{19}OFeCl_4$	5.345	9.430	2.110	1.690	0.935	48.14	4.42	31.62	12.22	0	+0.15	+0.03	-0.22
	6.050	10.930	2.365	1.915	1.065	48.21	4.37	31.65	12.31	+0.07	+0.10	+0.06	-0.13
C=48.14; H=4.27; Cl=31.59; Fe=12.44													

SUMMARY

It has been ascertained that carbon monoxide forms during the pyrolysis of halogen-containing organic compounds in an empty tube, and that this leads to low analytical figures for carbon.

For the simultaneous determination of carbon, hydrogen, and halogen it is preferable to perform the pyrolysis in the presence of a platinum catalyst. In the determination of halogen alone, the platinum catalyst is not needed.

LITERATURE CITED

[1] M. O. Korshun and N. S. Sheveleva, *Doklady Akad. Nauk SSSR*, **60**, 63 (1948).

[2] M. O. Korshun and N. E. Gel'man, *Novye Metody Elementarnogo Mikroanaliza* [New Methods of Elementary Microanalysis], p. 54, Goskhimizdat [State Publishing House of Chemical Literature], Moscow, 1949.

[3] R. Belcher and C. Spooner, *J. Chem. Soc.*, **1943**, 313.

[4] G. Ingram, *Analyst*, **73**, 548 (1948).

[5] A. F. Colson, *Analyst*, **73**, 541 (1948).

[6] G. L. Stragand and H. W. Safford, *Anal. Chem.* **21**, 625 (1949).

[7] R. Teston and F. E. McKenna, *Ibid.*, **19**, 193 (1947).

[8] A. Lacourt and Ch. T. Chang *Bull. soc. chim. Belg.*, **50**, 68, 115, 135 (1941); **52**, 157, 175 (1943).

[9] M. O. Korshun, *Zhur. Anal. Khim.*, **7**, 96, 101 (1952) M. O. Korshun and N. S. Sheveleva, *Ibid.*, 104.

[10] V. A. Klimova and M. O. Korshun, *Zhur. Anal. Khim.* **6**, 230 (1951).

CHAPTER 25

The Simultaneous Determination of Carbon, Hydrogen, and Sulfur in Organic Materials*

N. V. Sokolova

During simultaneous determination of carbon, hydrogen, and sulfur in organic compounds by means of Korshun and Sheveleva's method [1], after decomposition of the organic compound, the sulfur is converted into a mixture of oxides in an oxygen atmosphere. These oxides are quantitatively trapped with heated silver in the form of Ag_2SO_4. In these determinations the silver is placed in a separate, tared quartz apparatus which is connected by means of a ground glass joint to the end of the combustion tube.

We have used silver of 99.99% purity in the form of pieces of foil 0.06 mm thick. Tests showed that in this form, silver rapidly lost its activity and ceased to absorb the sulfur oxides quantitatively. E.b., 5.23g of silver foil was placed in the apparatus. 24 combustions were carried out with this filling. When it was used for more determinations, low results for sulfur were obtained. The silver was washed with boiling water until it gave a negative test for SO_4^{2-}. The wash liquor was found to contain 58.8 mg of SO_4 (as $BaSO_4$) which corresponded with 132 mg of Ag. Thus, of all the silver taken, only $0.132/5.23 \cdot 100 = 2.48\%$ Ag reacted with the sulfur oxides. In order to increase the efficiency of the silver we have used silvered pumice.

---

*Zhur. Anal. Kkim. **11**, 728-730 (1956).

For micro-analysis we have used silver deposited on pieces of unglazed clay plates [2], with small surfaces.

Preparation of Silvered Pumice

Pumice was ground to give particle sizes of 2-3 mm; it was sieved free of dust and treated by boiling with 10% nitric acid, followed by a few portions of 10% hydrochloric acid until the acid extracts of the pumice were colorless, indicating complete removal of impurities soluble in acid from the pumice; finally it was washed several times with boiling distilled water until a negative chloride reaction was obtained; it was then dried at 100°-120° and calcined in a crucible oven for 30 minutes at 800°.

The pumice was next covered in a porcelain crucible with a 20% solution of silver nitrate (2 ml of 20% $AgNO_3$/g of pumice). After 24 hours the solution was evaporated first on a hotplate and finally in a drying oven at 100-120°. After drying, the pumice was calcined in a crucible oven at 800° for 30-40 minutes until oxides of nitrogen disappeared. The treatment with silver nitrate was repeated once again. For this twofold treatment the total amount of silver nitrate used was 80% of the weight of pumice, or — calculated in terms of metallic silver — 50.4%. After treatment the pumice had a silvery grey color.

Determination of the Silver Content of the Preparation Obtained

Analysis was carried out as follows: an exact amount of ground pumice was treated with boiling 10% nitric acid, and, after cooling the nitric acid extract was titrated with 0.1 N ammonium thiocyanate in the presence of ferric alum. On repeating the nitric acid treatment of the pumice with nitric acid, no silver was found in the

TABLE 1

Determination of Sulfur in Organic Materials from Increase in Weight of Silver

Sample designation	Formula	Weight of sample in mg.	SO by wt. increase of silvered pumice	S, % Found	S, % Calc.	Deviation from theory
β-naphthalene sulfonic acid	$C_{10}H_7SO_3H.3H_20$	6.950	2.580	12.36	12.21	+0.15
		3.730	1.370	12.09		-0.12
		5.985	2.465	12.07		-0.14
		4.740	1.740	12.25		+0.04
		3.470	1.270	12.21		0
Sulfosalicylic acid	$HO-C_6H_3(COOH)(SO_3H).2H_20$	5.710	2.110	12.30	12.59	-0.29
		5.725	2.110	12.30		-0.29
Sulfanilic acid	$C_6H_4(SO_3H)(NH_2)$	4.530	2.480	18.28	18.48	-0.20
		4.280	2.330	18.17		-0.31
Thiourea	$CS(NH_2)_2$	3.910	4.920	42.00	42.10	-0.10
		5.750	4.735	42.14		+0.04

Mercaptobenzthiazole (Captax)	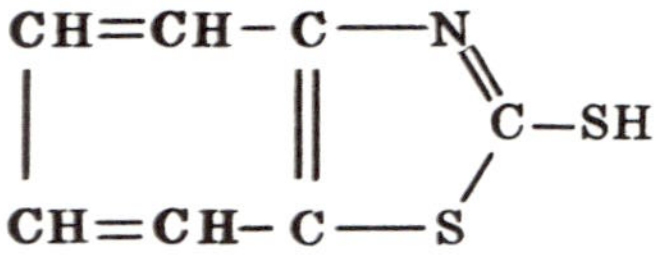	3.640	4.155	38.10	38.27	−0.17
		3.525	4.032	38.18		−0.09
Azobenzthiazole	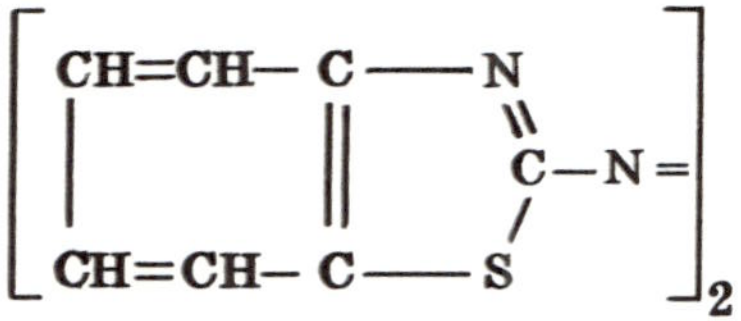	3.410	2.220	21.73	21.62	+0.01
		3.935	2.535	21.50		−0.12
		3.460	2.230	21.51		−0.11
Benzthiazole methylpyrazolone	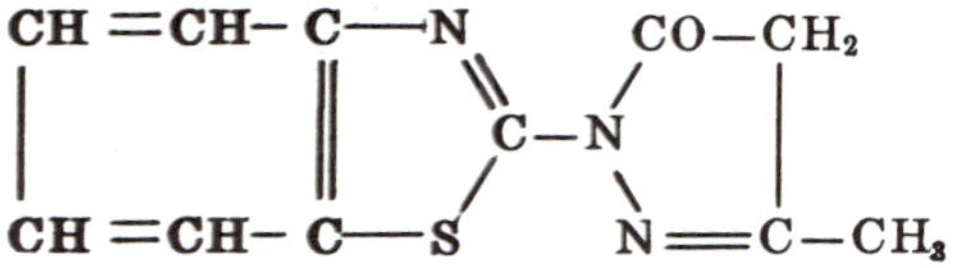	3.430	1.425	13.87	13.85	+0.02
		4.900	2.005	13.66		−0.19
		4.940	2.045	13.82		−0.03
	$C_{20}H_{22}O_5S_2$	3.385	1.525	15.04	15.75	−0.71
		5.840	2.695	15.40		−0.35
	$C_{29}H_{34}O_{10}S_2$	3.300	1.060	10.97	10.56	+0.41
		3.700	1.190	10.74		+0.18
	$C_{30}H_{34}O_{11}S_2$	3.210	0.980	10.23	10.00	+0.23
		3.730	1.115	9.98		−0.02
	$C_{16}H_{24}O_8S_2$	3.250	1.520	15.61	15.68	−0.07
		4.025	1.905	15.79		+0.11

extract. The preparation obtained as described above contained 40.32% Ag.

The Use of Silvered Pumice for Quantitative Determination of Sulfur

Pumice which has been silvered by the method described above has been used for trapping sulfur oxides during determination of sulfur in organic compounds by Korshun and Sheveleva's method [1]. Into the quartz apparatus which is connected by means of a ground-glass joint to the combustion tube was placed 2.42 g of silvered pumice containing 975.7 mg of Ag. This amount of silver can be used for combining with 434.3 mg of SO_4^{2-} (at temperatures from 650°-700°).

Determination was carried out exactly as laid down by Korshun and Sheveleva. When combustion was complete, the apparatus used for absorbing carbon dioxide and water was disconnected, the oven used for heating the apparatus containing silvered pumice was removed, and the pumice allowed to cool in an oxygen stream for 10 minutes; the apparatus containing pumice was then disconnected, stoppered with a ground-glass stopper, wiped with chamois, passed through a burner flame, kept for 10 minutes outside the balance, then for 10 minutes in the balance, and finally weighed on an MA-20 micro-balance. The absorption apparatus was weighed before analysis and a blank carried out.

Results for determination of the sulfur content of various organic compounds are given in Table 1.

This table only contains the results for sulfur, though in all cases carbon and hydrogen were also determined at the same time.

With this one charge of silvered pumice, 96 combustions in all were carried out of organic compounds whose sulfur content varied from 6.0 to 65%. As a result of absorption of sulfur oxides, the gain in weight of the silvered pumice was 281.7 mg; consequently, the percentage efficiency of the silver on pumice was 64.86%.

During subsequent combustions the apparatus increased in weight, but quantitative absorption of sulfur oxides did not occur, and in most cases the results obtained were low, e.g., during successive determinations of the sulfur content of mercaptobenzothiazole, the theoretical sulfur content of which is 38.3%, the results obtained were 34.34, then 32.80, and finally 18.20%.

In order to check on the amount of sulfur held by the silver, the pumice from the apparatus was washed with small portions of boiling distilled water until a negative reaction for SO_4^{2-} was obtained ($\sim$ one and one half liters of water); sulfate ions were then determined by precipitation with barium nitrate in nitric acid solution. 770 mg of $BaSO_4$ was found, which corresponded to 317 mg SO_4^{2-} and 713 mg of combined silver. The amount of unused silver remaining on the pumice was thus 262 mg. The percentage efficiency of silver as an absorber for sulfur oxides was therefore 64.86%. After washing off the silver sulfate, the pumice was again treated with silver nitrate and put back into the apparatus.

By using silvered pumice, owing to its porosity the metallic silver is distributed over a very large surface, which favors more complete use of it for reacting with sulfur oxides (64.8% instead of 2.5%).

Because of the small specific gravity of pumice ($\sim$ 0.4), an apparatus filled with silvered pumice weighs less than one containing metallic silver, thereby increasing the accuracy of determination of small amounts.

After use, the silvered pumice is readily freed from Ag_2SO_4 and can be used again for deposition of silver on it. Silver foil after use becomes brittle and loses its activity.

SUMMARY

The use of silvered pumice for absorption of sulfur oxides during gravimetric determination of sulfur in organic compounds, by combustion in a quartz apparatus in an oxygen stream, has several advantages over metallic silver.

LITERATURE CITED

[1] M. O. Korshun and N. S. Sheveleva, *J. Anal. Chem.* **7 2** 104 (1952).

[2] M. O. Korshun and N. E. Gel'man, *New Methods of Elemental Microanalysis* (Moscow-Leningrad State Chem. Press, 1949), pp. 29, 103.

CHAPTER 26

Micro and Semimicro Determination of Nitrogen by Hydrogenation of Organic Materials*

N. E. Gel'man and M. O. Korshun

Determination of nitrogen as ammonia obtained as the result of hydrogenating organic compounds has been known for more than 25 years; nevertheless the literature devoted to this method is very scant [1-5], and the method has not been widely used. The method is known in macro, semimicro, and micro modifications and consists of slow heating of an aliquot of test material mixed with nickel; the vapors formed are then passed in a slow stream of hydrogen (1 ml/minute) over a nickel catalyst heated to 300°. The ammonia formed is absorbed with acid, excess of which is titrated with alkali. The necessity for slow decomposition and low gas flow rate has been emphasized by all authors.

The introduction into elemental analysis of new principles of decomposition of organic materials, and the successful application of these principles to reductive decomposition as well [6-7], led us to develop a micro determination of nitrogen, by hydrogenation of organic materials, using a high gas flow rate and rapid pyrolytic decomposition of samples [8].

The method we have developed reduces in essence to the following: the aliquot of nitrogen-containing organic material is subjected to rapid thermal decomposition in a hydrogen atmosphere, and the decomposition products

Zhur. Anal. Khim.* **12, 128-33 (1957).

mixed with excess hydrogen are then passed over an iron catalyst at a gas flow rate of 35-40 ml/minute. Under these conditions, the nitrogen present in the organic material is quantitatively converted into ammonia, which is absorbed by a standard solution of potassium biiodate and determined iodometrically.

Method of Decomposition. All analysts working with hydrogenation methods use techniques for decomposing materials such that a mixture of the vapors of the test material and its decomposition and hydrogenation products arrives at the catalyst, while the basic mass of the material does not undergo any fundamental changes before it gets to the catalyst. With this method of decomposition which we shall call "distillation" we should like to contrast pyrolytic decomposition; in the latter, it is not the vapors of the test material which pass over the catalyst, but very reactive decomposition products formed from the test material at high temperatures.

During hydrogenation of various organic compounds using the technique of pyrolytic decomposition, we succeeded, in a matter of minutes, in achieving complete removal of nitrogen as ammonia (Table 1). When the "distillation" method was used, the results obtained were low.

Another essential advantage of pyrolytic decomposition over "distillation" is more complete hydrogenation of material to gaseous products, with consequent drop in the amount of carbon and resin formed during hydrogenation. When any carbon does happen to form, it settles out only on the inner surface of the quartz test tube in which the aliquot has been placed, while the combustion tube, and what is particularly important, the catalyst, both remain clean. The catalyst showed no signs of poisoning after 100-130 experiments had been carried out with pyrolytic

TABLE 1

Determination of Nitrogen by Rapid Hydrogenization of Samples

Material	Sample weight in mg	Duration of test in min. Decomposition only	Duration of test in min. Decomposition and purging	Nitrogen % Found	Nitrogen % Difference
Methylbenzamide	3.023	3.5	7.	10.24	-0.12
	3.965	4.5	7.5	10.38	+0.02
p-Aminoacetophenone	3.497	4	7	10.58	+0.22
	3.100	4	7	10.55	+0.19
Diphenylurea	5.083	5	12	13.26	+0.06
Triphenylguanidine	5.570	6	15	14.63	+0.01
	4.180	7	15	14.65	+0.03
Ammonium oxalate	3.030	5	14	19.55	-0.16
Oxamide	3.490	8	20	31.98	+0.11
	3.210	5	10	8.25	-0.13
Carbazole	3.575	6	12	8.45	+0.07

decomposition of samples under hydrogen. When the "distillation" technique was used, on the other hand, the catalyst was rapidly covered with a layer of carbon, and it had to be regenerated after a few analyses.

Catalyst. Instead of the classical catalyst for hydrogenation reactions — nickel — we used the specific catalyst for ammonia synthesis—iron; the optimum operating temperature of the latter is 400°, which is approximately 100° higher than that of nickel catalysts. The possibility of carrying out destructive hydrogenation at 400° instead of 300° is an essential advantage during the analysis of thermostable materials such as carbazole (Table 2).

The same lot of catalyst can be used for over 100 analyses without any loss of activity.

Determination of Ammonia. Ammonia formed as the result of hydrogenation was determined iodometrically,

TABLE 2
Determination of Nitrogen in Carbazole
(Sample weight 3-5 mg; nitrogen content 8.38%)

Catalyst	Operating temperature of the catalyst	Nitrogen in %	
		Found	Difference
Nickel	320	7.64	-0.74
		7.70	-0.68
	375	7.63	-0.75
		7.19	-1.19
		7.71	-0.67
Iron	300	7.39	-0.99
		7.50	-0.88
	400	8.33	-0.05
		8.38	-0.10
		8.40	+0.02

TABLE 3
Titration of Potassium Biiodate with Alkali
(5 ml of 0.02 N $KH(IO_3)_2$ and 1 ml of 10% KI solutions used)

	KOH used in ml		Deviation for one titration in ml
	5.000		0.000
	5.005		0.005
	5.005		0.005
	5.000		0.000
	5.000		0.000
	5.000		0.000
	4.995		0.005
	4.995		0.005
	5.000		0.000
	5.005		0.005
	4.995		0.005
	4.995		0.005
Mean	5.000	Mean	0.003

TABLE 4
Titration of Potassium Biiodate with Thiosulfate
(5 ml of 0.02 N $KH(IO_3)_2$ and 1 ml of 10% KI solutions used)

$Na_2S_2O_3$ used in ml		Deviation for one titration in ml	
	5.000		0.000
	5.000		0.000
	5.000		0.000
	5.000		0.000
	4.995		0.005
	4.995		0.005
	5.000		0.000
	4.995		0.005
	4.995		0.005
	5.000		0.000
	5.000		0.000
	5.000		0.000
Mean	5.000	Mean	0.002

using a standard solution of potassium biiodate for absorbing the ammonia. Potassium biiodate is well known in analytical practice [9]; it is readily prepared in a pure form, and keeps well both in the form of crystals, and also in dilute solutions. Excess biiodate can be determined either by titrating the H^+ of the acid salt with alkali, or determination of the iodine liberated on addition of KI; the iodine, the amount of which is equivalent to the amount of hydrogen ions in the acid salt, is back-titrated with sodium thiosulfate. Comparison of results of acidimetric and iodometric titration of 0.02 N solutions of potassium biiodate showed that the latter gives more accurate results (Tables 3 and 4).†

† We have also successfully used iodometric titration of excess biiodate, after absorption of ammonia, for micro- and semimicro Kjeldahl determination of nitrogen.

TABLE 5

Determination of Nitrogen in Samples of Different Weight

Test material	Weight in mg	0.02 N $KH(IO_3)_2$ used in ml	Nitrogen % Found	Nitrogen % Difference
Ammonium oxalate	3.710[a]	2.605	19.67	-0.04
$C_2H_8O_4N_2H_2O$; 19.71% N	5.259[a]	3.690	19.66	-0.05
Ethyl oxamate	3.224[a]	1.380	12.02	+0.05
$C_4H_7O_3N$; 11.97% N				
p-Phenylenediamine	3.836[a]	3.530	25.78	-0.13
$C_6H_8N_2$; 25.91% N	3.556[a]	3.290	25.92	+0.01
p-Aminoacetophenone	2.908[a]	1.075	10.36	—
C_8H_9ON; 10.36% N	4.103[a]	1.510	10.31	-0.05
Triphenylguanidine	3.352[a]	1.740	14.54	-0.08
$C_{19}H_{17}N_9$; 14.62% N	3.831[a]	1.980	14.48	-0.14
Methylbenzamide	3.405[a]	1.260	10.37	+0.01
C_5H_9ON; 10.36% N	3.570[a]	1.315	10.32	-0.04
α-Aminopyridine	4.382[a]	4.620	29.54	-0.23
$C_5H_6N_2$; 29.77% N	3.270[a]	3.460	29.64	+0.13
Dipyridylurea	3.685[a]	3.445	26.19	+0.03
$C_{11}H_{10}ON_4$; 26.16% N	4.085[a]	3.795	26.02	-0.14
6,6-Dicarbethoxy-5, 6, 13, 14-tetrahydro-1,2-benzophenanthridine	2.971[a]	4.070	3.84	+0.09
$C_{23}H_{23}O_4N$; 3.73% N				
Carbazole	3.753[a]	1.120	8.36	-0.02
$C_{12}H_9N$; 8.38% N	4.248[a]	1.275	8.41	+0.03
Anthranilic acid	10.39[b]	3.78	10.19	-0.03
$C_7H_7O_2N$; 10.21% N	8.55[b]	3.13	10.25	+0.04
p-Aminoacetophenone	11.65[b]	3.34	10.44	+0.08
C_8H_9ON; 10.36% N	11.71[b]	3.36	10.41	+0.05
Ethyl oxamate	12.58[b]	5.33	11.88	-0.09
$C_4H_7O_3N$; 11.97% N	11.40[b]	4.82	11.85	-0.12
Methylbenzamide	10.79[b]	4.02	10.40	+0.04
C_8H_9ON; 10.36% N	13.57[b]	5.10	10.53	+0.17
Diphenylurea	10.33[b]	4.92	13.34	+0.14
$C_{13}H_{12}ON_2$; 13.20% N	11.31[b]	5.31	13.15	-0.05

TABLE 5—*continued*

Test material	Weight in mg	0.02 N $KH(IO_3)_2$ used in ml	Nitrogen % Found	Difference
Carbazole	12.19[b]	3.64	8.37	-0.01
$C_{12}H_9N$; 8.38% N	12.94[b]	3.86	8.35	-0.03
Ammonium oxalate	8.27[b]	5.83	19.75	+0.04
$C_2H_8O_4N_2H_2O$; 19.71% N	7.65[b]	5.35	19.60	-0.11
o-Toluenesulfamide	3.792[a]	1.085	8.02	-0.16
$C_7H_9O_2NS$; 8.18% N	3.463[a]	0.995	8.05	-0.13
	4.113[a]	1.250	8.17	-0.01
Michler's thioketone	5.030[a]	1.795	10.00	—
$[C_6H_4(CH_3)_2N]CS$; 10.00% N	3.080[a]	1.120	10.19	+0.19
p-Toluenesulfamide	5.324[a]	1.518	7.99	-0.19
$C_7H_9O_2NS$; 8.18% N	4.456[a]	1.273	8.00	-0.18
The amide of 2,5-dimethyltetra-	3.403[a]	0.825	6.79	+0.10
hydro-1-thiopyrane-4-hydroxy-4-	3.019[a]	0.745	6.91	+0.02
acetic acid				
$C_9H_{17}O_2NS$; 6.89% N				
The anilide of 2,5-dimethyl-1-thio-	5.339[a]	0.942	4.94	+0.10
pyrane-4-hydroxy-4-acetic acid	5.462[a]	0.975	5.01	+0.17
$C_{15}H_{17}O_2NS$; 4.84% N				

[a]Weighed on a micro balance with an accuracy of 0.002 mg.
[b]Weighed on a semi-micro balance with an accuracy of 0.01 mg.

Field of Application of the Method. The method is applicable to the analysis of amino compounds, heterocyclic compounds, and also to individual representatives of various classes of nitrogen-containing materials (Tables 5 and 6).

We did not confine ourselves to analysis of materials containing C, H, O, and N, but investigated the possibility of determining nitrogen in the presence of sulfur. Ter Meulen [1] observes that in the case of analysis of sulfur compounds, the nickel catalyst must be replaced after each determination. We established that poisoning of iron catalysts does not occur immediately, but only after about 15 mg of hydrogen sulfide has passed over the catalyst; this amount of hydrogen sulfide corresponds to that formed on hydrogenation of about 25 samples of sulfur compounds containing about 15% sulfur, for an average weight of 4 mg per sample. If, however, the hydrogen sulfide is held back in the decomposition zone by mixing test sample with five times its amount of reduced nickel,‡ one can eliminate the poisoning effect of hydrogen sulfide on the iron catalyst. Actually, when aliquots of sulfur compounds, mixed with reduced nickel, were subjected to hydrogenation in a tube with a platinum catalyst, and the sulfur determined in the gaseous decomposition products by means of our method [6, 10], we found that hydrogen sulfide is practically quantitatively held back by the nickel, and only traces succeed in getting to the absorbers. Consequently, when materials containing sulfur are decomposed after addition of reduced nickel to the sample, only negligible amounts of hydrogen sulfide pass over the catalyst, and poisoning only occurs after a very long time. Thus, we carried out 23 nitrogen determinations, in materials containing sulfur, with the same catalyst. The results for determination of nitrogen in the presence of sulfur which are given in Table 5 were also obtained with one lot of catalyst. Thus, this method is also applicable to materials containing C, H, O, N, and S.

This method is applicable to samples differing widely in weight, of materials containing any amount of nitrogen; this distinguishes it from the Dumas gasometric micro determination, in which the weight of sample taken depends on the nitrogen content. Without modifying the

‡ Reduced iron was not used for mixing with the sample, since such iron is pyrophoric and clinkers readily. The nickel is reduced in a stream of hydrogen and is stored under carbon dioxide.

TABLE 6

Substance	mg. Sample	ml 0.05N IO_3^-	%N Found	%N Diff.
p-Aminoacetophenone C_8H_9ON;	16.70	2.49	10.43	+0.07
10.36% N	14.30	2.14	10.48	+0.12
Methylbenzamide C_8H_9ON;	14.40	2.15	10.46	+0.10
10.36% N	12.20	1.82	10.45	+0.09
Diphenylurea $C_{13}H_{12}ON_2$;	14.65	2.79	13.34	+0.14
13.20% N	13.75	2.63	13.40	+0.20
o-Aminopyridine $C_5H_6N_2$;	8.85	3.76	29.76	-0.01
29.77% N	12.10	5.17	29.92	+0.15
Dipyridylurea $C_{11}H_{10}ON_4$;	9.85	3.70	26.31	+0.15
26.16% N	9.60	3.59	26.19	+0.03
p-Phenylenediamine $C_6H_8N_2$;	11.45	4.27	26.11	+0.20
25.91% N	12.45	4.62	25.99	+0.08
Phenylurea $C_7H_8ON_2$;	12.90	3.81	20.69	+0.11
20.58% N	8.40	2.49	20.76	+0.18
Ammonium oxalate $C_2H_8O_4N_2H_2O$;	14.55	4.11	19.78	+0.07
19.77% N	9.85	2.79	19.84	+0.13
1,2-Dimethyl-4-ethynyl-4-hydroxydeca-	13.05	1.26	6.76	—
hydroquinoline $C_{13}H_{21}ON$; 6.76% N	13.85	1.33	6.73	-0.03
1-Ethyl-2-methyl-4-piperidone	9.45	1.33	9.89	-0.06
$C_8H_{15}ON$; 9.92% N	7.25	1.03	9.95	+0.03

Samples weighed on an analytical balance with accuracy of 0.05 mg.

method at all, one can work with sample weights of 2.5 to 20 mg. For high nitrogen contents and large sample weights, it is recommended that 0.05 N solutions be used instead of 0.02 N. Results for semimicro determinations of nitrogen, carried out on a micro apparatus using a semimicro balance and an analytical balance, are given in Tables 5 and 6.

Apparatus. The apparatus was the same as that used for determination of sulfur by hydrogenation [6, 11, 12].

It consists of a Kipp generator for hydrogen, two wash bottles, a screw clip (better still, one fitted with a micrometric screw) or a tap with threads for regulating the gas flow, a quartz tube for hydrogenation fitted with a ground glass outlet and two glass receivers also connected to it via ground glass joints, an open electric oven, type MA2-20, a gas burner, an aspirator, and a 1 liter measuring cylinder. In addition the following apparatus is necessary: 2 micro burets with a capacity of 10 ml, several 100 ml conical flasks with ground glass stoppers, a tube with a long handle for taking samples during establishment of the titer, and quartz tubes 65 mm long for the samples.

The following reagents are required: ferric nitrate, nickel nitrate or formate, crystalline potassium biiodate, potassium biiodate 0.02 N solution, 0.02 N sodium thiosulfate, 10% solution of potassium iodide, and 1% starch solution.

Preparation of Titrant Solutions. Potassium biiodate solution was prepared from the salt which had been recrystallized three times. 7.799 g of the biiodate was taken for preparation of 1 liter of 0.02 N solution. Thiosulfate solution was prepared by diluting 0.1 N thiosulfate.

Standardization. Crystalline potassium biiodate is the most suitable material for standardizing the thiosulfate.

When standardization is carried out in a neutral medium, it is permissible to take large amounts of salt (50-60 mg) and thereby eliminate weighing errors. Thiosulfate can also be standardized by means of potasium iodate (5-6 mg) or sublimed iodine (18-20 mg). In the latter case, the weights were taken in micro bottles with ground glass lids, containing 1-2 drops of a concentrated aqueous solution of potassium iodide. Standardization by all three methods gave results in good agreement with each other.

Potassium biiodate was standardized with the thiosulfate. The thiosulfate was checked every 5-7 days.

Preparation of Catalyst. Ferric nitrate is fused in its water of crystallization, the water is then removed, and the material is heated to dryness, next it is calcined until fumes of nitric oxide cease to be evolved; it is finally ground in a mortar to give lumps of 1-3 mm in size, which are sieved free of dust, and placed in the combustion tube to give a layer 10 cm long, such that the layer in the tube is in the middle of the electric oven; the layer is held in position by two plugs of asbestos fiber. In the remaining, empty part of the tube, between the catalyst layer and the quartz outlet, is placed a quartz or glass rod, whose diameter is slightly less than that of the combustion tube. Heated gases, on coming out of the catalyst layer, then proceed along the tube at a high velocity, and only a negligible amount of condensation of moisture on the tube walls is observed.

The catalyst is reduced at 475-500° in a rapid stream of hydrogen (gas flow about 100 ml/minute [13]). Reduction of catalyst takes 10-12 hours. Reduction is complete when no more moisture condenses on the cold parts of the tube. When not in use, the tube plus catalyst are left under a hydrogen pressure.

Analytical Procedure. The layer of catalyst in the decomposition tube is heated with the electric oven to 400-420° (thermocouple control); 5-7 ml of 0.02 N $KH(IO_3)_2$ is run into the receiver from a micro buret, and the receiver connected via the ground glass joint to the outlet of the combustion tube, while the other end of the receiver is connected to the aspirator. The sample, in the usual quartz test tube, is placed in the combustion tube at a distance of 2-3 cm from the electric oven, while the aspirator stopcock is closed and the Kipp apparatus stopcock is open. The tube is then closed, and the aspirator stopcock opened slowly (in order not to blow any liquid out of the receiver), the hydrogen flow is adjusted to 30-40 ml/minute, and a burner placed under the open end of the quartz test tube containing the sample. Decomposition is carried out for an average of 7 to 10 minutes. 200-400 ml of hydrogen is then passed through the tube, while the burner is moved along to the closed end of the tube.

While purging is being carried out (not more than 10 minutes), the part of the tube between the quartz test tube and the furnace is ignited; finally, all the tube is heated from the point of hydrogen entry up to the electric furnace; the aspirator tap is then turned off, the receiver is disconnected, and in its place is fitted a second receiver filled ready for the next determination; the whole setup can then be left until one is ready to do the next analysis.

The contents of the receiver are transferred to a 100 ml Erlenmeyer flask fitted with a ground glass stopper, the receiver is then rinsed three times with cooled distilled water, the wash liquors being added to the liquid already in the conical flask; 1 ml of 10% potassium iodide solution is added to the contents of the flask, the flask is let stand for 5 minutes in ice water and the iodine liberated titrated with thiosulfate. One determination usually takes about 30 minutes.

SUMMARY

1. Micro and semicro methods have been developed for determination of nitrogen by hydrogenation of nitrogen-containing organic materials; these are based on rapid pyrolysis of the material in a hydrogen atmosphere, whereupon the nitrogen is converted into ammonia over an iron catalyst, the ammonia being determined iodometrically.

2. The method can be used for analysis of amino com-

pounds and heterocyclic compounds containing C, H, O, N, and S. One determination takes about 30 minutes.

LITERATURE CITED

[1] H. ter Meulen and J. Heslinga, *New Methods of Organic Chemical Analysis* (1931).

[2] H. ter Meulen, *Rec. trav. chim.* **43**, 643 (1924); **44**, 271 (1925); **49**, 396 (1930).

[3] A. Lacourt, *Bull. soc. chim. Belg.* **43**, 82 (1934).

[4] P. M. Heertjes, *Chem. Weekblad* **34**, 827 (1937);

[5] A. Lacourt and Ch. T. Chang, *Bull. soc. chim. Belg.* **49**, 167 (1940).

[6] M. O. Korshun and N. E. Gel'man, *New Methods of Elemental Microanalysis* (State Chem. Press, Moscow, 1949).

[7] N. E. Gel'man, Candidate's Thesis, Institute of Organic Chemistry (Acad. Sci. USSR, 1948).

[8] N. E. Gel'man and M. O. Korshun, *Doklady Akad. Nauk SSSR* **72**, 895 (1950).

[9] I. M. Kolthoff, *Volumetric Analysis* (Moscow, 1932), Vol. II, p. 117.

[10] N. E. Gel'man, *Zavodskaya Lab.* **8**, 673 (1939).

[11] M. O. Korshun and N. E. Gel'man, *Apparatus for Elemental Quantitative Organic Microanalysis* (Acad. Sci. USSR Press, Moscow-Leningrad, 1947).

[12] M. O. Korshun and N. E. Gel'man, *Zavodskaya Lab.*, **12**, 754 (1946).

[13] S. S. Lachinov, *Zhur. Priklad. Khim.* **10**, 19 (1937).

CHAPTER 27

Simultaneous Determination of Carbon, Hydrogen, Silicon, and Sulfur in Sulfur-containing Organosilicon Compounds*

V. A. Klimova and E. G. Bereznitskaya

Simultaneous determination of carbon, hydrogen, silicon, and sulfur in sulfur-containing organosilicon compounds would appear to be a logical extension of the methods we have developed earlier for simultaneous determination of carbon, hydrogen, and silicon, and of carbon, hydrogen, silicon, and halogen [1, 2].

Thermal decomposition of sulfur-containing organosilicon compounds under rapid combustion conditions in a rapid stream of oxygen occurs with formation of some silicon carbide; i.e. the process is similar to that which happens in the case of compounds containing carbon, hydrogen, and silicon. Consequently, during analysis of such compounds, decomposition should be carried out in the presence of chromium sesquioxide on asbestos, as indicated for compounds containing C, H, and Si [1]. Chromium sesquioxide does not prevent formation of sulfur oxides and the latter can accordingly be trapped and determined by the Korshun-Sheveleva method [3], by means of silver heated to 700° in a special apparatus. Finely dispersed silica is quantitatively retained by the asbestos catalyst in the capsule used for decomposition. Nitrogen, when present, does not interfere. The method is simple and convenient. Results of determinations are given in the table.

Zhur. Anal. K im.* **12, 424-5 (1957).

The accuracy of determination for each element lies within the limits of ±0.3%.

SUMMARY

A method has been developed for simultaneous determination of C, H, Si, and S in sulfur-containing organosilicon compounds.

Experimental error for each element is within the limits of ±0.2 to 0.3%.

LITERATURE CITED

[1] V. A. Klimova, M. O. Korshun, and E. G. Bereznitskaya, *Zhur. Anal. Khim.* 9, 223 (1956).

[2[V. A. Klimova and E. G. Bereznitskaya, *Zhur. Anal. Khim.* **9**, 292 (1956).

[3] M. O. Korshun and N.S. Sheveleva, *Zhur. Anal. Khim.* **7**, 104 (1952).

Test compound	C in % Found	C in % Calculated	Δ C in %	H in % Found	H in % Calculated	Δ H in %
α-Thiocyanoethyltrimethylsilane $(CH_3)_3SiCH(CH_3)SCN$	45.29	45.23	+0.06	8.33	8.22	+0.11
	45.35		+0.12	8.33		+0.11
β-(Trimethylsilylmethylmercapto)-propionitrile $(CH_3)_3SiCH_2SCH_2CH_2CN$	48.65	48.50	+0.15	8.87	8.72	+0.15
	48.42		−0.08	8.77		+0.05
	48.55		+0.05	8.67		−0.05
α-Thiocyanomethyldiphenylmethylsilane $CH_3(C_6H_5)_2SiCH_2SCN$	67.09	66.88	−0.21	5.67	5.60	+0.07
	66.95		+0.07	5.61		+0.01
1,2-Di(α-thiocyanoethyl)tetramethyldisiloxane $O(Si(CH_3)_2CH_2SCN)_2$	34.80	34.75	+0.05	5.94	5.83	+0.11
	34.80		+0.05	5.82		−0.01
	34.63		−0.12	5.76		−0.07
β-(Methyldibutylsilymethylmercapto) propionitrile $CH_3(C_4H_9)_2SiCH_2SCH_2CH_2CN$	60.98	60.63	+0.35	10.47	10.57	−0.10
	60.73		+0.10	10.57		0.00
	60.40		−0.23	10.79		+0.22
Bis-(α-thiocyanomethyl) dimethylsilane $NCSCH_2Si(CH_3)_2CH_2SCN$	35.69	35.61	+0.08	5.02	4.98	+0.04
	35.63		+0.02	4.95		−0.03
	35.65		+0.04	5.01		+0.03
α-(Thiocyanomethyl) methyldipropylsilane $CH_3(C_3H_7)_2SiCH_2SCN$	53.57	53.66	−0.09	9.62	9.51	+0.11
	53.56		−0.10	9.53		+0.02
α-Thiomethyldiethylmethylsilane $CH_3(C_2H_5)_2SiCH_2SH$	48.86	48.58	+0.28	10.89	10.87	+0.02
	48.79		+0.21	10.99		+0.12
β-(Trimethylsilyl-γ-propylmercapto) propionitrile $(CH_3)_3Si(CH_2)_3SCH_2CH_2CN$	53.64	53.66	−0.02	9.56	9.51	+0.05
	53.87		+0.21	9.63		+0.12
β-(Methyldiethylsilymethylmercapto) propionitrile $CH_3(C_2H_5)_2SiCH_2SCH_2CH_2CN$	53.68	53.66	−0.02	9.44	9.51	−0.07
	53.63		−0.03	9.47		−0.04
γ-(Thiocyanopropyl) trimethylsilane $(CH_3)_3SiCH_2CH_2CH_2SCN$	48.40	48.50	−0.10	8.49	3.72	−0.23
	48.56		+0.06	8.75		+0.03

Test compound	Si in % Found	Si in % Calculated	Si Δ in %	S in % Found	S in % Calculated	Δ S in %	Sample weight, in mg
α-Thiocyanoethyltrimethylsilane $(CH_3)_3SiCH(CH_3)SCN$	17.67	17.63	+0.04	19.80	20.13	−0.33	9.31
	17.42		−0.21	20.06		−0.07	11.83
β-(Trimethylsilylmethylmercapto)-propionitrile $(CH_3)_3SiCH_2SCH_2CH_2CN$	16.16	16.20	−0.04	18.80	18.50	+0.30	10.12
	16.28		+0.08	18.66		+0.16	8.93
	16.46		+0.26	18.79		+0.29	11.44
α-Thiocyanomethyldiphenylmethylsilane $CH_3(C_6H_5)_2SiCH_2SCN$	10.15	10.42	−0.27	11.94	11.90	+0.04	12.39
	10.28		−0.14	12.03		+0.13	10.41
1,2-Di(α-thiocyanoethyl)tetramethyldisiloxane $O(Si(CH_3)_2CH_2SCN)_2$	20.02	20.31	−0.29	23.21	23.16	+0.05	11.22
	20.35		+0.04	23.35		+0.19	10.61
	19.95		−0.36	23.59		+0.43	11.12
β-(Methyldibutylsilylmethylmercapto) propionitrile $CH_3(C_4H_9)_2SiCH_2SCH_2CH_2CN$	10.67	10.91	−0.24	12.23	12.45	−0.22	9.42
	10.57		−0.34	12.40		−0.05	11.63
	10.75		−0.16	12.42		−0.03	10.22
Bis-(α-thiocyanomethyl) dimethylsilane $NCSCH_2Si(CH_3)_2CH_2SCN$	13.70	13.88	−0.18	31.99	31.69	+0.30	9.62
	—			31.54		−0.14	12.83
	13.61		−0.27	31.57		−0.12	13.06
α-(Thiocyanomethyl) methyldipropylsilane $CH_3(C_3H_7)_2SiCH_2SCN$	14.15	13.95	+0.20	15.77	15.92	−0.15	10.27
	14.13		+0.18	16.05		+0.13	12.27
α-Thiomethyldiethylmethylsilane $CH_3(C_2H_5)_2SiCH_2SH$	19.01	18.93	+0.08	21.58	21.62	−0.04	9.02
	19.16		+0.23	21.79		+0.10	11.73
β-(Trimethylsilyl-γ-propylmercapto) propionitrile $(CH_3)_3Si(CH_2)_3SCH_2CH_2CN$	13.82	13.95	−0.13	16.02	15.92	+0.10	9.56
	13.87		−0.08	16.15		+0.23	9.20
β-(Methyldiethylsilylmethylmercapto) propionitrile $CH_3(C_2H_5)_2SiCH_2SCH_2CH_2CN$	13.81	13.95	−0.14	15.69	15.92	−0.23	10.32
	13.63		−0.32	15.87		−0.05	10.56
γ-(Thiocyanopropyl) trimethylsilane $(CH_3)_3SiCH_2CH_2CH_2SCN$	16.43	16.20	+0.23	18.52	18.50	+0.02	14.35
	16.27		+0.07	18.61		+0.11	7.84

CHAPTER 28

Microdetermination of Carbon and Hydrogen in Organofluorine Compounds*

N. E. Gel'man, M. O. Korshun, and N. S. Sheveleva

Determination of carbon and hydrogen in organofluorine compounds has been studied by many research workers to date. Investigations in elemental microanalysis have been carried out within the framework of the classical methods of Pregl and Friedrich, and have been directed towards overcoming the difficulties associated with the quantitative oxidation of organofluorine compounds and complete absorption of fluorine-containing products inside the combustion tube. Various reagents which have formed part of the packing of the combustion tube in addition to the usual oxidizing agents and platinum contacts have been tried as absorbers for fluorine [1-6]. During analysis of highly fluorinated hydrocarbons and carbon fluorides, it has been recommended that combustion temperature be increased, and that combustion should be carried out in a stream of moist oxygen [2, 7-9]. The latter makes determination of hydrogen impossible; in this connection, determination of hydrogen in organofluorine compounds has been especially considered, but has not hitherto been solved satisfactorily [4, 8, 10, 11].

In developing a method for determining carbon and hydrogen in materials containing fluorine, we tried to make it as reliable as the method used for analysis of the

Zhur. Anal. Khim. **12**, 526-33 (1957).

usual materials. The basis of our method was pyrolytic combustion in a rapid stream of oxygen [12, 13]; this technique has been successfully used in many microanalytical laboratories in Russia, and recently abroad [14, 14a]. During pyrolytic combustion of organofluorine compounds the fluorine can be quantitatively held back in the decomposition zone of the test aliquot by an "internal absorbent", which is packed into the quartz tube used for decomposition of the material [15]. The "internal absorbent" may be a material which is stable at about 1000° and which can be converted into a non-volatile, high melting fluoride and a readily decomposed carbonate. Magnesium oxide is the material which satisfies these requirements most closely:

	MgO	MgF_2	$MgCO_3$
Melting point	2800°	1125°	Starts to decompose at 300°
Boiling point	3600°	2260°	

Pyrolytic combustion in the presence of magnesium oxide for absorption of fluorine opens up various possibilities of simultaneous determination of carbon, hydrogen, and fluorine [15, 16]. The simplest method would be to determine the fluorine content approximately on the basis of the change in weight of the tube containing magnesium oxide in which pyrolytic decomposition had been carried out. This variation is applicable to materials which are free from other elements which are trapped by magnesium oxide. In accordance with the equation $MgO + 2F = MgF_2 + \frac{1}{2}O_2$, absorption of 38 weight units of fluorine would lead to a change in weight of the absorbent of only 22 weight units; e.g. during combustion of 7.6 mg of material containing 50% fluorine, an error in weighing the tube containing magnesium oxide of 22 μg would lead to an error in fluorine determination of 0.5% (absolute). In practice the fluorine content found may deviate from the theoretical value by as much as a few percent. Nevertheless, in a number of cases, a knowledge of the approximate fluorine content may be important; consequently, in the present article, in addition to determination of carbon and hydrogen, we have also described the variation in which all three elements are determined on one aliquot.

TABLE 1[a]

Name and formula of test material	Weight in mg	Found C	Found H	Difference % C	Difference % H
2,4-Dinitrophenylhydrazone of	5.025	48.76	2.88	-0.16	-0.13
trifluoropropiophenone	5.432	48.79	2.98	-0.13	-0.03
C 48.92 %; H 3.01%, $C_{15}H_{11}F_3N_4O_4$					
The amide of α-hydroperfluoro-	5.960	24.81	2.06	-0.01	-0.03
propionic acid. C 24.82%;	5.060	24.84	2.10	+0.02	+0.01
H 2.09%, $C_3H_3F_4NO$					
The anilide of α-hydroperfluoro-	4.307	44.23	2.60	-0.06	0.00
isobutyric acid. C 44.29%;	5.100	44.35	2.65	+0.06	+0.04
H 2.60%, $C_{10}H_7F_6NO$					

[a]Analyses carried out by A. S. Lapteva.

Another variation for the simultaneous determination of carbon, hydrogen, and fluorine which is based on the pyrohydrolysis of the magnesium fluoride formed during determination of carbon and hydrogen [16] seems to us to have greater possibilities.

The expediency of using magnesium oxide for absorption of fluorine during elemental analysis is confirmed by the work of other investigators, who have used magnesium oxide, or preparations based on it, under the conditions used in classical microanalysis as a component part of the packing used in the combustion tube; it has been used in conjunction with cupric oxide and metallic silver [6], with lead dioxide, cupric oxide, metallic silver [3], or with metallic silver, lead oxide, lead dioxide, in the form of eight alternate layers [5]. In every instance it can be stated that magnesium oxide is a better absorbent for fluorine than anything else which has been used previously for this purpose in elemental analysis; it is superior, for example, to lead oxides, lead chromate, sodium fluoride, potassium fluoride, etc.

The method, in essence, consists of sprinkling the sample together with magnesium oxide into a quartz capsule, and subjecting it to pyrolytic combustion in an empty combustion tube in an oxygen stream. The fluorine is thereupon trapped by the magnesium oxide, while carbon and hydrogen are determined as usual [12, 13]. The magnesium oxide must be conditioned beforehand. When other halogens are present in the test material they are absorbed by metallic silver which is placed in the combustion tube and heated to 550-600° [17]; nitrogen oxides are trapped outside the combustion tube by means of dry absorbents [18]; sulfur, phosphorus, boron, silicon, and metals (apart from the alkali and alkaline earth metals) do not interfere with determination of carbon and hydrogen, but remain in the magnesium oxide layer with the fluorine. In substances containing only C, H, O, N, and F, the latter can be determined approximately at the same time as carbon and hydrogen from the changes in weight of the tube containing the magnesium oxide.

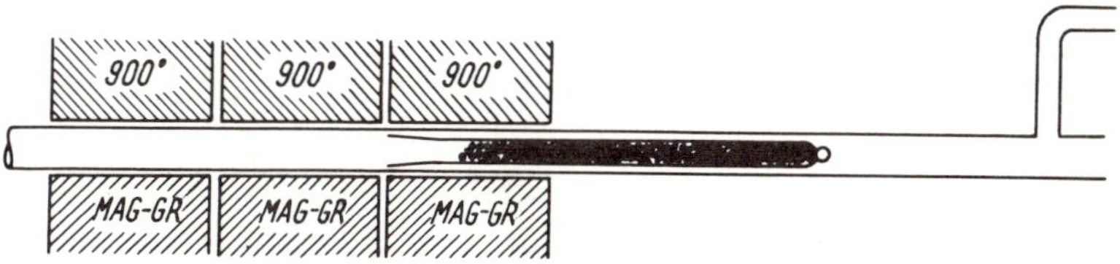

TABLE 2

Name and formula of test material	Weight in mg	Found % C	Found % H	Found % F	Difference % C	Difference % H	Difference % F
Methyl α-hydroperfluoroisobutyrate $C_5H_4O_2F_6$	8.986	28.64	1.96	55.96	+0.06	+0.04	+1.69
C=28.58%; H=1.92%; F=54.27%	5.015	28.55	1.91	60.88	-0.03	-0.01	+6.61
Ethyl α-hydroperfluoroisobutyrate $C_6H_6O_2F_6$	11.670	32.12	2.76	51.50	-0.03	+0.06	+0.63
C=32.15%; H=2.70%; F=50.87%	8.021	32.15	2.83	50.40	0.00	+0.13	-0.47
Methyl β-hydroperfluoroisobutyl ether	7.342	25.84	1.63	68.45	-0.03	-0.11	+2.95
$C_5H_4OF_8$(b.p.70°)	10.374	25.81	1.78	63.84	-0.06	+0.04	-1.66
C=25.87%; H=1.74%; F=65.50%							
Methyl perfluoroisobutenyl ether $C_5H_3OF_7$ (b.p.90°)	11.532	28.38	1.61	60.83	+0.07	+0.18	-1.88
C=28.31%; H=1.43%; F=62.71%	4.542	28.55	1.65	64.68	+0.24	+0.21	+1.97
The amide of α-hydroperfluoropropionic acid	8.117	24.78	2.35	52.86	-0.04	+0.27	+0.47
$C_3H_3ONF_4$	8.641	24.67	2.16	51.47	-0.15	+0.08	-0.98
C=24.82%; H=2.08%; F=52.39%	8.178	24.75	2.11	54.85	-0.07	-0.03	+2.46
The diethylamide of α-hydroperfluoroisobutyric acid	11.708	38.41	4.41	46.33	+0.16	0.00	+0.95
$C_8H_{11}ONF_6$(m.p.51°)	8.508	38.43	4.55	47.09	+0.18	+0.14	+1.71
C=38.25%; H=4.41%; F=45.38%							
The anilide of α-hydroperfluoroisobutyric acid	5.730	44.45	2.66	43.46	+0.16	+0.06	+1.42
$C_{10}H_7ONF_6$	6.062	44.59	2.67	43.34	+0.33	+0.07	+1.30
C=44.29%; H=2.60%; F=42.04							
α-Hydroperfluoropropionitrile C_3HNF_4(b.p.41°)	9.680	28.59	0.84	61.02	+0.23	+0.05	+1.20
C=28.36%; H=0.79%; F=59.82%	8.411	28.30	0.94	—	-0.06	+0.15	—
Diethylperfluoroisobutenylamine $C_6H_{10}NF_7$	6.795	38.00	3.98	56.20	+0.05	0.00	+3.67
C=37.95%; H=3.98%; F=52.53%	16.234	37.78	4.12	52.67	-0.17	+0.14	+0.14

On the basis of the results for combustion of non-volatile high melting fluorine-containing compounds, we became convinced that fluorine is absorbed in a thin layer of magnesium oxide during pyrolytic decomposition of the test sample. Table 1 contains the results of analyses carried out in a tube 6.5 cm long containing a layer of magnesium oxide 5 cm long. Depending on the volatility of test material, decomposition starts at various distances from the bottom of the tube, so that the length of the magnesium oxide layer required for quantitative trapping of the fluorine changes accordingly. Consequently, on carrying out series of analyses of materials whose structure and properties are unknown, it is necessary to increase the length of the magnesium oxide layer appreciably. According to our observations, even during combustion of liquids boiling at 35-40°, a 12-13 cm layer of MgO is quite sufficient. When it is necessary to carry out an approximate determination of fluorine along with the determination of carbon and hydrogen, combustion of the test material is carried out in a long quartz capsule (15 cm) filled with magnesium oxide, and placed in an empty combustion tube (Fig. 1). In this way it is possible to analyze various materials including low boiling liquids containing more than 70% fluorine (See Table 2). When it is necessary to determine only carbon and hydrogen, another variation is better; in this variation combustion is carried out in a short tube filled with magnesium oxide, this tube in turn being placed in a hollow quartz sleeve (with sintered bottom) (Fig. 2) containing the same absorbent. The magnesium oxide in the sleeve is placed there with the object of trapping small amounts of fluorine which are not trapped in the tube itself; this additional magnesium oxide can be used without changing it during the whole of a working day, while that in the tube itself should be changed for

TABLE 3

Name and formula of test material	Weight in mg	Found % C	Found % H	Difference % C	Difference % H
1,1,1-Trifluoro-2-bromo-3-trichloroacetoxypropane $C_5H_3BrCl_3F_3O_2$	10.370	17.80	0.87	+0.05	−0.02
C=17.75%; H=0.89%; F=16.84%	12.640	17.88	0.91	+0.13	+0.02
Perfluorodiethylmethylamine $C_5F_{13}N$ (b.p. 45°)	9.334	18.69	0.20	−0.02	+0.20
C=18.71%; F=76.93%	8.685	18.59	0.24	−0.12	+0.24
Perfluorotriethylamine $C_6F_{15}N$ (b.p. 70°)	10.360	19.57	0.17	+0.15	+0.17
C=19.42%; F=76.80%	11.063	19.69	—	+0.24	—
Ethyl trifluoroacetate $C_4H_5F_3O_2$ (b.p. 70°)	9.223	34.00	3.66	+0.19	+0.11
C=33.81%; H=3.55%; F=40.11%	10.697	33.74	3.57	−0.07	+0.02
Diethyl-β-fluoroethylphosphinate $C_6H_{14}FO_4P$	6.970	36.00	7.03	0.00	−0.02
C=36.00%; H=7.05%; F=9.49%	6.290	35.93	6.97	−0.07	−0.08
o-Ethyl-S-β-fluoroethyl ester of methylphosphinic acid $C_5H_{12}FOPS$	6.060	32.29	6.53	+0.04	+0.03
C=32.25%; H=6.50%; F=12.85%	6.818	32.46	6.66	+0.21	+0.16
β-Fluoroethyl *p*-nitrophenyl ester of methylthionophosphinic acid	7.698	38.58	3.98	−0.13	+0.01
$C_9H_{11}FO_4NPS$ C=38.71%; H=3.97%; F=6.80%	5,816	38.50	4.06	−0.21	+0.09
2,2-Diphenylenedioxonium boron fluoride $C_{18}H_{13}BF_4O$	9.444	65.14	3.97	+0.02	+0.02
C=65.12%; H=3.95%; F=22.88%	8.328	65.31	4.08	+0.19	+0.13
Diphenylchloronium boron fluoride $C_{12}H_{10}BClF_4$	7.981	52.09	3.71	−0.03	+0.06
C=52.12%; H=3.65%; F=27.49%	7.793	52.15	3.90	+0.03	+0.25
Diphenylbromonium boron fluoride $C_{12}H_{10}BBrF_4$	8.560	44.97	3.12	+0.05	−0.03
C=44.92%; H=3.15%; F=23.68%	6.460	44.93	2.95	+0.01	−0.20
β-Chlorovinylphenyliodonium boron fluoride $C_6H_7BClF_4I$	7.490	27.26	2.02	−0.02	+0.02
C=27.28%; H=2.00%; F=21.57%	7.666	27.13	1.99	−0.15	−0.01

each determination. Using this sleeve it is possible to carry out determinations considerably more quickly than when a long tube on its own is used, and far less magnesium oxide has to be used. Results for determinations in which this second variation was used are given in Table 3.

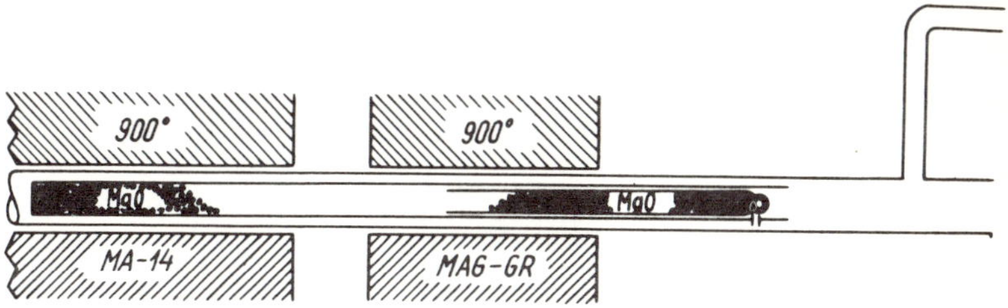

Fig. 2

Apparatus

A standard apparatus‡ for microdetermination of carbon and hydrogen is necessary for carrying out an analysis [13]; the following are required in addition:

1. When carbon and hydrogen are to be determined in fluorine-containing materials:

a) a hollow quartz sleeve with a sintered bottom;

b) quartz tubes for the samples.

The sleeve should fit readily into the combustion tube, while the quartz sample capsule, in turn, should easily fit into the sleeve, so that there is a gap of about 2 mm between tube and sleeve. This gap is necessary so that during combustion enough oxygen can get into the sleeve. When a standard combustion tube is used, the external diameters of sleeve and tube should be 8-8.5 and 5-6 mm

‡ The complete apparatus is produced by the "Laborpribor" factory in Klin.

respectively. All the same it is more difficult to carry out pyrolytic combustion in a narrow tube than in the usual type of tube; consequently, we recommend the following dimensions for the combustion tubes, sleeves, and tubes for the sample, as being more convenient:

	Length in mm	Diameter in mm	
		External	Internal
Combustion tube	500	13.5	11.5
Sleeve	200	10	8.5
Tube for sample	65-90	7	5.5

c) an MA-14 open electric oven 14 cm long, operating at 900°.

2. When carbon, hydrogen, and fluorine are to be determined simultaneously:

a) quartz sample capsules 150 mm long and 8-9 mm in diameter, fitted with ground glass stoppers [1]. The surface of the sample tube should be flawless;

b) quartz scoops for introducing aliquots of solid test material into the sample tube (Fig. 3);

c) quartz rods for pushing capillaries containing test material into the sample tubes when the material is a liquid (Fig. 4);

d) two MAG-6R electric ovens operating at 900°, length 6 cm;

3. For conditioning magnesium oxide:

a) a gas holder and the usual purifying system for oxygen§;

b) ampoules made from high melting sample tubes for storing the conditioned magnesium oxide;

c) an MA-O2-20 oven operating at 1150°, length 20 cm.

---

§ If the apparatus used for C and H determinations is only used periodically, then use can be made of its purifying system; the latter is connected to the tube used for conditioning after the combustion tube, or connected in parallel with it.

Fig. 3

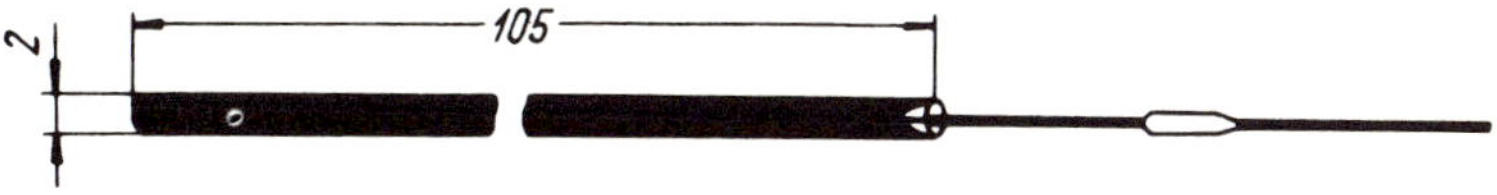

Fig. 4

Conditioning the Magnesium Oxide

Air-dry magnesium oxide contains considerable amounts of water and carbon dioxide at room temperature, and it gradually gives these up on heating. Magnesium oxide loses its hygroscopicity after calcining at 1600° and higher; at the same time, however, it loses its chemical activity [19], and such a preparation becomes inert even towards fluorine. In the temperature range we are interested in, namely 800-1000°, the change in the absorptive properties of magnesium oxide on increasing the temperature becomes so significant that it might be the source of large errors in the elemental analysis. This makes it essential to condition any magnesium oxide to be used as an internal absorbent for fluorine.

We tested magnesium oxide in the form of dust, granules, and preparations deposited on a carrier. The best method proved to be deposition of magnesium oxide on pumice, which is readily strewn into the sample tube, hardly corrodes quartz at all and has no tendency to form dust. The pumice used was material which had been ground, freed from dust, washed with hydrochloric acid and ignited in a muffle [20]. It is convenient to use two fractions; a fine fraction down to 0.5 mm, and a coarse fraction 1-1.5 mm in cross section. The ignited pumice whose granules have a light brown color is placed in a porcelain beaker, and is lightly moistened with distilled water (so that the granules do not stick together); it is then mixed with powdered magnesium oxide. 0.5g of magnesium oxide is taken per gram of pumice. The mixture obtained is dried to a friable state when it is sieved free from dust. An old combustion tube can be used for igniting the preparation; the end of such a tube is cut and is ground at an angle of 45°. A lump of ashless filter paper is put into this tube, the purpose of the filter being to keep the granules in place when they are loaded into the tube; the dried preparation is loaded into the tube and the oxygen rate adjusted to 30-40 ml per minute; the tube is placed in a cold furnace with the filter outside the furnace limits and all the magnesium oxide inside. As the furnace begins to heat, too rapid combustion of the filter should be avoided; otherwise some of the preparation may be expelled from the oven. The required temperature (1100-1150°) is reached in the course of 20-30 minutes. After the filter paper has been burned, the furnace is drawn over the preparation so that all of it is ignited at a temperature of not less than 1100°; 1.5 hours after commencing the heating, the oven is removed, and the ignited preparation transferred to glass ampoules with long necks which are then sealed off. One ampoule containing about 1.5 g of the preparation can be used for several determinations; the ampoule is sealed off immediately after some of the preparation has been taken from it. The preparation obtained in this way has a white color with a slight yellow tinge. The oxygen used for this conditioning is purified in the same way as that used for actual combustion during analysis.

When carbon, hydrogen, and fluorine are determined simultaneously, granulated magnesium oxide without a carrier is used; it is conditioned immediately before analysis. The ignited preparation is strewn into a long quartz capsule, which is then placed in the combustion tube; it is ignited with three electric furnaces at 900°; then ignited two times again for 10 minute intervals with 10 minute intervals for cooling between each ignition.

Analytical Procedure

a) Determination of Carbon and Hydrogen

The setup normally used for carbon and hydrogen is mounted [12, 13], the electric furnaces are switched on, and absorption apparatus containing Anhydrone and Ascarite connected; some of the coarse fraction of the preparation of magnesium oxide on pumice is poured through a funnel with a long stem into the sleeve until ⅓ of its length is packed with the preparations; the sleeve as a whole is slipped into the combustion tube while the aspirator stopcock is closed; some of the fine fraction of the preparation is then loaded into the quartz sample tube, and the latter placed in the sleeve and its position adjusted inside the sleeve such that all the magnesium oxide is within the zone heated to 900° by MA-14 and MAG-6R ovens. Two capsules are used for each determination; magnesium oxide is ignited in one of them prior to combustion; the other takes the sample. A blank is carried out in the course of 15-20 minutes at an oxygen rate of 20 ml/minute; the absorption apparatus is disconnected and prepared for weighing (wiping and discharging any electrical charges); the end of the combustion tube is covered and the oven used for heating the sleeve is drawn back. The absorption apparatus is weighed and connected to the setup, the aspirator stopcock, however, still being kept closed; an aliquot of test sample is introduced into a quartz sample capsule. The sleeve is moved on to the open end of the combustion tube, the tube containing the ignited magnesium oxide is withdrawn and some of the latter is strewn over the sample in the sample capsule; the latter is then replaced in the sleeve. The sleeve and the sample in its capsule are moved into the combustion tube to occupy the same position used for the blank; the magnesium oxide in the sleeve is heated with an MA-14 oven, the oxygen flow is started (20 ml/minute) and pyrolytic decomposition of sample is commenced by placing a gas burner or electric furnace under the open end of the sample capsule; the burner is then moved forward in a direction opposite to the oxygen flow. The arrangement of the sleeve and sample capsule in the combustion tube is shown in Fig. 2. During combustion of liquids, the sample is introduced in a sealed capillary into the empty sample capsule; magnesium oxide is then strewn into the tube and the fine hair end of the capillary broken against the bottom of the sample tube by pressing on the other end of the capillary which sticks out with tweezers. When the liquid is nonvolatile, there is no need to seal the capillary.

Most compounds liberate carbon during pyrolysis, so that the analyst, by observing the formation of a dark coating on the magnesium oxide, can readily regulate the rate of combustion. Normally, combustion takes 10-15 minutes, but volatile liquids which decompose with charring must be burned more slowly (15-20 minutes). During analysis of materials boiling below 50°, it is recommended that the part of the sample capsule where the sample itself is located be cooled with dry ice after introduction of the sample.

After combustion it is essential to ignite the sleeve along the whole of its length with furnaces MA-14 and MAG-6R for 10 minutes, in order to decompose the magnesium carbonate formed, after which the absorption apparatus is disconnected from the setup and weighed. A fresh portion of magnesium oxide is strewn into a clean sample capsule, the used sample capsule is removed from the sleeve and replaced by the new one. While weighing is in progress, the sleeve is heated along the whole of its length by electric furnaces for 15 minutes, oxygen being passed through the apparatus during the heating. The absorption apparatus is connected to the setup, the furnaces are drawn back, the next test aliquot is taken and the analysis of the fresh sample carried out as already described. The satisfactory operation of the setup is checked by carrying out combustion of a pure fluorine-containing compound, and the respective corrections applied for the weight of CO_2 and H_2O. If the test material contains other halogens in addition to fluorine, the former are not held back by magnesium oxide completely [16]; accordingly, in such a case, a platinum contact 3-4 cm long is inserted between the sleeve and the outlet of the combustion tube; the platinum contact is heated with the same furnace as that used for heating the sleeve. After the platinum contact a layer of metallic silver 4-5 cm is inserted; this is heated with an electric furnace to 550-650°. Sulfur, boron, phosphorus, silicon, and metals (apart from the alkali and alkaline earths) in addition to fluorine, are held back in the sample capsule by magnesium oxide and do not interfere with the determination. One determination takes 60-70 minutes.

b) Simultaneous Determination of Carbon, Hydrogen, and Fluorine

When the blank is carried out, a long quartz sample capsule containing conditioned magnesium oxide without carrier is introduced into an empty combustion tube; it is then ignited with three electric furnaces for two ten-minute periods as indicated above. The absorption apparatus is disconnected, after which the sample capsule plus magnesium oxide are withdrawn; it is stoppered with a ground glass stopper and placed on a long metal block near the balance. The absorption apparatus is weighed; it is then connected immediately to the setup, the aspirator stopcock being kept closed. The sample capsule plus magnesium oxide are weighed 30 minutes after their removal from the combustion tube. When the sample is a solid, it is introduced by means of a small quartz scoop into the sample capsule; the scoop is tapped gently so that the sample falls far enough into the capsule for the latter to be stoppered after combustion. Liquid test samples are introduced in capillaries into the sample capsule by means of a quartz rod. If the capillary has had to be fused, then the fused end has to be broken against the bottom of the sample tube. Sample capsule plus sample are finally placed in the combustion tube, the stopcock of the aspirator is opened and a furnace is drawn over the open end of the sample capsule fit on so that not less than 3-4 cm of MgO is within the heating zone. Two other electric furnaces are drawn over the capsule in addition to the first; towards completion of combustion the whole of the magnesium oxide layer is heated with the furnaces. When combustion is complete, the capsule is ignited as described above, and then the weighing carried out. When using magnesium oxide in a quartz apparatus, it must be borne in mind that sample and the quartz surface are readily electrified during movement of sample capsules along the combustion tube. It is recommended that electric charges be removed by passing those parts of the apparatus likely to pick up such charges through a burner flame.

The authors wish to thank E. N. Merkulova for permission to use some of her results.

SUMMARY

1. It has been established that during rapid pyrolytic combustion at 900° of organo-fluorine compounds in a layer of magnesium oxide contained in a quartz apparatus, quantitative oxidation of highly fluorinated compounds and complete retention of fluorine can be achieved using only a short length of magnesium oxide.

2. A microanalytical method has been developed for determination of carbon and hydrogen in organo-fluorine compounds (solid and liquid, including volatile liquids) containing nitrogen, halides, sulfur, boron, phosphorus, silicon, and metals not forming carbonates which are difficult to decompose.

3. Another variation is suggested for analysis of materials containing carbon, hydrogen, oxygen, nitrogen, and fluorine, whereby, in addition to determining carbon and hydrogen, it is possible to get an approximate value for the fluorine content.

LITERATURE CITED

[1] N. S. Nikolaev, *Izvest. Akad. Nauk SSSR., Otdel. Khim. Nauk*

[2] R. Belcher and R. Goulden, *Mikrochen.* 36/37, 679 (1951).

[3] W. M. Trockmorton and G. H. Hutten, *Anal. Chem.* **24**, 2003 (1952).

[4] W. Bodenheimer and M. Goldstein, *Bull. Res. Council Israel* **3**, 53 (1953).

[5] C. A. Rush, S. S. Cruikshank, and E. J. Rhodes, *Mikrochim. Acta*, 1956, 858.

[6] R. N. McCoy and E. L. Bastin, *Anal. Chem.* **28**, 1776 (1956).

[7] O. J. Milnes, *Anal. Chem.* **22**, 315 (1950).

[8] *Fluorine and Its Compounds*, Edited by J. Simons, [Russian Edition] Vol. 2, Moscow, 1956, p. 143.

[9] S. J. Clark, *Quantitative Methods of Organic Microanalysis*, London, 1956, p. 52.

[10] *Ibid.* p. 66.

[11] J. F. Miller, H. Hunt, H. B. Hass, and E. T. McBee, *Anal. Chem.* **19**, 146 (1947).

[12] M. O. Korshun and V. A. Klimova, *Zhur. Anal. Khim.* **2**, 274 (1947); **3**, 176 (1948); **4**, 292 (1949).

[13] M. O. Korshun and N. E. Gel'man, *New Methods of Elemental Microanalysis* Moscow, 1949.

[14] J. A. Kuck and E. C. Grim, *Mikrochim. Acta,* **1954**, 201.

[14a] J. A. Kuck and P. L. Altieri, *Mikrochim. Acta,* **1956**, 1550.

[15] N. E. Gel'man and M. O. Korshun, *Doklady Akad. Nauk SSSR* **89**, 685 (1953).

[16] M. O. Korshun, N. E. Gel'man, and K. I. Glazova, *Doklady Akad. Nauk SSSR* **111**, 1255 (1956).

[17] M. O. Korshun and N. S. Sheveleva, *Zhur. Anal. Khim.* **7**, 104 (1952): **11**, 376 (1956).

[18] V. A. Klimova and M. O. Korshun, *Zhur. Anal. Khim.* **6**, 230 (1951).

[19] G. V. Kukolev, *Chemistry of Silicon and Physical Chemistry of Silicates*, p. 275.

[20] M. O. Korshun, E. A. Terent' eva, and V. A. Klimova, *Zhur. Anal. Khim.* **9**, 275 (1954).

CHAPTER 29

The Analysis of Fluoroorganic Compounds. Simultaneous Microdetermination of Fluorine and Nitrogen in Organic Compounds*

N. E. Gel'man, M. O. Korshun, M. N. Chumachenko, and N. I. Larina

In earlier published work it was established [1, 2] that in the elementary analysis of fluoroorganic compounds, magnesium oxide is a reliable reagent for quantitative binding of fluorine liberated in the decomposition of organic substances. It was shown at that time that fluorine absorbed by magnesium oxide could be quantitatively liberated from the layer of absorbent in the form of HF by hydrolytic decomposition of the magnesium fluoride by steam at high temperature [3]. The high temperature hydrolysis, or pyrohydrolysis goes by the following scheme:

$$\text{Metal-}F_2 + H_2O \rightarrow \text{Metal oxide} + 2HF.$$

TABLE 1

Burning in a Layer of Pure Nickel Oxide

Substance	Weight g	F % Calc.	F % Found	F % Diff.
α-Hydroperfluoroisobutyric anilide	7.000	42.04	39.63	- 2.41
$C_{10}H_7F_6NO$	4.790	42.04	41.06	- 0.98
Tris(*m*-nitrophenyl) oxonium borofluoride $C_{18}H_{15}BF_4N_3O_7$	5.200	16.20	15.74	- 0.46
1-Bromo-2(trifluoromethyl)-1,1,3,3-3-pentafluoropropane* C_4HBrF_8	6.720	54.09	28.09	-26.00
	4.690	54.09	21.67	-32.42

*B. p. 56°

Doklady Akad. Nauk SSSR* **123, 468-470 (1958).

TABLE 2

Determination of Fluorine and Nitrogen in One 3.5-8 mg Sample

Substance	N %			F %			Remarks
	Calc.	Found	Diff.	Calc.	Found	Diff.	
β-(p-fluorophenyl)-β-alanine	7.64	7.67	+0.03	10.37	10.12	-0.25	
$C_9H_{10}FNO_2$	7.64	7.55	-0.09	10.37	10.39	+0.02	
Trifluoroacetone 2,4-dinitrophenylhydrazone	19.18	19.10	-0.08	19.52	19.71	+0.19	
$C_9H_7F_3N_4O_4$	19.18	19.20	+0.02	19.52	19.88	+0.36	
β-Trifluoromethyl-γ-piperidino-							
trifluoropropenylbenzene	5.53	5.35	-0.18	30.02	29.99	-0.03	Pyrohy-
$C_{11}H_{15}F_4NO$	5.53	5.38	-0.15	30.02	29.69	-0.33	drolysis
N-(α-Trifluoromethyl-β-phenyl							carried
1-trifluoropropenyl) piperidine	4.34	4.43	+0.09	35.29	35.66	+0.37	out in a
$C_{15}H_{15}F_6N$	4.34	4.32	-0.02	35.29	35.60	+0.31	nickel tube
α-Hydroperfluoroisobutyric anilide	5.16	5.08	-0.08	42.04	42.14	+0.10	
$C_{10}H_7F_6NO$	5.16	5.14	-0.02	42.04	42.24	+0.20	
p-Toluidine salt of phenylfluorophosphinic acid	5.24	5.15	-0.09	7.10	7.15	+0.05	
$C_{13}F_{15}FNO_2P$	5.24	5.26	+0.02	7.10	6.94	-0.16	

Difluorochloroacetanilide	6.81	6.94	+0.13	18.49	18.39	-0.10	
$C_8H_6ClF_2NO$	6.81	6.84	+0.03	18.49	18.69	+0.20	
Tris(*m*-nitrophenyl)oxonium borofluoride	8.89	8.76	-0.13	16.20	16.09	-0.11	Pyrohydrolysis carried out in a platinum tube
$C_{18}H_{15}BF_4N_3O_7$	8.89	8.82	-0.07	16.20	16.51	+0.31	
α,β,β,β-Tetrafluoropropiophenone 2, 4-dinitrophenylhydrazone	14.51	14.65	+0.14	19.43	19.85	+0.42	
$C_{15}H_{10}F_4N_4O_4$	14.51	14.70	+0.19	19.43	19.57	+0.14	
α-Hydro-β-methoxyperfluoropropionitrile *	10.15	10.13	-0.02	40.98	41.02	+0.04	
$C_4H_3F_3NO$	10.15	10.28	+0.13	40.98	40.97	-0.01	
α-Hydroperfluoroisobutyric acid piperidide	5.32	5.27	-0.15	43.31	43.19	-0.12	
$C_9H_{11}F_6NO$	5.32	5.36	+0.04	43.31	43.09	-0.22	

* B. p. 106°

On the basis of this fact, we considered it possible to work out a simultaneous determination of fluorine and nitrogen from the same sample by placing magnesium oxide in the zone of decomposition of the substance in the Dumas determination of nitrogen, taking up the fluorine by the magnesium oxide, and then pyrohydrolyzing the magnesium fluoride which was formed. For this purpose we used the modification of the microdetermination of nitrogen by the Dumas method which was worked out by M. O. Korshun and M. N. Chumachenko and was used for the analysis of hard-to-burn compounds; in this they used pyrolytic burning of the sample in a layer of nickel oxide.

This variation was chosen because pyrolytic burning according to M. O. Korshun [4], carried out in a quartz apparatus, not only assures quantitative combustion, but permits easy replacement of the preparation which absorbs the fluorine after each experiment. The presence of nickel oxide as well as magnesium oxide in the zone of decomposition does not hinder the pyrohydrolytic determination of fluorine, since nickel fluoride is hydrolyzed at a lower temperature than is magnesium fluoride [5]. At the same time, nickel oxide itself at high temperatures does not hold back fluorine completely. In Table 1 are given the results of determining fluorine when the sample was burned in a layer of pure nickel oxide. Combustion was carried out at 900° in an electric furnace 6 cm long.

For simultaneous determination of fluorine and nitrogen a 3-8 mg sample was placed in a 9 cm long quartz combustion tube filled with a granulated preparation of nickel oxide which contained 15-20 weight % MgO, and this tube was inserted in a larger combustion tube; combustion was carried out at 900-950° in a CO_2 atmosphere. Liquid samples were taken up in thick-walled quartz capillaries which were also placed in the tube and covered with oxidizer. At the end of the combustion of the substance and after the nitrogen had been driven into the azotometer,† the contents of the quartz tube were poured into the pyrohydrolytic tube, designed by N. E. Gel'man and K. I. Glazova, through which steam was passed at 1000° at a rate of 0.5 ml of condensate per minute. The pyrohydrolysis lasted 20-25 minutes. Fluorine was titrated in the condensate by thorium nitrate [6] and nitrogen was determined by the volume collected in the azotometer. Either a platinum or a nickel tube was used for the pyrohydrolysis. In work with the nickel tube we introduced an empirical correction for the loss of fluorine in the apparatus during pyrohydrolysis.

Nitrogen was determined with an accuracy of 0.2%, fluorine to 0.5%. The results of the analyses of pure substances are given in Table 2.

The determination of fluorine and nitrogen in one sample of organic substance has been carried out first by us.

LITERATURE CITED

[1] N. E. Gel'man and M. O. Korshun, *Doklady Akad. Nauk SSSR* **39, 4,** 685 (1953).

[2] N. E. Gel'man, M. O. Korshun, and N. S. Sheveleva *Zhur. Anal. Khim.* **12, 4,** 526 (1957).

[3] M. O. Korshun N. E. Gel'man, and K. I. Glazova, *Doklady Akad. Nauk SSSR* **111, 6,** 1255 (1956).

[4] M. O. Korshun and V. A. Klimova, *Zhur. Anal. Khim.* **2, 5,** 274 (1947); **3, 3,** 176 (1948); **4, 5,** 292 (1949); M. O. Korshun and N. E. Gel'man, *New Methods of Elementary Microanalysis*, Moscow, 1949.

[5] L. Domange, *Ann. Chim.* **7,** 225 (1937).

[6] M. O. Korshun, V. A. Klimova, and M. N. Chumachenko, *Zhur. Anal. Khim.* **10, 6,** 358 (1955).

† Translation of Russian — Publisher's note (Means nitrogen meter).

CHAPTER 30

The Simultaneous Microdetermination of Carbon, Hydrogen, and Halogen in Organic Compounds*

M. O. Korshun, N. E. Gel'man, and N. S. Sheveleva

In previous articles we have described methods for the simultaneous microdetermination of carbon, hydrogen, and halogens by pyrolytic combustion in an empty tube [1-2], and also described apparatus in which we have incorporated several modifications [3]; in addition, the characteristics of the elemental analysis of halogen-containing materials have been considered [4]. It was established that the presence of halogens, or of halogen-containing materials, inhibits the oxidation of carbon to carbon dioxide, as a result of which, the gases issuing from the combustion tube have been found to contain carbon monoxide. In order to ensure quantitative oxidation of carbon to CO_2 during the analysis of halogen-containing compounds, it was found necessary to introduce platinum contacts into the oxidizing zone. At the same time it was shown that, in those cases where absorption of the halogen was carried out immediately after the products of pyrolytic combustion had left the quartz capsule, and the gases freed of halogen had passed into the oxidizing zone, quantitative conversion of carbon into CO_2 could be achieved without using a platinum catalyst. Nevertheless, as pointed out already [4], such an experimental technique would preclude simultaneous determination of C, H, and halogen.

Zhur. Anal. Khim.* **13, 695-701 (1958).

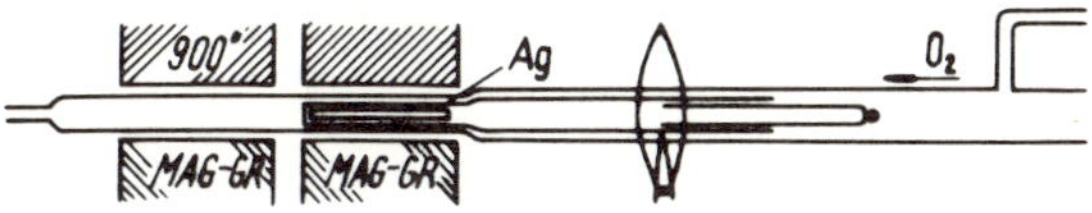

Fig. 1. Apparatus for the simultaneous determination of carbon, hydrogen, and halogens.

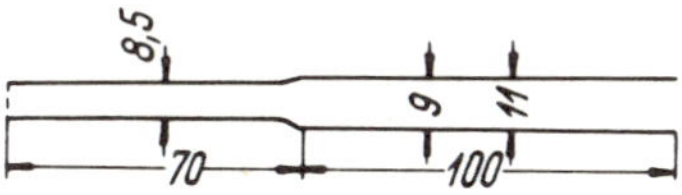

Fig. 2. Sleeve 8.5 and 11 mm external diameter; 9 mm internal diameter.

Recently a technique has been developed in our laboratory, in which, by carrying out the combustion in a so-called sleeve [5], it has been found possible to introduce essential improvements into the determination of carbon, hydrogen, and halogen. In this new variation the tube containing the sample, silver, and the oxidizing material contact the gas flow in the combustion tube in the said order. For this purpose the silver and the tube containing the sample are placed in a hollow thin-walled quartz sleeve (Fig. 2), which, after combustion, is removed from the tube and weighed. In this case, as mentioned already [4], there is no need to place platinum contacts in the oxidizing zone, but, at the same time it is possible to weigh the silver by using the sleeve. The weight of the sleeve is almost half that of the apparatus used previously, which included a heavy ground joint; accordingly, it is possible to achieve considerably better reproducibility in weighing the silver, and, thereby, essentially improve the accuracy of halogen determination. In addition, there is no further need to use a combustion tube with ground joints, and the life of the tube itself is increased (Fig. 3), since it is not corroded by the silver which is located in the sleeve.

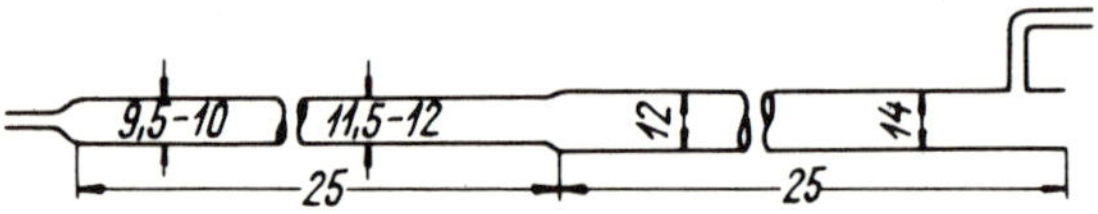

Fig. 3. Combustion tube. 9.5—10 mm and 12 mm internal diameter; 11.5—12 mm and 14 mm external diameter.

TABLE 1
Absorptive Capacity of Metallic Silver at 575°

Weight of silver in mg	Determinations carried out with same aliquot	Halogen caught by silver in mg	Absorptive capacity of silver for halogen in mg of halogen per g
1.45	42	88	61
1.18	32	77	65
1.18	25	76	64
0.80	18	47	59
1.36	40	81	60
1.62	32	65	40
Mean 1.27		72	58

The halogens are absorbed by pure metallic silver in the form of foil, mesh, or wire,‡ or by silver precipitated on a support. When metallic silver is used in the sleeve, about 1.5 g is introduced into the sleeve; this is approximately half the amount used in the ground-joint absorption apparatus [4]. The layer of silver 5-6 cm long is heated with the MAG-6R electric furnace to 550-600°. This amount of silver can be used for 30-40 determinations; this corresponds to an increase in weight as a result of absorption of halogen of about 70 mg. Table 1 contains results for

‡ We had at our disposal silver foil 0.04 mm thick, and gauze made from 0.1 mm diameter wire.

the absorptive capacity of metallic silver expressed in milligrams of halogen per g of silver. Comparison of our results with those of Mitsui [6], which he obtained during a detailed study of the determination of halogens in organic compounds, confirm a conclusion which we had already come to previously, that the absorptive capacity of metallic silver depends not only on its surface area, but also on the temperature [4]. In the case under discussion, increasing the temperature from 425-575° gave a several-fold increase in absorptive capacity.

Our experiments also confirmed results obtained by Mitsui which showed that silver preparations with a large surface area have a good absorptive capacity at temperatures which are too low for use with silver mesh or foil. Of particular interest in this connection is a preparation of silvered pumice, since, according to our observations, silver deposited on a porous carrier absorbs halogen very well, and corrodes the walls of the quartz tube to a considerably lesser extent than metallic silver. This is a consequence of the fact that the main bulk of the silver halide is held by the well-developed surface and does not fall on the quartz walls. For the simultaneous determination of C, H, and halogen, we tried granulated, silvered pumice prepared according to Sokolova's method [7], and became convinced of its complete suitability for our purpose. The silvered pumice was tested at 425°, i.e. at a temperature which is lower than the melting point of the silver halide, which is 455°, 434°, and 588° for AgCl, AgBr, and AgI, respectively. In the course of these tests we established that the absorptive capacity of this preparation was almost double that of the electrolytic silver proposed by Mitsui (Table 2).

In addition to a more efficient use of silver, another advantage of working with silvered pumice is that, thanks to absorption of the halogen at a comparatively low temperature (425-575°), corrosion of the sleeve is so insignificant that the latter can be used repeatedly with fresh lots of material.

When metallic silver is used, the silver halide spills over the sleeve, which, as a rule, is taken out of service after the first series of experiments.

The silvered pumice should not completely fill the cross section of the sleeve, since, if this is done, there will be some resistance to the gas flow, which will be sufficient to force part of the gaseous combustion products into bypassing the pumice layer.

TABLE 2

Absorptive Capacity of Various Silver Preparations

	Metallic silver in the form of mesh, foil, or wire		Electrolytic silver	Silvered pumice
	Mitsui's results	Author's results	Mitsui's results	Author's results
Temperature to which the silver is heated	425°	575°	425°	425°
Capacity expressed in mg halogen per g of silver	3.7	55	162	292

TABLE 3

Determination of Carbon and Chlorine in Hexachloroethane

Test material	C % Found	C % Difference	Cl % Found	Cl % Difference	% Total C and Cl	Notes
Hexachloro-	7.68	−2.46	82.85	−6.99	90.53	Combustion carried
ethane C_2Cl_6	9.59	−0.55	85.72	−4.14	95.31	out with gas burner
C 10.14%	10.25	+0.11	89.66	−0.20	99.91	Combustion carried
Cl 89.86%	10.22	+0.08	89.59	−0.27	99.81	out with an electric furnace

Experimental Procedure. Before starting an analysis, a blank is carried out with the furnaces heating. For this purpose, the sleeve containing the silver is inserted in the combustion tube, and the furnace drawn over the part where the silver is located; when metallic silver is used, the furnace temperature is raised to 575° while when silvered pumice is used, the temperature is adjusted to 425°. The absorption apparatus is connected to the combustion tube and 500 ml of oxygen passed through at a rate of 35-40 ml per min. When the blank test is finished, the absorption apparatus is disconnected, and, after wiping it and removing the electric charge on it, it is transferred to the balance. The electric furnace is drawn back from the sleeve and the latter is then cooled in a weak oxygen stream in the course of 7-8 minutes. The sleeve is removed from the combustion tube by means of a hook, and is then electrically discharged by passing it over the oxidizing flame of a burner, and placed on a long block near the balance. The absorption apparatus is weighed 15 and 20 minutes after wiping. After weighing the absorption apparatus, the sleeve is weighed 25 minutes after wiping the apparatus. The weighed apparatus is connected to the setup and left under oxygen pressure with the aspirator stopcock closed. An aliquot of test material is weighed into a capsule and the latter placed in the sleeve; the whole is then placed in the combustion tube. The sample capsule is inserted approximately halfway into the sleeve. In some cases the analyst may find it more convenient to use a capsule 9 cm long; the latter is placed at a position ⅓rd the length of the sleeve. Long capsules are used for pyrolysis of readily volatile compounds (b.p. 40-50°) and for organosilicon compounds which contain halogens. Combustion of fully halogenated hydrocarbons is best carried out with a MAG-6R electric furnace at 900° rather than with a gas burner.

TABLE 4

Simultaneous Determination of Carbon, Hydrogen, and Halogen

Test material	Amount in mg	C % Found	C % Diff.	H % Found	H % Diff.	Halogen % Found	Halogen % Diff.
1, 1, 1, 2-Tetrachloro-3-*p*-tolylpropane $C_{10}H_{19}Cl_4$	6.558	44.16	+0.01	3.78	+0.07	52.20	+0.06
C = 44.15%; H=3.71%; Cl=52.14%	6.884	44.14	-0.01	3.80	20.09	52.22	+0.08
Hexachloroethane C_2Cl_6	8.145	10.22	+0.08	—	—	89.53	-0.33
C = 10.14%; Cl = 89.86%	9.900	10.30	+0.16	—	—	89.71	-0.15
p-Chlorocinnamic acid $C_9H_7ClO_2$	7.444	59.13	-0.06	3.78	-0.08	19.21	-0.21
C = 59.19%; H = 3.86%; Cl = 19.42%	9.767	59.09	-0.10	3.94	+0.08	19.25	-0.17
Bromobenzoic acid $C_7H_5BrO_2$	6.015	42.01	+0.19	2.61	+0.10	39.87	+0.12
C = 41.82%; H = 2.51%; Br=39.75%	6.860	41.95	+0.13	2.66	+0.15	39.61	-0.14
α, β-Dibromoethyl-2-propoxybenzene $C_{11}H_{14}Br_2O$	8.387	41.28	+0.17	4.50	+0.12	49.74	+0.11
C = 41.01%; H = 4.38%; Br=49.63%	6.112	41.10	+0.09	4.49	+0.11	49.49	-0.14
Acetovinyltrimethylammonium chloride	6.615	51.27	-0.10	8.64	+0.02	21.87	+0.20
$C_7H_{14}CNO$ C = 51.37%; H = 8.62%;	8.050	51.30	-0.07	8.66	+0.04	21.64	-0.03
Cl = 21.67%							
Triphenyloxonium iodide $C_{18}H_{15}IO$	6.960	57.91	+0.14	4.03	-0.01	34.31	+0.39
C = 57.77%; H = 4.04%; I = 33.92%	7.778	57.99	+0.22	4.03	-0.01	34.07	+0.15
2-Phenyldihydroquinolizine perchlorate	6.526	59.00	+0.07	4.04	+0.08	11.55	-0.05
$C_{15}H_{12}ClNO_4$	6.410	59.07	+0.14	4.07	+0.11	11.73	+0.13
C=58.93%; H = 3.96%; Cl = 11.60%							

Table 3 contains some comparative results which show that analysis of hexachloroethane by means of a gas burner leads to incomplete combustion. Results of analysis of halogen-containing materials carried out by the technique just described are given in Table 4.

The same operational procedure and apparatus can also be used for the simultaneous determination of C, H, and S; the only difference in this case is that the silvered pumice is kept at a temperature of 750°. It is necessary to use different capsules for the determination of sulfur and halogen. Results for the determination of C, H, and S are given in Table 5.

As is evident from the results given in Tables 5, 6, and 7, in those cases where the test material contains halogen or sulfur, the technique involving the use of a sleeve is applicable to methods developed earlier for the simultaneous determination of several elements on one aliquot of test material, namely, to the determination of carbon, hydrogen, halogen, and silicon [8-11], carbon, hydrogen, sulfur, and phosphorus [12-14], carbon, hydrogen, halogen, or sulfur and metal [4]; the latter is possible only when, as the result of combustion, nonvolatile metal oxides or salts which are stable at a temperature of 900° are obtained. Determinations which we have carried out of carbon, hydrogen, halogen, and inorganic residue on one aliquot of test material which contained in addition to C, H, and I, one atom each of phosphorus and copper, showed that under the conditions of pyrolytic combustion, the weight of inorganic residue corresponds to copper pyrophosphate, and that it is possible on the basis of this residue to calculate the content of copper and phosphorus in materials of a similar type (Table 7). The results given in Tables 4, 5, 6, and 7 illustrate the possibilities which are opened up by the use of pyrolytic combustion during the analysis of organic compounds of complex composition.

TABLE 5
Simultaneous Determination of Carbon, Hydrogen, Sulfur, and Some Other Elements

Test material	Amount in mg	C % Found	C % Diff.	H % Found	H % Diff.	S % Found	S % Diff.	P % Found	P % Diff.	Fe %[a] Found	Fe %[a] Diff.
Sulfanilic acid	8.741	41.84	+0.23	4.21	+0.14	18.51	0.00				
$C_7H_7NO_3S$ C = 4.07%; H = 41.61%; S = 18.51%	7.700	41.61	0.00	4.22	+0.15	18.47	-0.04				
Phenylsemicarbazone of O,O-diethyl-S-acetonyldithio-	6.790	58.25	-0.10	8.02	-0.14	8.65	0				
phosphate $C_{16}H_{30}N_2O_4S$ C = 58.35%; H = 8.16%; S = 8.65%	5.848	58.29	-0.06	8.01	-0.15	8.69	+0.04				
Phenylsemicarbazone of O,O-diethyl-S-acetonyldithiophosphate	5.578	44.82	+0.04	5.73	-0.18	17.14	+0.06	8.19	-0.06		
$C_{14}H_{22}N_3O_3PS_2$ C = 44.78%; H = 5.91%; S = 17.08%; P = 8.25%	5.761	45.00	+0.22	6.01	+0.10	17.06	-0.02	8.19	-0.06		
Ferrocenylthiocyanomethane	7.045	55.95	-0.10	4.32	+0.01	12.18	-0.29	—	—	21.88	+0.16
$C_{12}H_{11}FeNS$ C = 56.05%; H = 4.31%; S = 12.47%; Fe = 21.72%	5.682	55.87	-0.18	4.45	+0.14	12.01	-0.45	—	—	21.54	-0.18

[a] Calculated from the weight of Fe_2O_3

TABLE 6

Simultaneous Determination of Carbon, Hydrogen, Chlorine, and Silicon

Substance	Mg.Sample	%C	Δ	%H	Δ	%Cl	Δ	%Si	Δ
Methyldodecyldichlorosilane	5.890	55.32	+0.21	9.93	+0.03	24.70	-0.33	9.57	-0.33
$C_{13}H_{28}Cl_2Si$ C = 55.11%;	5.670	55.28	+0.17	9.90	-0.06	25.22	+0.19	9.80	-0.10
H = 9.96%; Cl = 25.03%; Si = 9.90%									
Decyltrichlorosilane $C_{10}H_{21}Cl_3Sl$	7.725	43.68	+0.12	7.79	+0.11	38.21	-0.37	10.58	+0.40
C = 43.56%; H = 7.68%;	5.802	43.71	+0.15	7.71	+0.03	38.03	-0.55	10.55	+0.37
Cl = 38.58%; Si = 10.18%									

TABLE 7

Simultaneous Determination of Carbon, Hydrogen, Iodine, and Inorganic Residue

Test material	Amt. mg	C % Found	C % Diff.	H % Found	H % Diff.	I % Found	I % Diff.	Cu % Found	Cu % Diff.	P % Found	P % Diff.
Complex formed between copper iodide and dipropyl phosphinate	7.718	34.47	-0.12	4.65	+0.05	30.77	+0.31	15.35	+0.10	7.49	+0.05
$C_{12}H_{19}CuIO_2P$ C = 34.59%; H = 4.60%; I = 30.46%; Cu = 15.25%; P = 7.44%;	7.296	34.58	-0.01	4.73	+0.13	30.25	-0.21	15.36	+0.11	7.49	+0.05
Complex formed between copper iodide and diethyl phenylphosphinate	7.720	30.81	-0.09	3.91	+0.02	33.08	+0.42	16.35	—	7.97	—
$C_{10}H_{15}CuIO_2P$ C = 30.90%; H = 3.89%; I = 32.66%; Cu = 16.35%; P = 7.97%	6.130	30.55	-0.35	3.83	-0.06	33.25	+0.59	16.18	-0.17	7.89	-0.08

*Calculated on the basis of a residue of $Cu_2P_2O_7$.

SUMMARY

An improved modification of the method for the simultaneous determination of C, H, and halogen has been developed. Thanks to the use of a sleeve, the accuracy of determination of the halogen is high, and there is no need to use platinum contacts, while a combustion tube of the normal type without any ground-glass joints can be used.

Silvered pumice is used for absorbing the halogen, and it has been demonstrated that this material can be used at 425°; it has been shown that at this temperature the absorptive capacity of silver precipitated on pumice is considerably greater than that of electrolytic silver.

The method described is also applicable to the determination of C, H, and S, and for the determination of several elements on one aliquot of test material, in accordance with methods developed previously.

LITERATURE CITED

[1] M. O. Korshun and N. S. Sheveleva, *Doklady Akad. Nauk SSSR* **60**, 63 (1948).

[2] M. O. Korshun and N. E. Gel'man, *New Methods of Elemental Micro Analysis*, Moscow, (1949).

[3] M. O. Korshun and N.S.Sheveleva, *Zhur. Anal. Khim.* **7**, 104 (1952).

[4] M. O. Korshun and N.S.Sheveleva, *Zhur.Anal.Khim.* **11**, 376 (1956).

[5] N. E. Gel'man, M. O. Korshun, and N. S. Sheveleva, *Zhur. Anal. Khim.* **12**, 526 (1957).

[6] T. Mitsui and H. Sato, *Mikrochimica Acta* **1956**, 1603.

[7] N. V. Sokolova, *Zhur. Anal. Khim.* **11**, 728 (1956).

[8] V. A. Klimova M. O. Korshun and E. G. Bereznitskaya, *Doklady Akad. Nauk, SSSR* **84**, 1175 (1952).

[9] V. A. Klimova, M. O. Korshun and E. G. Bereznitskaya, *Doklady Akad. Nauk SSSR* **96**, 81 (1954).

[10] V. A. Klimova and E. G. Bereznitskaya, *Zhur. Anal. Khim.* **11**, 223 (1956).

[11] V.A. Klimova and E. G. Bereznitskaya, *Zhur. Anal. Khim.* **11**, 292 (1956).

[12] V. A. Klimova, M. O. Korshun and E. G. Bereznitskaya, *Doklady Akad. Nauk SSSR* **96**, 287 (1954).

[13] M. O. Korshun, E. A. Terent'eva and V. A. Klimova, *Zhur. Anal. Khim.* **9**, 275 (1954).

[14] M. O. Korshun, E. A. Terent'eva and V. A. Klimova, *Doklady Akad. Nauk SSSR* **100**, 707 (1955).

CHAPTER 31

New Variation of the Dumas Method of Determining Nitrogen*

V. A. Klimova and I. F. Dubinina

In the determination of nitrogen by the Dumas method it is essential to bring about the quantitative oxidation of the compound and at the same time obtain all the nitrogen in the form of elementary nitrogen. This is attained by the use of incandescent copper oxide as oxidant and incandescent copper as reducing agent for oxides of nitrogen. In this method both reduced copper and copper oxide are used in great excess. We consider that the use of such an excess of oxidizing and reducing agents is necessary only because the method of combustion is not satisfactory: the combustion is carried out by slow distillation of the substance into the oxidation and reduction zones, and incompletely burned fragments of the compound or unreduced oxides of nitrogen may be carried away in the convection of the hot gases through the filling of the tube. Also, products of the incomplete oxidation of the substance may condense in a thick layer in the still cold part of the filling and, in some cases, react chemically with the filling, e.g., in the analysis of nitriles, in which cuprous cyanide may be formed and may not yield the whole of its nitrogen in the analytical process.

It is not possible to determine nitrogen in nitriles and in silver and alkali-metal nitrates; unreliable and variable results may be obtained also in the analysis of polynitro

Izvest. Akad. Nauk SSSR, Otdel. Khim. Nauk **1958**, 129-32.

compounds, heterocyclic compounds, condensed-ring compounds, compounds with angular methyl groups, methylated amines, etc.

With the object of determining nitrogen in compounds of this sort, several modifications of the micro-Dumas method have been proposed recently [1-5]. Almost all these methods demand considerable complication of the apparatus and procedure, and even then they do not always give reliable results. It seemed to us quite probable that the actual method of combustion is the main cause of failure in the analysis of some types of nitrogen-containing compounds, and that the use of enormous excesses of oxidizing and reducing agents in the Dumas method is based merely on an attempt to ensure adequate contact between the compound being analyzed and the incandescent part of the filling. If we could carry out the combustion in such a way as to ensure complete and adequately long contact of the compound with a small amount of oxidizing agent, and of the oxides of nitrogen formed with a small amount of reducing agent, then not only would the cumbersome filling of the combustion tube be unnecessary, but it might be possible to obtain quite acceptable results for compounds that cannot be analyzed at present by the Dumas method.

Martin [6] has carried out some interesting work in this direction. He proposed to free a part of the tube from filling. In the empty part of the combustion tube he placed a small capsule containing the weighed sample of substance covered with a layer of copper oxide, and he carried out the combustion in this capsule, gradually moving the burner from its open end to its closed end. Martin states that with such a method of combustion only gases pass to the filling. This method of decomposing the substance must be regarded as quite satisfactory. However, while removing part of the filling from the tube, Martin left the second half of the tube unchanged, i.e., there were still two layers of copper oxide with an intermediate layer of copper. Thus, there remained a layer of oxidizing agent to deal with possible incomplete combustion of the substance in the test tube; also, as previously, reduction of oxides of nitrogen was effected in the combustion tube with a large excess of reducing agent.

EXPERIMENTAL

We concluded that the contact between the substance and the oxidizing agent was already adequate in the proposed technique of combustion and there was therefore no need for a supplementary layer of copper oxide in the combustion tube. Also, it would be better to carry out the reduction of oxides of nitrogen in the same capsule. We therefore went further than Martin by freeing the tube from all filling apart from a short layer of Hopcalite I, which was placed at the nitrometer end of the combustion tube. A layer of Hopcalite 70 mm in length is quite sufficient for the complete oxidation of carbon monoxide at almost any possible rate of passage of gas through the apparatus. In order to make quite sure of the oxidation of difficultly combustible compounds, we added a little nickel oxide to the copper oxide. Hence, in the method that we propose, the combustion of the substance and the reduction of oxides of nitrogen are carried out in the sample capsule, which is placed in an empty combustion tube. The oxidant used is a mixture of 100 parts of wire-form copper oxide and 1 part of nickel oxide powder; the reductant is copper wire previously reduced in a stream of hydrogen at 400-500°. As compared with the Dumas method, the amount of copper oxide is reduced to $\frac{1}{15}$th, and the amount of reduced copper to $\frac{1}{4}$th of the previous amount.

By the proposed method some of the defects of the generally accepted micro-Dumas method can be eliminated.

Thus, in combustion by our method the gases leave the capsule gradually, after the combustion and reduction of oxides of nitrogen. The experimental results reported below show that we succeeded in effecting quite adequate contact with small amounts both of oxidant and of reductant. Also, the fact that in the combustion the furnace moves to meet the vapor of the substance, results in conditions of combustion that are on the whole more favorable, because the vapor of the substance and the thermal-decomposition and combustion products do not pass into a cold layer of filling, but into a hot layer, with which they react immediately. In the first place this accelerates combustion and, in the second, it eliminates the possibility of the distribution of the substance along the tube, which sometimes occurs in the Dumas method. This latter behavior probably plays an important part in the analysis of nitriles. In combustion in the capsule cuprous cyanide either is not formed or is destroyed quantitatively in the heating. The absence of large excesses of oxidant and reductant naturally eliminates the errors associated with these excesses; it makes it possible to manage with a small amount of copper oxide and it protects the quartz tube from attack.

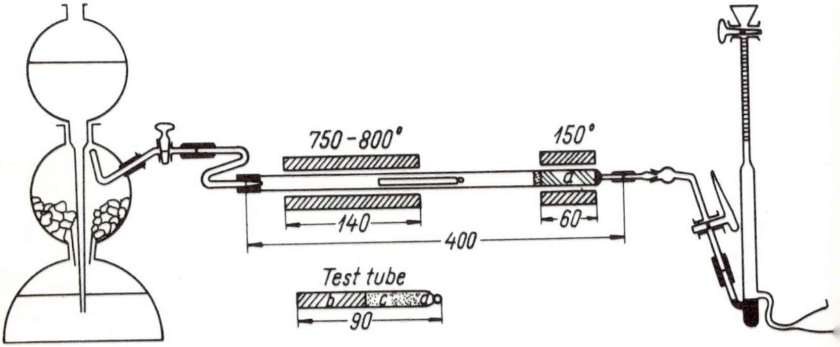

Fig. 1. Scheme of apparatus (in mm): a) Hopcalite, b) copper, c) copper oxide, d) copper oxide and nickel oxide with reducing agent.

The proposed method was checked with numerous compounds. Good results were obtained in all cases; some of them are given in the table.

Nitrogen Content %

No.	Substance	Calculated	Found by proposed method	Diff.	Found by Dumas method	Diff.
1	Silver nitrate	8.23	8.20	−0.03	7.20	−1.03
		8.23	8.24	+0.01	6.39	−1.03
2	2-(Dimethylaminomethyl)-N,N-dimethylcyclohexylamine dipicrate	17.45	17.49	+0.04	12.74	+0.29
		17.45	17.42	−0.03	17.50	+0.05
3	Decahydro-4-quinolinol hydrochloride	7.37	7.56	+0.17	7.94	+0.57
		7.37	7.41	+0.02	8.12	+0.75
4	1[2-(2-thienyl)ethyl]piperidine methiodide	4.38	4.47	+0.14	3.89	−0.54
		4.38	4.26	−0.07	3.86	−0.47
5	3-[1-(bromoethynyl)cyclohexyloxy]propionitrile	5.47	5.61	+0.14	–	–
		5.47	5.56	+0.09	–	–

Our assembly (see Figure) consists of a Kipp apparatus for the generation of carbon dioxide, a combustion tube 400 ml in length and 10 mm in internal diameter, in the nitrometer end of which is a layer of Hopcalite 70 mm in length, a stopcock, a nitrometer containing 50% potassium hydroxide, and two furnaces. A furnace, 60 mm in length, maintains the layer of Hopcalite at 150° throughout the working day. A second furnace, 140 mm in length, effects the combustion at 800°.

The sample was placed in a quartz capsule, 90 mm in length and 5 mm in internal diameter. A 100:1 mixture of copper and nickel oxides was sprinkled on the sample; the tube was shaken in order to mix the sample with the first portions of the mixture, and it was then rather more than half filled with mixture. The rest of the test tube was filled with reduced copper.† A given portion of copper

† The reduction of copper oxide was carried out at 400-500° in a stream of hydrogen. The copper so prepared was sufficiently active.

can be used several times, since in practice only a small layer of it is oxidized. However, repeated heating results in a change in the structure of its surface and a reduction in its chemical activity. Hence, if the copper acquired a metallic luster it was necessary to replace it by a fresh portion.

The filled capsule was placed in the middle of the combustion tube with its open end toward the Kipp apparatus. Air was then displaced from the system by carbon dioxide in the way recommended in the micro-Dumas method. The stopcock of the Kipp apparatus was then closed and a combustion furnace, heated to 750-800°, was applied to the combustion tube in such a way that the half of the capsule directed toward the Kipp apparatus was within the furnace. The nitrometer contained a solution of alkali. Wet cottonwool or ice was placed on the part of the combustion tube containing the closed end of the capsule. Such cooling is particularly useful in the analysis of volatile or readily decomposed substances. The combustion furnace was gradually moved toward the closed end of the test tube, i.e. toward the sample of the substance.

The rate of combustion was regulated by the movement of the furnace as in the Dumas method, i.e., so that one bubble separated in the nitrometer every two minutes. When the furnace had been moved to such an extent that the closed end of the test tube was in the hottest section and the evolution of bubbles in the nitrometer had stopped, displacement of the gases into the nitrometer was begun. The combustion furnace was removed only when this displacement was complete. The further operations and the calculation of results were the same as in the micro-Dumas method.

SUMMARY

1. A new variation of the micro-Dumas method has been developed.

2. It has been shown that there is no need to have any filling in the combustion tube apart from a short layer of Hopcalite if the combustion is carried out in the sample capsule; a short length of filling in the test tube can deal adequately with the complete oxidation of the compound and the reduction of oxides of nitrogen.

3. It was shown that good analytical results can be obtained by the proposed method (accurate within ±0.2%), even with compounds that do not readily undergo complete combustion.

LITERATURE CITED

[1] W. Kirsten, *Mikrochemie* **40**, 121 (1952).

[2] G. Ingram, *Mikrochim. Acta* **1953**, 131.

[3] J. Unterzaucher, *Mikrochemie* **36-37**, 706 (1951).

[4] P. D. Sternglanz, R. C. Thompson, W. L. Savell, *Anal. Chem.* **23**, 1027 (1951).

[5] T. D. Parks, E. L. Bastin, E. J. Agazzi, F. R. Brooks, *Anal. Chem.* **26**, 229 (1954).

[6] F. Martin, *Bull. soc. chim. France* **1953**, C65.

CHAPTER 32

Simultaneous Determination of Carbon, Hydrogen, and Nitrogen*

V. A. Klimova and G. F. Anisimova

After carbon and hydrogen, nitrogen is the element that must be determined most frequently in the analysis of organic compounds. There can be no doubt about the expediency of determining it simultaneously with carbon and hydrogen, but there is no rapid and reliable method of doing this. The method that we propose for the simultaneous determination of C, H, and N consists in the pyrolytic decomposition of the substance in a rapid stream of oxygen in an empty combustion tube. Carbon and hydrogen are then quantitatively oxidized to carbon dioxide and water [1], but the nitrogen is obtained partly in the form of oxides and partly as elementary nitrogen [2, 3]. The water, nitrogen oxides, and carbon dioxide formed are absorbed, respectively, with Anhydrone, a solution of potassium dichromate in sulfuric acid carried on silica gel, and Ascarite, and are weighed [4, 5]. Elementary nitrogen mixed with a large excess of oxygen passes over a prepared copper in which oxygen is absorbed, and nitrogen is then displaced by carbon dioxide into a nitrometer, in which it is collected over KOH solution and measured. The nitrogen content of the substance is made up of the amount obtained as oxides and the amount obtained as elementary nitrogen.

Pure oxygen is prepared by the catalytic decomposition of hydrogen peroxide in presence of cupric oxide. Oxygen is collected in a gas holder over saturated brine. For one combustion 500 ml is used. A control experiment with

**Izvest. Akad. Nauk SSSR, Otdel. Khim. Nauk* 1958, 791-2.

this amount of oxygen gives 0.032-0.037 ml of unabsorbed gas. The rate of passage of oxygen is 20-25 ml/minute.

For the preparation of the copper reagent, 300 ml of 40% NaOH solution was added to a solution of 150 g of $CuCl_2 \cdot 2H_2O$ in 1200 ml of water heated to 60°. To the reaction mixture we added 15 g of silica gel dried at 200° for 2.5 hours; the mixture was stirred and left overnight. The mass was then washed by decantation until it ceased to have an alkaline reaction (test with phenolphthalein); it was filtered. The dark-colored mass was dried with periodic stirring in a drying oven at 200° for 6-7 hours; it broke down into grains and was reduced in this form in a stream of hydrogen at 250-280°. The working temperature of the preparation was 250-280°; 1 g of it absorbed about 100 ml of oxygen.

The apparatus is an arrangement for the rapid microdetermination of carbon and hydrogen [1] to the terminal tube of which a three-way stopcock has been attached; the stopcock is connected to a Mariotte vessel and to a glass tube filled with the prepared copper. The tube containing

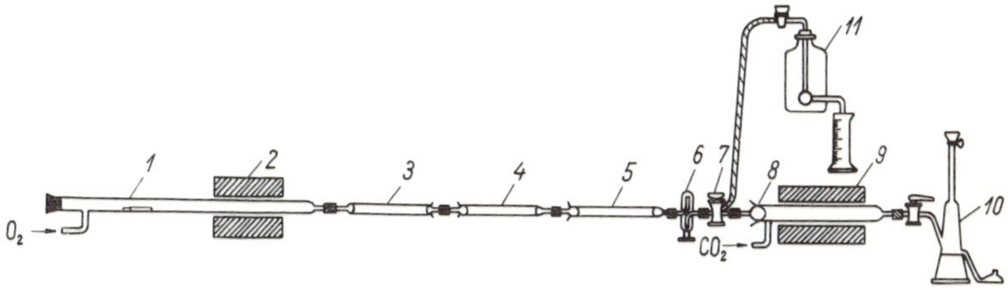

Fig. 1. Apparatus for the simultaneous determination of C, H, and N: 1) combustion tube; 2) electric furnace; 3) 4) 5) absorption tubes for water, nitrogen oxides, and carbon dioxide; 6) guard tube; 7) three-way stopcock; 8) tube containing prepared copper; 9) electric furnace; 10) nitrometer; 11) aspirator.

TABLE 1

No.	Formula of substance	C % Found	C % Calc.	C % Δ	H % Found	H % Calc.	H % Δ	N % Found	N % Calc.	N % Δ
1	$C_7H_{15}ON_2$	53.36	53.50	0.14	9.39	9.55	0.16	26.81	26.75	0.06
		53.46	53.50	0.04	9.39	9.55	0.16	26.73	26.75	0.02
2	$C_6H_3(OH)(NO_2)_3$	31.42	31.44	0.02	1.25	1.31	0.06	18.20	18.34	0.09
		31.20	31.44	0.22	1.25	1.31	0.06	18.25	18.34	0.14
3	$N=C(C_2H_5)C_3H_7$	51.30	51.43	0.13	5.63	5.71	0.08	20.03	20.00	0.03
	$\vert$	51.60	51.43	0.17	5.67	5.71	0.04	19.83	20.00	0.17
	$NH-C_6H_3(NO_3)_2$									
4	$N=C_{13}H_{20}$	61.41	61.29	0.12	6.53	6.45	0.08	15.20	15.05	0.15
	$\vert$ $NH-C_6H_3(NO_2)_2$	61.44	61.29	0.15	6.57	6.45	0.13	15.23	15.05	0.18
5	$C_3H_{15}ON_3$	59.51	59.67	0.16	8.25	8.29	0.04	23.38	23.20	0.18
		59.75	59.67	0.08	8.35	8.29	0.06	23.08	23.20	0.12

the prepared copper has a side tube connected to a Kipp apparatus, and the end, which is drawn out, is connected to the stopcock of the nitrometer (see Figure).

For analysis, a 2-8 mg sample of the substance to be analyzed is placed in a small capsule. The capsule is placed in the combustion tube, and 400-500 ml of oxygen is passed through the apparatus to remove air introduced with the capsule and as a result of diffusion into the absorption apparatus during weighing. The tube containing copper is now connected to the Kipp apparatus and checked for absence of air (presence of microbubbles in the nitrometer). The stopcock of the Kipp apparatus is then closed, the guard tube is connected to the tube containing the prepared copper, and the combustion (pyrolysis [1]) is started. Combustion is continued for 20-25 minutes. When it is complete, the absorption tubes are weighed and nitrogen-containing substances are analyzed by the proposed method. The time required for analysis is 60-65 min. The results of the analyses are given in the Table.

SUMMARY

It was shown that carbon, hydrogen, and nitrogen can be determined simultaneously by the pyrolytic decomposition of the organic substance in a stream of oxygen in an empty combustion tube. Carbon and hydrogen are determined in the usual way, and nitrogen is determined from the sum of the amounts present as nitrogen oxides and as elementary nitrogen.

LITERATURE CITED

[1] M.O. Korshun and V.A. Klimova, *Zhur.Anal.Khim.* **2**, 274 (1947).
[2] M.O. Korshun and V.A. Klimova, *Zhur.Anal.Khim.* **3**, 176 (1948).
[3] A. Friedrich, *Mikrochemie* **23**, 129 (1937).
[4] V.A. Klimova and M.O. Korshun, *Zhur.Anal.Khim.* **6**, 230 (1951).
[5] V. A. Klimova and K. S. Zabrodina, *Izvest. Akad. Nauk SSSR., Otdel. Khim. Nauk* **1959**, 582

CHAPTER 33

Simultaneous Microdetermination of Carbon, Hydrogen, and Sulfur*

E. I. Margolis and N. F. Egorova

A large number of macro-, as well as micromethods, exist for the quantitative determination of sulfur in organic compounds.

The method developed by M. O. Korshun and N. S. Sheveleva [1, 2] for the gravimetric determination of carbon, hydrogen and sulfur in a sample of substance weighing 4-7 mg consists of decomposing the substance in a combustion tube in a rapid stream of oxygen (30-40 ml/min), with subsequent absorption of the oxidation products by suitable absorbents. Here the sulfur oxides are trapped using metallic silver (as fine ribbon 2-3 mm wide) ,which is contained in a special quartz tube heated to 650-800°. The precision of determining all three elements is of the order of ±0.2-0.3%. With all its merits, the method possesses a serious disadvantage: the quartz tube with the silver corrodes badly at high temperature and has to be replaced frequently.

In 1954 P. N. Fedoseev and R. M. Lagoshnaya [3] proposed a method for the simultaneous macrodetermination of carbon, hydrogen and sulfur, in which the sulfur oxides were trapped using metal oxides and silicates. The authors came to the conclusion that sodium silicate is the best absorbent. With this method, sulfur-containing organic compounds are decomposed in the presence of amorphous chromium oxide (coating the weighed sample of substance being analyzed) in a rapid stream of oxygen at

**Vestnik. Moskov. Univ. Ser. Mat., Mekh., Astron., Fiz., Khim.* 13, 209-13 (1958)

800°; here the sulfur is determined from the weight increase of the absorbent, contained in the combustion tube in porcelain boats, heated to 700-750°. The same authors proposed a method for the quantitative determination of sulfur oxides, in which the silicates of sodium, potassium, strontium and barium were used as the absorbents [4].

In the present paper we investigated the possibility of using the silicates of sodium and strontium as absorbents for sulfur oxides in the joint microdetermination of carbon, hydrogen and sulfur in organic compounds.

In order to make the above described method suitable for microanalysis, we had to make a number of important changes in it. Our study revealed that sodium silicate, used successfully by P. N. Fedoseev and R. M. Lagoshnaya [3] when 0.1-0.15 g samples of sulfur-containing organic compounds were analyzed, showed grave disadvantages when attempts were made to use it in microelemental analysis. As was already indicated [3], particles of sodium silicate are carried out by the oxygen stream during analysis, as a result of which low analytical results are obtained for sulfur. In addition, in order to remove the carbon dioxide and water absorbed from the air by the sodium silicate, we had to resort to high temperatures, which led to vitrification and coalescence, and consequently to a reduction in the active surface area of the absorbent. If sodium silicate is deposited on pumice in order to prevent particle loss, then no vitrification is observed unless the amount of sodium silicate below 10% of the weight of the support, but with such a low content of sodium silicate it proves necessary to change the absorbent very often.

Strontium silicate does not suffer from the above indicated limitations, and in addition, it is easily pressure-molded, has a high porosity, a low density, and releases the carbon dioxide and water absorbed from the air with comparative ease.

To obtain strontium silicate, we added an excess of saturated sodium silicate solution with stirring to a saturated solution of strontium nitrate. The precipitate was washed well with warm distilled water, at first by decantation, and then on a Buchner funnel, until the test with diphenylamine for NO_3^- ion was negative, after which the product was dried and calcined in a stream of oxygen at 650-700°.

A quantitative absorption of sulfur oxides by strontium silicate occurred when certain conditions (worked out in the present study) were observed, thus making it possible to use strontium silicate in the microelemental analysis of sulfur-containing organic compounds.

The experiments made to find the optimum conditions for using strontium silicate were run in a modified apparatus for the microdetermination of carbon, hydrogen and sulfur [2]. A platinum catalyst (about 1.5 g of platinum wire as a spiral coil) was placed in the oxidation zone. The combustion tube was connected by means of a ground joint to the quartz tube (length 22 cm) containing the strontium silicate, which had been previously calcined at 600-650° to constant weight.†

By means of special experiments it was established that the optimum temperature for heating the absorbent lies between 600 and 700°, at an oxygen flow rate of 15-20 ml/min. The carbon dioxide and water were absorbed by Ascarite and Anhydrone, respectively.

Based on our study, the following method of analysis was adopted.

The weighed quartz tube with strontium silicate is connected to the combustion tube, the whole is heated in a split [sectional] electric furnace at 600°, then a stream of oxygen is passed through the whole system for 5-7 min at

† Increasing the temperature above 700° leads to a reduction in the activity of the absorbent.

the rate of 35-50 ml/min, after which the weighed tubes for the absorption of water and carbon dioxide are connected, and the oxygen flow rate is reduced to 15-20 ml/min. After this, the weighed sample (4-8 mg) of substance under analysis in a capsule is placed in the combustion tube, and then burned. Oxygen is then passed at the same rate as before for 8-10 min, the split furnace is removed from the apparatus (to permit cooling), and a stream of oxygen is passed at the rate of 35-50 ml/min for another 12-15 min. All of the changes in the oxygen flow rate were made by adjusting the clamps on the rubber tubing through which the oxygen was admitted.

At the end of the test the tubes were disconnected and weighed after 10 min, the tube containing the strontium silicate being weighed last.

The precision of the determinations for all three elements was ±0.2-0.3%.

The data obtained in analyzing some pure sulfur-containing compounds with the indicated method are given in Table 1.

For the calculations we used the formula:

$$\% \text{ of element} = \frac{a \times F \times 100}{b}$$

where a is the increase in the weight of the absorbent, b is the weight of the substance, and F is the conversion factor.

To calculate the sulfur content we used the factor:

$$F_S = \frac{S}{SO_3}; \log F = 1.60254$$

If only the carbon and hydrogen in a sulfur-containing organic compound have to be determined quantitatively, then the strontium silicate can be placed in the combustion tube, and the analysis run in the standard apparatus for

TABLE 1

Name of compound and empirical formula	Wt. of sample mg.	Wt. mg			C%		H%		S%		Difference		
		CO_2	H_2O	SO_3	Found	Calc.	Found	Calc.	Found	Calc.	C%	H%	S%
Dibenzyl sulfide,	4.498	12.959	2.686	1.669	78.63	78.45	6.68	6.58	14.86	14.96	+0.18	+0.10	−0.10
$C_{14}H_{14}S$	3.970	11.403	2.319	1.470	78.38	78.45	6.54	6.58	14.89	14.96	−0.07	−0.04	−0.13
Thianthrene,	4.498	10.974	1.490	3.305	66.59	66.63	3.70	3.78	29.42	29.65	−0.04	−0.08	−0.23
$C_{12}H_8S_2$	5.368	13.121	1.847	3.943	66.71	66.63	3.85	3.78	29.41	29.65	+0.08	+0.07	−0.24
Didecyl sulfide,	3.965	11.101	4.785	1.001	76.48	76.35	13.50	13.45	10.11	10.19	+0.13	+0.05	−0.08
$C_{20}H_{42}S$	5.859	11.410	4.892	1.048	76.47	763.5	13.45	13.45	10.30	10.19	+0.12	−0.00	+0.11
Thionaphthol,	4.770	13.128	2.121	2.406	75.11	74.96	4.98	5.03	20.18	20.01	+0.15	−0.05	+0.17
$C_{10}H_8S$	5.859	16.122	2.641	2.915	75.09	74.96	5.04	5.03	19.92	20.01	+0.13	+0.01	−0.09
Sulfanilic acid,	6.600	10.112	2.459	3.021	41.81	41.68	4.17	4.07	18.33	18.51	+0.13	+0.10	−0.18
$C_6H_7NO_3S$	7.952	12.162	2.922	3.711	41.74	41.68	4.11	4.07	18.69	18.51	+0.06	+0.04	+0.18
3-Phenyl-2-thiazolidinecarboxylic acid,	5.637	11.800	2.648	2.124	57.13	57.39	5.31	5.29	15.19	15.39	−0.26	+0.02	−0.20
$C_{10}H_{11}NO_2S$	4.974	10.430	2.350	1.914	57.22	57.39	5.28	5.29	15.41	15.89	−0.17	−0.01	+0.02

TABLE 2

Compound	mg. sample	mg. CO_2	mg. H_2O	% C Found	% C Calc.	% H Found	% H Calc.	Δ % C	Δ % H
Decyl mercaptan,	6.174	15.600	6.963	68.95	68.89	12.62	12.72	+0.06	−0.10
$C_{10}H_{22}S$	5.951	15.011	6.673	68.84	68.89	12.55	12.72	−0.05	−0.17
Thiophenol,	4.421	10.600	2.125	65.41	65.41	5.38	5.48	0.00	−0.10
C_6H_6S	4.853	11.662	2.349	65.53	65.41	5.43	5.48	+0.12	−0.06
2-Acetyl-3-*p*-tolylthiazolidine,	4.159	9.929	2.536	65.15	65.12	6.82	6.83	+0.03	−0.01
$C_{12}H_{15}ONS$	4.223	10.058	2.667	65.01	65.12	7.06	6.83	−0.11	+0.23
2-Benzoyl-3-*p*-tolylthiazolidine,	4.411	11.673	2.453	72.20	72.05	6.22	6.05	+0.15	+0.17
$C_{17}H_{17}ONS$	5.795	15.328	3.212	72.19	72.05	6.20	6.05	+0.14	+0.15
Sulfanilic acid,	4.400	6.705	1.647	41.59	41.68	4.18	4.07	−0.09	+0.11
$C_6H_7NO_3S$	4.990	7.634	1.838	41.75	41.68	4.12	4.07	+0.07	+0.05
Dibenzyl sulfide,	4.946	14.240	2.940	78.56	78.45	6.65	6.58	+0.11	+0.07
$C_{14}H_{14}S$	6.249	17.983	3.708	78.52	78.45	6.64	6.58	+0.07	+0.06

the microdetermination of carbon and hydrogen. 1 g of strontium silicate, previously calcined in a stream of oxygen at 600-650°, is placed in two quartz boats located at the end of the tube, and then heated with a second electric furnace at 600-650°. Here the oxygen flow rate during the time of combustion should be kept at 10-12 ml/min.

When these conditions are observed, the precision of determining carbon and hydrogen is ±0.2% (Table 2).

As a result, it was established that strontium silicate can successfully replace silver in the quantitative microanalysis of sulfur-containing organic compounds. The use of strontium silicate permits the determination of carbon and hydrogen in the presence of sulfur, and also the simultaneous determination of carbon, hydrogen and sulfur in one sample. The practicality of using strontium silicate for the quantitative absorption of sulfur oxides finds additional support in the fact that the quartz tube in which the analysis is run does not corrode, and the apparatus can be used for a long time.

SUMMARY

With strontium silicate used as the absorbent for sulfur oxides, a method was developed for the simultaneous microdetermination of carbon, hydrogen and sulfur in organic compounds.

LITERATURE CITED

[1] M. O. Korshun, and N. S. Sheveleva, *Doklady Akad. Nauk SSSR* **60**, 63 (1948).

[2] M. O. Korshun, and N. S. Sheveleva, *Zhur. Anal. Khim.* **7**, 104 (1952).

[3] P. N. Fedoseev and R. M. Lagoshnaya, *Ibid.*, **9**, 224 (1954).

[4] P. N. Fedoseev and R. M. Lagoshnaya, *Sov. Patent 101,313* (1955).

CHAPTER 34

Microdetermination of Carbon and Hydrogen*

N. F. Egorova and A. S. Zabrodina

The microdetermination of carbon and hydrogen in organic compounds by combustion in an empty tube in a stream of oxygen was first proposed by Belcher and Spooner [1]. They ran the combustion in a tube with an internal diameter of 9 mm, and at an oxygen flow rate of 50 ml/min.

The method of Belcher and Spooner has some important advantages over combustion methods that employ a tube with a filling, including rapidity of combustion, no need to use a filling, absence of pressure regulation, etc. However, a serious disadvantage of the method indicated is the fact that if during combustion the compound either flares up or a carbon-containing smoke is formed, then low values are obtained for carbon owing to incomplete combustion.

To obtain complete combustion it has been proposed [2] that the length of the tube be substantially increased by making it spiral in the heating zone. Ingram [3] increased the diameter of the tube to 12 mm, inserted a plug of quartz wool [wadding] inside it to trap carbon-containing smoke, and then ignited the plug. Later Belcher and Ingram [4] burned the substance in a tube with a diameter of 9 mm, again using quartz wool to trap the carbon-containing smoke, and then passed the combustion products through a vertical tube with a diameter of 25 mm and having baffles, which tube was heated at 900°. All

**Vestnik Moskov. Univ. Ser. Mat., Mekh., Astron. Fiz., Khim.* 13, 235-8 (1958).

of the above mentioned authors used a boat to hold the sample.

M. O. Korshun and V. A. Klimova [5] worked out a method for combustion in an empty tube that was fundamentally different from the method of Belcher and Spooner. As proposed by Korshun and Klimova, the combustion is run in a quartz tube with an internal diameter of 9-10 mm; the sample is first subjected to thermal decomposition in a capsule with practically no oxygen present, after which the pyrolysis products are passed through a zone having a length of 20 cm, heated to 850-950°, where they are oxidized completely to water and carbon dioxide. With such a combustion technique it is necessary to run the pyrolysis slowly, as otherwise the vapors of the substance can flash outside the capsule and thus escape complete oxidation.

Later it was proposed [6] that the gaseous pyrolysis products escaping from the capsule be burned by means of a special attachment, avoiding in this way explosion in the heated zone of the tube.

In the micromethod proposed by us for the rapid determination of carbon and hydrogen the combustion of the sample is run in a stream of oxygen (30-40 ml per min), in a wide, empty tube with an internal diameter of 19-20 mm (see Figure 1). Doubling the diameter of the tube, as com-

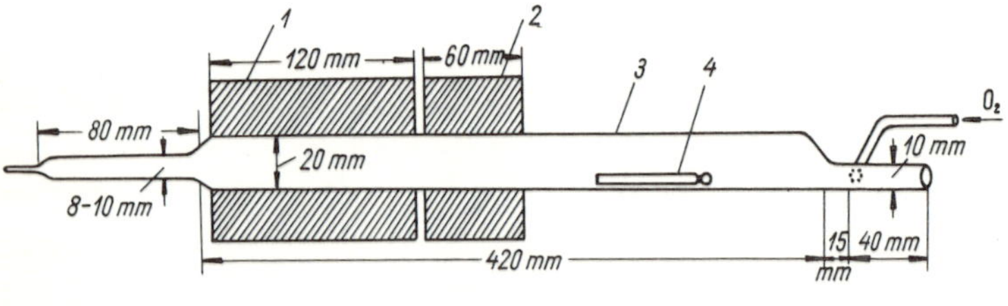

Fig. 1. Diagram of the apparatus for the microdetermination of carbon and hydrogen. 1) stationary electric furnace; 2) traveling electric furnace; 3) combustion tube; 4) capsule.

pared to that usually used, leads to a 4-fold increase in the volume of the oxidation zone, which assures an excess of oxygen sufficient for complete oxidation of the decomposition products even with a very rapid combustion of the substance. For this reason an ejection of the decomposition products from the capsule and their burning in the tube are not only safe, but they are even especially induced in order to accelerate the combustion process. As a result, because of the large excess of oxygen present, the possibility of incurring an error in the determination owing to incomplete combustion is excluded.

METHOD OF ANALYSIS

According to our method, the substance is burned in a wide, transparent quartz tube, heated by two electric furnaces to 850-900°. The first furnace, 6 cm long, is movable, while the second furnace, 12 cm long, is stationary. The remainder of the apparatus is the same as the standard apparatus used for the determination of carbon and hydrogen by the method of M. O. Korshun and V. A. Klimova [5].

After the previously weighed absorption tubes have been connected to the combustion tube, the weighed sample (3-9 mg), contained in a quartz capsule, is inserted in the combustion tube to a distance of 5-6 cm from the first electric furnace. The method used to burn the sample, at an oxygen flow rate of 30-40 ml per min, depends upon the nature of the substance. If the substance being analyzed is a solid, then in order to melt it, the burner is placed directly under the substance for several seconds, after which it is placed under the capsule at a distance of 2-2.5 cm from the closed end. When the substance under analysis is a liquid, the burner is placed in such manner that it heats half of the capsule at the open end. Violent decomposition of the substance occurs in a matter of a few seconds, accompanied

TABLE 1

Name and formula of compound	Weight of substance, mg.	Combustion time, min.	Amount of C %			Amount of H %		
			Found	Calc.	Difference	Found	Calc.	Difference
2,2,4,4-Tetramethylpentane C_9H_{20}	3.122	4	84.44	84.28	+0.16	15.64	15.72	−0.08
	4.427	4	84.39	84.28	+0.11	15.75	15.72	+0.03
Naphthalene, $C_{10}H_8$	7.214	2	93.81	93.71	+0.10	6.39	6.29	+0.10
	6.330	2.5	93.86	93.71	+0.15	6.43	6.29	+0.14
Cinnamic acid, $C_9H_8O_2$	5.851	3	72.90	72.96	−0.06	5.60	5.44	+0.16
	4.369	3	72.77	72.96	−0.19	5.50	5.44	+0.06
1-Propionyl-2,5-endoethylene-3-cyclohexene, $C_{10}H_{14}O$	7.950	3.5	79.95	79.96	−0.01	9.46	9.39	+0.07
	10.365	5	79.91	79.96	−0.05	9.54	9.38	+0.15
3,5-Dimethoxyphenyl ester of 2-chromonecarboxylic acid, $C_{16}H_{14}O_6$	7.185	4	66.35	66.25	+0.10	4.29	4.32	−0.03
	5.431	4	66.33	66.25	+0.08	4.29	4.32	−0.03
2-Methyl-5-ethylpyridine $C_8H_{11}N$	6.090	5	79.26	79.29	−0.03	9.17	9.15	+0.02
	4.018	4.5	79.33	79.29	+0.04	9.19	9.15	+0.04
Picric acid $C_6H_3N_3O_7$	5.391	4.5	31.44	31.45	−0.01	1.45	1.32	+0.13
	4.608	4	31.39	81.45	−0.06	1.52	1.32	+0.20

m-Dinitrobenzene,	4.582	3	43.05	42.87	+0.18	2.46	2.39	+0.07
$C_6H_4N_2O_4$	6.620	3	42.99	42.87	+0.12	2.54	2.69	+0.15
Nitrocyclohexane,	6.397	4	55.84	55.80	+0.04	8.59	8.58	+0.01
$C_6H_{11}NO_2$	6.519	4	55.84	55.30	+0.04	8.61	8.58	+0.03
1-Benzyl-2-methyl-3-carbethoxy-5-	6.090	6.5	74.25	74.28	−0.03	6.61	6.54	+0.07
methoxyindole, $C_{20}H_{21}NO_3$	5.839	7	74.12	74.28	−0.16	6.58	6.54	+0.04
5-Chloro-3,5-dimethyl-3-ethyl-	10.125	4.5	69.16	69.26	−0.10	11.96	12.16	−0.20
heptane, $C_{11}H_{23}Cl$	6.865	4	69.21	69.26	−0.05	11.98	12.16	−0.18
1-Butyryl-2-chloro-1-cyclopentene	11.174	2.5	62.47	62.61	−0.14	7.57	7.59	−0.02
$C_9H_{13}ClO$	5.293	2	62.64	62.91	+0.03	7.42	7.59	−0.17
m-Bromoethylbenzene,	4.761	3	52.05	51.92	+0.13	4.84	4.90	−0.06
C_8H_9Br	6.150	3	51.98	51.92	+0.06	4.94	4.90	+0.04
1-Benzyl-2-methyl-5-methoxydi-								
methylethylamine methiodide	4.985	4	56.06	56.00	+0.06	6.08	6.04	+0.04
$C_{21}H_{27}IN_2O$	4.729	4	55.89	56.00	−0.11	6.14	6.04	+0.10
2-Amino-5-propionylthiazole	4.840	4	45.98	46.13	−0.15	5.10	5.16	−0.06
$C_6H_8N_2OS$	5.373	4	46.18	46.13	+0.05	5.21	5.16	+0.05
5-Thiocyanoindole,	4.790	13	62.16	62.05	+0.11	3.45	3.47	−0.02
$C_9H_6N_2S$	5.120	17	61.96	62.05	−0.09	3.55	3.47	+0.08
Dithienyl ketone,	7.350	3	55.62	55.64	−0.02	3.13	3.11	+0.02
$C_9H_6OS_2$	6.155	3	55.76	55.64	+0.12	2.96	3.11	−0.15

by the ejection of burning vapors, at times together with flakes of carbon. After this, the short electric furnace is used to burn the decomposition products that have settled on the walls of the tube and in the capsule. Because of the large excess of oxygen present, this process does not require more than 2-4 min. The combustion is ended by passing 250-300 ml of oxygen through the apparatus at the rate of 50-60 ml per min. As a result, the total time required for combustion and displacement requires 7-10 min.

During the time of combustion it is necessary to heat the exit end of the combustion tube and the capillary of the absorption tube in contact with it, in order to prevent the condensation of water vapors in them.

The combustion water and carbon dioxide are absorbed in Anhydrone and Ascarite, respectively, and are determined in the usual manner by the weight increase of the Pregl absorption tubes. Prior to weighing, the absorption tubes are kept near the balance for 5 min on metal blocks.

In analyzing halogen- and sulfur-containing compounds, metallic silver and platinum catalysts are placed in the combustion tube. The oxygen flow rate during combustion is fixed at 25-35 ml per min.

The authors ran a large number of analyses with the given method. The results of some of these analyses are given in the Table.

SUMMARY

A method was developed for the microdetermination of carbon and hydrogen in organic compounds, based on the combustion of the latter in a wide, empty tube with a large excess of oxygen present. This method permits running the combustion of samples very rapidly (2-4 min), and excludes the possibility of incurring error because of incomplete combustion. The order of accuracy for the determination of carbon and hydrogen is $\pm 0.2\%$.

LITERATURE CITED

[1] R. Belcher, and C. Spooner, *J. Chem. Soc.* **1943**, 313.

[2] A. F. Colson, *Analyst* **73**, 541 (1948).

[3] G. Ingram, *Ibid.*, **73**, 548 (1948).

[4] R. Belcher, and G. Ingram, *Anal. Chim. Acta* **4**, 118 (1950).

[5] M. O. Korshun and V. A. Klimova, *Zhur. Anal. Khim.* **2**, 274 (1947).

[6] G. de Vries and E. van Dalen, *Anal. Chim. Acta* **7**, 274 (1952)

CHAPTER 35

Microdetermination of Carbon and Hydrogen in Organic Compounds by Measurement of Electroconductivity*

N. E. Gel'man, M. O. Korshun, Ven-Yun Van, and N. A. Balashova

The recent tendency towards automation of analytical processes is reflected in papers on the elementary analysis of organic compounds, particularly the determination of C and H. These papers aim at accomplishing automatic control of combustion of the substance under conditions of classical microanalysis, i.e., in packed tubes and for strictly constant small rates of oxygen [1-8]. The analytical finish is gravimetric, and its automation is generally not considered. This type of automation, firstly, solves the problem only partially, since the gravimetric finish of the analysis requires no less time and attention of the analyst than combustion, and, secondly, does not solve the problem under the most favorable conditions for quantitative oxidation of the sample.

It seems to us that the correct route to automatic combustion is the establishment of conditions under which complete conversion of the substance to CO_2 and H_2O is ensured under standard experimental conditions within a short time, and which require neither regulation of the combustion rate, nor even observations thereof. The basis for such conditions may be the Korshun-Klimova pyrolytic combustion [9], which, after testing on many thousands of organic compounds, achieved fame as the most universal combustion method. A substantial increase in the excess

Doklady Akad. Nauk SSSR* **129, 1046-8 (1959).

of oxygen in pyrolytic combustion was achieved by us by decreasing the sample weight by approximately a factor of 5. The combustion of a 1 mg sample is carried out for 3-4 minutes, without observing its behavior. These conditions are in good agreement with those selected for non-gravimetric methods. The amount of CO_2 obtained from the combustion is measured by the change in electroconductivity of the absorbent solution. This method of measuring CO_2 is not new, and is widely used for the quantitative determination of small concentrations of CO in gases [10, 11, 12], for C in metals and alloys [13, 14, 17], for C in minerals [15, 16], etc.

In recent years, there have been proposed rapid, automated methods for determining O, C, and S in inorganic and also in organic substances, based on measurement of electroconductivity [12, 13, 14, 17, 18, 19]. The authors of these papers, dealing with the analysis of inorganic compounds, quite correctly point out the fruitfulness of also using this principle in inorganic analysis, and give a series of examples of the analysis of organic compounds. However, the determination of H by change of electroconductivity has not been worked out so far, and determination of C and H in one sample by a conductometric method has not been accomplished as yet. The present paper is dovted to this task.

To determine H, we subject the water obtained by combustion of the organic compound to decomposition on an incandescent contact layer of carbon in a stream of pure N_2 or argon. The O in the water is quantitatively converted to CO, which can be determined in various ways. This principle was used earlier in elementary analysis for the iodimetric determination of C and H [20]. We used it for the conductometric finish of the analysis.

The essence of the proposed method for determination of C and H is as follows: A weighed sample (ca. 1 mg) of organic substance is subjected to rapid pyrolytic combustion in a stream of O_2 moving at 30-35 ml per min. The water obtained is frozen with a dry ice-acetone mixture and the remaining CO_2 is passed into a conductivity cell containing the absorbent solution. The trap with the frozen-out water is then placed in an apparatus for determining O, the water is evaporated on to a contact layer of carbon black, the CO obtained is oxidized on CuO to CO_2, and this is also passed into a conductivity cell.

On the basis of conductometric apparatus described in the literature [11, 15, 16], we constructed a cell capable of determining not only small amounts, but any percentage content of C in a sample, by increasing the cell capacity to 55 ml. Complete stirring of the absorbent by the stream of gas is obtained within a few minutes; as a result, the duration of the experiment from initial to final recording is not more than 10 minutes. For this cell, the linear relationship between amount of C and change in conductivity is maintained up to a C content of about 1000μg in the sample (Fig. 1). In addition to the cell, the measuring part of the apparatus contains a ZG-10 oscillator as source of AC, an EO-7 oscillograph as null instrument, a capacity magazine ME-3, an MTV resistance bridge, connected as a Wheatstone bridge.

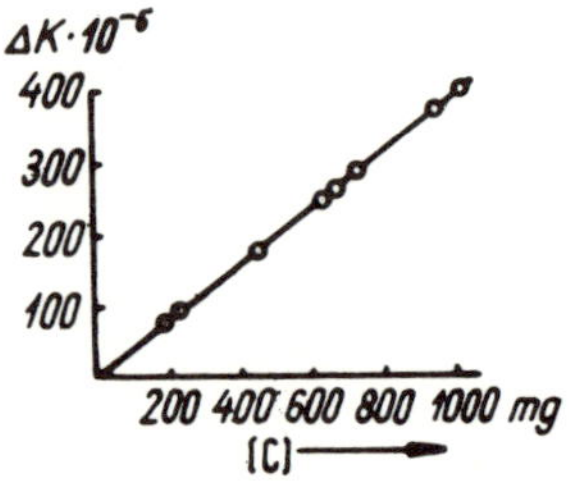

Fig. 1

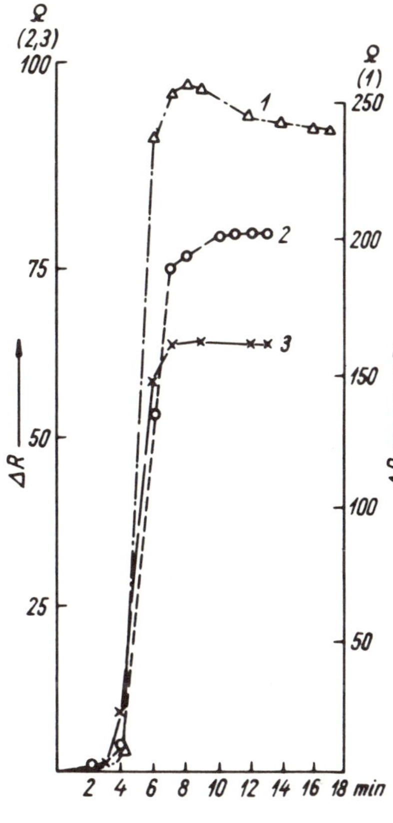

Fig. 2. Absorbent: 0.005N NaOH; amount of carbon 1028 μg. 2—0.01 N $Ba(OH)_2$ with 1047 μg. 3—0.01N NaOH with 1036 μg.

0.01N solutions of NaOH or $Ba(OH)_2$ are used as absorbents for CO_2. Comparison of the curves of change of resistance of these absorbents during the experiment shows that, for $Ba(OH)_2$, the resistance is constant at the end of the experiment, while for NaOH the curve goes through a maximum, after which the resistance begins to decrease appreciably (Fig. 2). Some authors also observed this phenomenon and explained it by a local increase in concentration of CO_3^{-2} in the absorbent solution, which disap-

TABLE 1

Compound	mg. sample	C content (%)			H content (%)		
		Calc.	Found	Diff.	Calc.	Found	Diff.
Hexachloroethane	2.221	10.14	10.50	+0.36	—	—	—
C_2Cl_6	2.662	10.14	10.22	+0.08	—	—	—
Oxalic acid	1.458	19.05	19.20	+0.15	4.76	4.66	–0.10
$C_2H_2O_4.2H_2O$	1.347	19.05	19.26	+0.21	4.76	4.57	–0.19
Picric acid	2.257	31.45	31.50	+0.05	1.32	1.35	+0.03
$C_6H_3N_3O_6$	1.652	31.45	31.67	+0.22	1.32	1.30	–0.02
Bromobenzoic acid	0.905	41.81	42.08	+0.27	2.43	2.53	+0.10
$C\text{-}H_5BrO_2$	1.220	41.81	41.61	–0.20	2.43	2.41	–0:02
Trans-α-decalone 2,4-dinitrophenyl-	1.054	57.83	57.83	0.00	6.02	6.14	+0.12
hydrazone	1.125	57.83	58.01	+0.18	6.02	6.06	+0.04
$C_{16}H_{20}C_4N_4$							
Benzotriazole	1.006	60.50	60.77	+0.27	4.23	4.33	+0.10
$C_6H_5N_2$	0.934	60.50	60.47	–0.03	4.23	4.24	+0.01
1-Hydroxy-9,11-dimethylperhydro-	1.002	75.00	74.67	–0.33	8.31	8.21	–0.18
12(14), 8(13)-phenanthradien-7-	0.954	75.00	75.10	+0.10	8.31	8.46	+0.15
one 1-acetate $C_{18}H_{24}O_3$							
Naphthalene	0.710	93.68	93.38	–0.30	6.32	6.60	+0.28
$C_{10}H_8$	0.988	93.68	93.56	–0.08	6.32	6.47	+0.15

pears when complete stirring is accomplished [13]. Under our conditions the reason for decreased resistance is, apparently, hydrolysis of the carbonate ions, due to which the number of hydroxyl ions in the solution increases and the resistance decreases. The absence of similar maxima when working with $Ba(OH)_2$ solutions, where because of formation of a precipitate of $BaCO_3$, there is practically no hydrolysis, confirms the correctness of this assumption. It should be noted that, for 0.005 N solutions of NaOH, the observed decrease in resistance is considerably greater than for 0.01 N solutions, which are therefore more convenient to work with (Fig. 2, 1).

Table 1 shows the results of C and H determinations for organic compounds with various C contents. All analyses showed good agreement of results after two combustions. The mean square error in determination was +0.14% abs. for C; +0.01% abs. for H.

The proposed method for microdetermination of C and H in organic compounds is, to a considerable extent, free from subjective errors in combustion and analysis, as shown by the better reproducibility of the results as compared with gravimetric determinations, despite small sample size. It permits automation of all stages of analysis, except taking the samples, and in this respect has promise. The method is also interesting for the fact that it can be used to study the combustion of substances under various conditions and to obtain information about the mechanism of decomposition.

LITERATURE CITED

[1] H. Reihlen, *Chem. Fabrik*, **1934**, 63.

[2] W. Zimmermann, *Mikrochem. vereinigt mit Mikrochim. Acta* **31**, 149 (1943).

[3] F. O. Fischer, *Anal. Chem.* **21**, 827 (1949).

[4] R. Levy, P. Mathieu, *Bull. soc. chim. France*, **1952**, 737.

[5] B. Bobransky, *Mikrochim. Acta* **1956**, 1735.

[6] T. T. White, V. A. Campanile,et al., *Anal. Chem.* **30**, 409 (1958)

[7] T. Kono, *Mikrochim. Acta* **1958**, 461.

[8] G. Kainz, *Z. anal. Chem.* **166**, 427 (1959).

[9] M. O. Korshun and V. A. Klimova, *Zhur. Anal. Khim.* **2**, 274 (1947); **3**, 176 (1948); **4**, 292 (1949). M. O. Korshun and N. E. Gel'man: "New Methods of Elementary Microanalysis" (Moscow, 1949).

[10] B. Bruns, et al., *Zhur. Anal. Khim.* **2**, 294 (1947).

[11] W. Schmidt, D. Baasch, *Forschungsberichte des Wirtschafts-und Verkehrsministeriums Nord-Rhein-Westfahlen*, No. 67 (1954).

[12] K. Drekopf, B. Braukmann, *Brennstoffchemie* **36**, 203 (1955).

[13] W. Koch, H. Malissa, *Arch. Eisenhüttenwes.*, **27**, 695 (1956).

[14] H. Malissa, *Mikrochim. Acta*, **1957**, 553.

[15] P. A. Kryukov and N. E. Shul'ts, *Gidrokhim. Materialy* **23**, 110 (1955).

[16] P. A. Kryukov and E. V. Rengarten, *Zhur. Anal. Khim.* **10**, 51 (1955).

[17] R. Boulin, *Chim. Anal.*, **40**, 72 (1958).

[18] W. Koch, S. Eckhard, H. Malissa, *Arch. Eisenhütenwes.*, **29**, 543 (1958).

[19] F. Ehrenberger, S. Gorbach, W. Mann, *Mikrochim. Acta*, **1958**, 778.

[20] J. Unterzaucher, *Chem.-Ing.-Tech.* **1950**,39; *Mikrochim. Acta.* **1957**, 448.

CHAPTER 36

Relation Between the Length of the Contact Layer and the Method of Decomposition During the Direct Determination of Oxygen*†

M. O. Korshun and E. A. Bondarevskaya

During the last fifteen years an important event in the development of microelemental analysis was the solution of the problem of the direct determination of oxygen. Published literature reviews illustrate how this problem, which was the object of an intense effort on the part of many chemists in the course of more than one hundred years [1-3], developed. The essence of the method generally accepted at the moment for the direct determination of oxygen is the thermal decomposition of the test material in a stream of pure nitrogen, the conversion of all the oxygen into CO over carbon which has been heated to 1100-1120° [4-6] or 1150° [2, 7, 9], and the subsequent oxidation of the CO into CO_2 which is then determined gravimetrically or titrimetrically.

Up to 1947 direct determination of oxygen was carried out only in Germany [4-6] and in the Soviet Union [2, 7, 9]. The method then widely entered into analytical laboratory practice in America [10-16] and Europe [17-22]. During recent years much work has been published on the direct determination of oxygen in various materials: in organic compounds [2, 4-7, 14, 19], coal [23-29]; in rubber, resin, gutta percha, petroleum products and asphalt [30-32, 11] and in inorganic materials [2, 4, 6, 33]. The method

* *Zhur. Anal. Khim.* **14**, 123-7 (1959).
† Communication 15, see J. Anal. Chem. 13, 695 (1958).

for the direct determination of oxygen has already found its place in the textbooks on quantitative organic microanalysis [9, 34-37].

Naturally, during such wide application, the method has been studied in detail by many analysts and many modifications and improvements have been made. Workers who have worked on improving the method have turned their attention to: 1) the preliminary purification of the inert gas used (nitrogen or helium) and to improvements in this technique [4-6, 9, 13, 14, 15, 21-22, 30]; 2) studying the sources of error connected with the quality of the quartz [14, 19-22, 34, 39-42]; the carbon [2, 4, 9-14, 18-19, 21, 43-47] and the iodine pentoxide [13, 38, 43-45, 48-51] used; and 3) looking for a better way of determining the CO [11-12, 14, 19, 21-22, 27-28, 30, 46, 52-56]. Nevertheless, questions touching on the relation between the optimum value of the contact layer and the method of decomposing the test material have not yet been considered. The present article is devoted to this question.

From the experience gained in our laboratory on the determination of carbon and hydrogen, and the simultaneous determination of C and H and other elements, it has been established that preliminary pyrolysis of the test materials considerably facilitates quantitative oxidation [9, 57-61] as well as reduction [9, 62-63) of the aliquot, as a result of which, it is not the vapors of the sample, but the active products of its decomposition, which are subjected to oxidation or reduction. Decomposition can be carried out in a similar manner during the determination of oxygen as well.

When the sample is placed in a quartz capsule and is pyrolytically decomposed (i.e., the capsule is heated along the direction from the open end towards the closed end with a furnace or with a burner) and the material is not distilled from the boat, as is usually done by all workers engaged in the direct determination of oxygen, then one might expect that the decomposition products issuing from the capsule will, undoubtedly, be much more reactive than the vapors of the undecomposed material, and will react considerably more readily than the latter with the contact layer of carbon at 1150°. Thanks to this, a much shorter layer of carbon will be sufficient than that hitherto used. In order to convince ourselves of the efficiency of such a method of decomposition for the determination of oxygen, we carried out pyrolytic decomposition of the sample within a quartz capsule which was used in an empty tube in the decomposition, pure and dry nitrogen being passed through at the rate of 10-12 ml per min. Since no carbon contact layer was present in the tube, the oxygen came off as water, carbon monoxide, and carbon dioxide. The carbon monoxide was oxidized to carbon dioxide over cupric oxide at 100° and the oxygen-containing products determined by weight. It was established in this way that during preliminary pyrolysis of an organic material carried out in a quartz capsule at 900-1000°, in an atmosphere of pure nitrogen, when an empty decomposition tube was used, the total percentage of oxygen calculated on the basis of the water and the carbon monoxide and dioxide amounts to 90% of the total content of oxygen in the analytical sample (Table 1). Accordingly, when the method described above is used for decomposition, a considerable breakdown of the organic substance can be achieved. The fact that some of the oxygen is not accounted for can probably be ascribed to the presence of oxygen-containing fragments of the molecule in the decomposition products.

Experiments were then carried out with a shortened layer of carbon in the decomposition tube. A layer of granulated carbon 5 cm long was placed in the quartz tube and heated to 1150° (±20°) with a tubular furnace 10 cm long. The sample was introduced into the tube in a quartz cap-

TABLE 1

Test material	mg. sample	% O Found as			Total % O Found	
		H_2O	CO	CO_2	absolute	relative
Saccharose	5.509	30.98	7.86	7.31	46.15	89
$C_{12}H_{22}O_{11}$, 51.42% O	5.983	29.76	7.08	7.60	44.44	86
Salicylic acid	4.876	4.83	20.32	4.01	29.16	84
$C_7H_6O_3$, 34.75% O	5.028	7.21	19.11	2.69	29.61	84
γ-Methyl glucoside	5.710	16.80	5.83	7.27	29.90	59
$C_7H_{14}O_6$, 49.44% O	5.959	17.96	4.58	8.24	30.78	61
Benzoic acid	4.450	2.57	22.25	0.17	24.99	95
$C_7H_6O_2$, 26.20 % O	4.437	2.32	21.56	0.55	24.43	93
Cis-4-ketocyclohexanedi-	4.689	2.45	20.31	15.58	38.44	89
carboxylic acid						
$C_8H_{10}O_5$, 42.97 % O	5.063	2.33	22.57	16.04	40.94	94

TABLE 2

Test material	mg. sample	% O found	% O diff.	Experimental conditions
Saccharose	12.240	51.28	−0.14	Length of the contact layer of granulated carbon 5 cm, temperature 1150°
$C_{12}H_{22}O_{11}$, 51.42 % O	12.134	51.53	+0.09	
Salicylic acid	13.745	34.56	−0.19	
$C_7H_6O_3$, 34.75 % O	11.371	34.76	+0.01	Nitrogen flow rate 10-12 ml/minute
Cis-4-ketocyclohexanedicarboxylic acid	4.025	42.87	−0.16	
$C_8H_{10}O_5$, 42.97 % O	3.471	43.09	+0.12	Decomposition of test material was carried out in a quartz capsule 7 cm long with a gas burner
Dimethyl ester of azobenzoic-3-3′-dicarboxylic acid	8.497	21.37	−0.10	
$C_{16}H_{14}O_4$, 21.47 % O	8.486	21.42	−0.05	
Urea	9.585	26.69	+0.03	
CH_4ON_2, 26.66 % O	11.491	26.60	−0.06	

sule 7 cm long; the latter was placed at a distance of 5-6 cm from the furnace. The capsule was initially heated with a gas burner at the open end in the same way as in the determination of carbon and hydrogen [9, 57-63]. The gaseous decomposition products, upon issuing from the capsule, were carried along with the nitrogen flow into the contact layer of granulated carbon heated to 1150°; here, all the oxygen contained in the vapor was quantitatively converted into carbon monoxide. The latter was oxidized to CO_2 by means of cupric oxide, the CO_2 being finally caught in Ascarite. The results obtained showed that the accuracy of the determination of oxygen in organic materials when the contact layer of carbon is shortened to 5 cm (Table 2) is adequate.

As our further experiments showed, determination of oxygen in organic compounds can also be carried out in an empty tube, when the contact layer of carbon is placed directly in the quartz capsule over the sample itself. A micro electric burner 6 cm long was used for heating the capsule at 1150 . By moving the burner slowly forward from the open end of the capsule towards the closed end, quantitative conversion of the oxygen in the sample to carbon monoxide was achieved in the capsule itself. The CO was subsequently oxidized as described above. Results of analyses carried out by this technique are given in Table 3.

These experiments showed that by adopting preliminary pyrolysis of the test material in a quartz capsule, there is no need for a long contact layer of carbon. When the above technique is used, a carbon layer 5 cm long can be used instead of a 16-20 cm long layer. When this method is adopted, the carbon can be placed either in the decomposition tube or directly in the capsule containing the test sample. Accordingly, there is no need for a long tube for carrying out the decomposition, and it is possible to work

TABLE 3

Test material	mg. sample	% O found	% O diff.	Experimental conditions
Saccharose	6,601	51.25	−0.17	The contact layer of granulated carbon 5 cm long was placed in a quartz test tube 9 cm long. Decomposition was effected with a micro electric burner 6 cm long at 1150°.
$C_{12}H_{22}O_{11}$, 51.42 % O	11.023	51.24	−0.18	
Benzoic acid	6.930	26.17	+0.03	
$C_7H_6O_2$, 26.20 % O	5.962	26.45	+0.25	
α, α, β-Methyl ester of naphthalenetricarboxylic acid	9.432	31.01	−0.14	
$C_{16}H_{20}O_6$, 31.15 % O	8.593	31.00	−0.15	
2, 2-Dimethyl-3-di-β-cyanoethylpyrone	8.397	13.49	−0.18	
$C_{13}H_{18}O_2N_2$,13.67% O	9.049	13.65	−0.02	
Uric acid	7.905	28.62	+0.05	
$C_6H_4O_3N_4$, 28.57 ¾ O	7.481	28.72	+0.15	

with a tube which is only 30 cm long instead of 50 cm. Experiments which we have carried out with an empty tube with carbon inside the decomposition capsule give a future possibility that the length of the contact layer can be reduced even further. The use of platinized carbon [63-65] made it possible for us to lower the temperature to 900°. A separate article will be devoted to this work.

SUMMARY

It has been established that during the pyrolytic decomposition of material in a nitrogen atmosphere in a quartz capsule placed in an empty quartz tube, almost quantitative decomposition of the sample occurs, during which up to 90% of the oxygen is liberated as H_2O, CO, and CO_2.

Preliminary pyrolysis of the sample considerably facilitates quantitative formation of carbon monoxide, so that it is possible to cut down the length of the contact layer, and, accordingly, also cut down the length of the reaction tube.

It has been shown that it is possible to determine oxygen with an empty combustion tube if the carbon layer is put in the quartz capsule with the sample.

LITERATURE CITED

[1] M. S. Skanavi-Grigor'eva, *Uspekhi. Khim.* **7**, 713 (1938).

[2] M. O. Korshun, Dissertation Inst. Org. Chem. USSR (1943); *Trudy Komissii Anal. Akad. Nauk SSSR* I (IV), 168 (1947).

[3] P. J. Elving and W. B. Ligett, *Chem. Rev.* **34**, 1299 (1944).

[4] M. Schütze, *Z. anal. Chem.* **118**, 244 (1939).

[5] W. Zimmermann, *Z. anal. Chem.* **118**, 258 (1939).

[6] J. Unterzaucher, *Ber.* **73B**, 391 (1940).

[7] M. O. Korshun. *Zavodskaya Lab.* **10**, 241 (1941).

[8] M. O. Korshun and N. E. Gel'man. *Apparatus for Quantitative Elemental Micro Analysis*, Moscow, 1947.

[9] M. O. Korshun and N. E. Gel'man, *New Methods of Elemental Micro Analysis*, Goskhimizdat, Moscow-Leningrad, 1949, p. 56.

[10] V. A. Aluise, T. R. Hall, F. C. Staats and W. W. Becker, *Anal. Chem.* **19**, 347 (1947).

[11] R. A. Dinerstein and R. W. Klipp, *Anal. Chem.* **21**, 545 (1949).

[12] A. O. Maglott and J. B. Lewis, *Anal. Chem.* **22**, 1051 (1950).

[13] C. C. Harris, D. M. Smith, and J. Mitchell, *Anal. Chem.* **22**, 1297 (1950).

[14] V. A. Aluise, H. K. Alber, H. S. Conway, C. C. Harris, W. H. Jones, *Anal. Chem.* **23**, 530 (1951).

[15] W. Kirsten, *Anal. Chem.* **25**, 75 (1953).

[16] C. L. Ogg, *Anal. Chem.* **26**, 116 (1954).

[17] C. E. Spooner, *Bull. soc. chim. France* **1947**, 115 ; *Metallurgia* **37**, 157 (1948).

[18] R. Berret and P. Poirier, *Bull. soc. chim. France* **1949**, 539.

[19] R. A. Deinum and A. C. Schouten, *Anal. Chim. Acta* **4**, 286 (1950); **3**, 353 (1949).

[20] R. Levy, *Bull. soc. chim. France* **1952**, 672.

[21] A. Lacourt, *Mikrochim. acta* **1954**, 735.

[22] A. F. Colson, *Analyst* **79**, 784 (1955).

[23] E. A. Druyan-Rempel', and S. A. Babushkin, *Zavodskaya Lab.* **16**, 238 (1950).

[24] S. A. Babushkin and E. A. Druyan-Rempel', *Zavodskaya Lab.* **17**, 1418 (1951).

[25] E. A. Druyan-Rempel', *Zavodskaya Lab.* **19**, 430 (1953).

[26] S. G. Aronov, L. L. Nesterenko, and L. P. Sologub, *Zavodskaya Lab.* **17**, 1415 (1951).

[27] H. Guerin and M. Bastick, *Bull. soc. chim. France* **1953**, 304.

[28] J. Garach and G. Weldener, *Chim. analyt.* **36**, 211 (1954).

[29] M. Bastick, *Bull. soc. chim. France* **1952**, 308.

[30] W. W. Walton and W. H. Smith, *J. Res. Nat. Bur. Standards* **40**, 493 (1947).

[31] P.N.Degens, Jr. and J.D.Kommer, N. N. Maatschappij. *Labor. Amsterdam Pepf.*. **11**, 746 (1947).

[32] P. Gouverneur, M. A. Schreuders and P. N. Degens, Jr. *Anal. Chim. Acta*, **5**, 293 (1951).

[33] O. Lourenco. *Anais assoc. quím. Brasil*, **10**. 57 (1951).

[34] F. Pregl-Roth, *Quantitative organische Mikroanalyse*, Wien, 1949 p. 73-84.

[35] A. Steyermark, *Quantitative Organic Microanalysis*. Philadelphia, 1951, p. 208.

[36] R. Belcher and A. L. Godbert, *Semimicro Quantitative Organic Analysis*, London, 1954.

[37 S. J. Clark, *Quantitative Methods of Organic Microanalysis*, London, 1956.

[38] J. Unterzaucher, *Mikrochem.* **36/37**, 706 (1951).

[39] W. Kirsten, *Mikrochem.* **34**, 151 (1949); **36/37**,609 (1951).

[40] D. Lindqist, *Acta chem. scand.* **2**, 177 (1948).

[41] R. Hey and H. J. Riley, *J. Chem. Soc.*, **1948**,1362.

[42] W. Otting, *Mikrochem.* **38**, 551 (1951).

[43] J. Unterzaucher, *Analyst* **77**, 584 (1952).

[44] J. Unterzaucher, *Bull. soc. chim. France* **1953**, 71.

[45] V. A. Campanile, J. H. Badley, et al., *Anal. Chem.* 23, 1421 (1951).

[46] J. Holowchak, G. F. C. Wear, *Anal. Chem.* **23**, 1404 (1951).

[47] M. Dandy and E. Stehr, *Anal. Chem.* **23**, 10, 1408)1951).

[48] W. H. Jones, *Anal. Chem.* **25**, 1449 (1953).

[49] H. A. Pieters, *Chem. Weekblad* **43**, 455 (1947).

[50] E. Huffman, *Anal. Chem.* **23**, 531 (1951).

[51] J. Unterzaucher, *Mikrochim. Acta* 4-6, 822 (1956).

[52] J. Unterzaucher, *Chemie-Ingenieur-Technik* **22**, 39 (1950).

[53] R. Hinkel and R. Raymond, *Anal. Chem.* **25**, 470 (1953).

[54] M. Schütze, *Ber.* **77B**, 484 (1944).

[55] D. Shepherd, *Anal. Chem.* **19**, 77 (1947).

[56] H. W. Deinum, *Brennstoff-Chem.* **35**, 334 (1954).

[57] M. O. Korshun and V. A. Klimova, *Zhur. Anal. Khim.* **2**, 274 (1947); **3**, 476 (1948); **4**, 292 (1949).

[58] M. O. Korshun and N. S. Sheveleva, *Doklady Akad. Nauk SSSR* **60**, 63 (1948); *Zhur. Anal. Khim.* **7**, 104 (1952); 11, 376 (1956).

[59] M. O. Korshun, *ibid.* **7**, 96, 101 (1952).

[60] M. O. Korshun, E. A. Terent'eva, and V. A. Klimova, *ibid.* **9**, 275 (1952); M. O. Korshun and E. A. Terent'eva, *Doklady Akad. Nauk SSSR* **100**, 707 (1955).

[61] V. A. Klimova, M. O. Korshun, and E. A. Bereznitskaya, *ibid.* **84**, 1175 (1952); 96, 81, 287 (1954); *Zhur. Anal. Khim.* **11**, 224 (1956); V. A. Klimova and E. A. Bereznitskaya, *ibid.* **11**, 292 (1956).

[62] N. E. Gel'man and M. O. Korshun, *Doklady Akad. Nauk SSSR* (1950); *Zhur. Anal. Khim.* **12**, 128 (1957).

[63] M. O. Korshun and E. A. Bondarevskaya, *Doklady Akad. Nauk SSSR* **110** (1956).

[64] J. Oita and H. Conway, *Anal. Chem.* **26**, 600 (1954).

[65] F. H. Oliver, *Analyst* **80**, 593 (1956).

CHAPTER 37

A Rapid Micromethod for the Determination of Halogens in Organic Compounds*

N. V. Sokolova, T. A. Orestova, and N. A. Nikolaeva

The problem which confronted us was to search for a simple micromethod for determining halogens, both in monomers and polymers, which would require simple apparatus and would be suitable for compounds containing other elements in addition to halogens.

Many microanalytical methods are known for the determination of halogens in organic compounds: catalytic oxidation by combustion of test material in an oxygen stream using a platinum catalyst [1]; the Carius method [2]; fusion of the test material with potassium metal [3]; combustion of the organic material in a stream of oxygen, the halogen being caught by silver metal [4] with simultaneous determination of carbon, hydrogen, and halogen. All of these methods require fairly complicated apparatus, and, with the exception of the last method, it is necessary to transfer the test halogen into another vessel.

Of interest is the rapid micro method of determining halogens suggested by Schöniger [5] which is carried out in a simple apparatus. Combustion of the organic test material is carried out in a flask filled with oxygen, the halogen liberated being then titrated in the same flask.

Combustion of a material in vessels filled with oxygen was first suggested in 1892 by Hempel [6] for the macro-

Zhur. Anal. Khim.* **14, 472-7 (1959).

determination of sulfur in coal and other organic materials; the method was subsequently improved by Graefe [7] and Marcusson and Doscher [8]. The latter determined halogen also by this method. The method, however, did not find wide application since combustion was carried out in flasks of too large a volume (i.e., from 6 to 10 liters); these often cracked. In 1922 Votocek [9-11] extended this method further and used it for the determination of chlorine in organic materials. Later, Mikl and Pech [12, 13] determined halogen and sulfur, both independently, and in the presence of each other, in organic materials.

All of the variations mentioned above for the determination of halogen and sulfur by combustion in a vessel filled with oxygen were used for macro samples of 0.5-0.1 g. Schöniger [5], for the microdetermination of halogen, reduced the weight of sample to 4-8 mg and the capacity of the flask to 300 ml. Chlorine and bromine were determined acidimetrically by Vieböck's method [14] using mercury cyanate.

For halogen determination we modified the construction of the flask, since the flasks used by Schöniger have some essential drawbacks. We used conical or flat bottomed flasks with a side arm and a stopcock for connecting to the atmosphere. The ground glass stopper had a drawn out end so that it was possible to fuse a platinum wire† into it.

Chlorine and bromine were determined by titrating with mercuric nitrate or perchlorate, while iodine was determined as iodate by Leipert's method [15] using sodium thiosulfate as titrant.

---

† A flask without a side arm (as used by Schöniger) is not suitable to work with. It is sometimes impossible to remove the stopper on account of the vacuum which develops in the flask after absorption of the combustion products. The use of a stopper with a drawn out end, as in our construction, makes it easy to fuse the platinum wire into it without damaging the stopper.

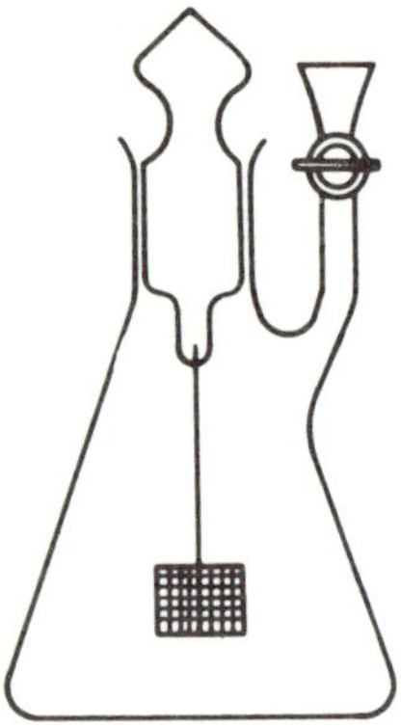

Fig. 1

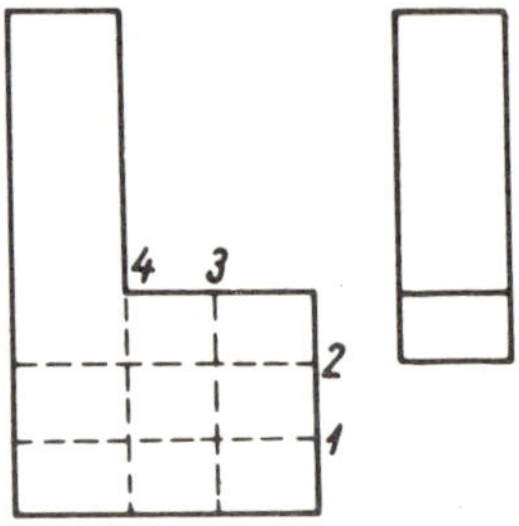

Fig. 2

A sample of the halogen-containing organic material is wrapped up in ashless filter paper and the latter fastened to the platinum wire fused into the drawn out end of the stopper; the latter is rapidly ignited in the flask filled with oxygen. The halogen thereupon is quantitatively converted to the hydrogen halide which is absorbed by a solution of alkali. The halogen is titrated in the same flask.

Apparatus. Determination is carried out in a conical or flat bottom flask made of heat-resistant glass, and with a capacity of 200-350 ml; the flask is fitted with a ground glass stopper and is provided with a side arm. The latter is provided with a stopcock and terminates in a funnel (Fig. 1). The bottom of the stopper is drawn out, and into the drawn-out end is fused a platinum wire 70-80 mm long and 0.7-0.8 mm in diameter, depending upon the height of the flask used. The platinum wire should not get closer than 20-30 mm to the bottom of the flask. To the end of the wire is fastened a platinum grid 20 x 30 mm in size; the grid is attached to the wire either by bending the latter or by welding it on.

Ashless filter paper made by the D. I. Mendeleev Moscow Chemico-Technological Institute was used for combustion of the test samples; the shape and size of the filter papers are shown in Fig. 2. For analysis of liquids, capillary tubes with a small expanded portion in the middle, or without such an expanded portion, were used, the latter being fused at one end (Fig. 3).

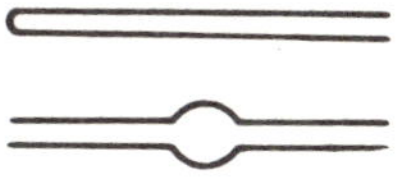

Fig. 3

Experimental Procedure. A sample of solid standard (3-15 mg, in certain cases up to 50 mg depending on the halogen content), is weighed out by difference from a weighing tube onto the center of the filter paper; the latter is then carefully wrapped up by folding along the dotted lines (Fig. 2).

Nonvolatile liquids are taken in capillaries. It is useful to seal one end of such capillaries with a small piece of paraffin wax, since the latter helps complete conversion of the halogen into the hydrogen halide and also prevents loss.

The capillary plus liquid sample are also wrapped up in filter paper in such a way that the open end of the capillary is covered by several layers of paper.

The sample is wrapped up in the filter paper as if it were a little parcel. The parcel is fitted into the platinum grid, so that the long narrow end of the filter paper is pointing upwards (Fig. 4). The lower part of the platinum grid is bent upwards slightly so that the sample is held securely in position and its complete combustion is ensured.

Determination of Chlorine and Bromine. 10 ml of distilled water, 1 ml of 2 N potassium hydroxide, and 3 drops

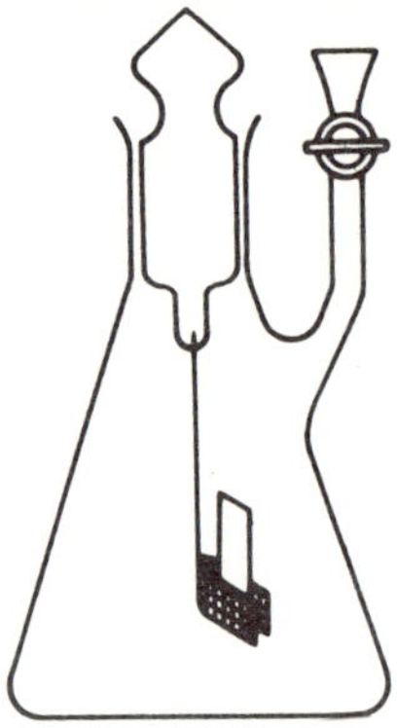

Fig. 4

of hydrogen peroxide are placed in the flask for absorbing the halogen. In order to expel the air from the flask, oxygen is passed into it from a cylinder, through a flask containing 50% potassium hydroxide, for 2-3 minutes at the rate of 100-120 ml per minute. The upper end of the filter paper is then ignited and the stopper rapidly inserted in the flask. During the combustion process, which lasts 15-20 seconds, the flask should be left at rest, the stopper must, however, be firmly kept in position by hand in order to prevent the gases formed from forcing it out.

When combustion is complete, in order to accelerate absorption of the combustion products, the flask is vigorously shaken for 6-10 minutes until the cloud over the liquid in the flask has disappeared completely.

Unless the flask is shaken, absorption of the combustion products may take 45-60 minutes. Before opening the flask, a small amount of distilled water is poured into the side arm and into the neck surrounding the stopper. The vacuum in the flask is destroyed by opening the stopcock in the side arm, the stopper is then easily lifted out. The stopper and platinum wire grid are carefully washed with distilled water over the flask.

TABLE 1

Compound	mg. sample	% Cl Found	% Cl Calc.	△
Triphenyllead chloride	4.025	7.50	7.50	0.00
$(C_6H_5)_3$ PbCl	5.200	7.29	7.50	−0.21
Ditosylxylitane chlorahydrin	5.345	7.67	7.69	−0.02
$C_{19}H_{21}S_2O_7Cl$	4.805	7.79	7.69	+0.10
Diphenyllead dichloride	4.045	16.07	16.40	−0.33
$(C_6H_6)_2PbCl_2$	4.375	16.27	16.40	−0.13
Sodium *p*-chlorobenzenesulfonate	3.920	16.23	16.52	−0.29
$C_6H_4ClSO_3Na$	5.640	16.30	16.52	−0.22
Dinitrochlorobenzene	4.550	17.38	17.50	−0.12
$C_6H_5Cl(NO_2)_2$	24.365	17.31	17.50	−0.19
p-Chlorophenylmethacrylamide	3.050	18.33	18.12	+0.11
$ClC_6H_4NHCOC_3H_5$	4.620	18.38	18.12	+0.16
p-Nitrobenzyl chloride	4.390	20.97	20.66	+0.31
$C_6H_4CH_2NO_2Cl$	4.860	20.90	20.66	+0.24
p-Thiocyanochlorobenzene	5.640	21.17	20.90	+0.27
C_6H_4ClCNS	5.275	21.09	20.90	+0.19
	3.560	20.90	20.90	0.00
Aminoxylitane hydrochloride		21.11	20.91	+0.20
$C_5H_{12}O_3NCl$		21.21	20.91	+0.30
p-Chloronitrobenzene	2.785	22.59	22.53	+0.06
$C_6H_4NO_2Cl$	4.265	22.69	22.53	+0.16
p,p'-Aminochlorodiphenyl	5.380	17.55	17.39	+0.16
$(C_6H_4)_2NH_2Cl$	4.705	17.20	17.39	−0.19
Chlorobenzoic acid	7.640	22.87	22.64	+0.23
ClC_6H_4COOH	6.095	22.93	22.64	+0.29
p-Chlorophenol	3.365	27.66	27.62	+0.04
$C_6H_4(OH)Cl$	3.940	27.78	27.62	+0.16
p-Chloroaniline	3.950	27.59	27.93	−0.34
$ClC_6H_4NH_2$	3.670	27.58	27.93	−0.35
Semicarbazine hydrochloride	4.280	31.84	31.78	+0.06
$H_2NCONHNH_2HCl$	5.055	31.80	31.78	+0.02

TABLE 2

Test material	Sample weight, mg.	Chlorine content found, %		
		in a flask filled with O_2	in O_2 current	by fusion with potassium
Copolymer of isoprene and chlorostyrene, Sample IVS* No. 13	5.290	13.42	—	—
	4.440	—	13.44	—
The same No. 25	2.910	24.55	—	—
	34.590	—	—	24.21
The same No. 32	5.670	11.05	—	
	7.860	—	10.98	
The same No. 34	3.515	11.12	—	—
	3.020	11.10	—	—
	5.460	—	10.96	—
The same No. 35	5.425	17.51	—	—
	5.060	17.82	—	—
	6.285	—	17.82	—
The same No. 39	5.490	22.82	—	—
	5.295	22.93	—	—
	3.245	—	22.63	—
	4.455	—	22.92	—
Three-dimensional polymer Mark KFUKh, Sample* No. 25	4.400	9.45	—	—
	4.520	9.57	—	—
	27.845	—	—	9.25
The same No. 1	2.685	10.61	—	—
	26.670	—	—	10.61
The same No. 2	4.140	10.80	—	—
	3.115	10.93	—	—
	36.940	—	—	11.19
The same No. 4	4.225	10.50	—	—
	5.060	—	10.73	—
Polychlorovinyl	4.450	56.35	—	—
	3.470	56.43	—	—
	3.115	—	56.42	—
	3.305	—	56.23	—
Chloride residue in treated chitin	39.960	0.23	—	—
	51.310	0.18	—	—
	35.435	—	—	0.28
	28.290	—	—	0.22
The same No. 32	40.010	0.33	—	—
	44.730	0.34	—	—

* A sample prepared in the Institute of High Polymer Compounds.

TABLE 3

Test material	Sample weight mg.	Bromine content % found	Bromine content % theoretical	Deviation %
Methylbromophthalimide $BrCH_3C_8H_2O_2NH$	5.000	33.24	33.21	+0.03
	4.435	33.03		-0.18
Dimethylbromostyrene $(CH_3)_2BrC_6H_2CH=CH_2$	2.745	37.81	37.91	-0.10
	2.550	37.62		-0.29
p-Bromonitrobenzene $BrC_6H_4NO_2$	3.500	39.58	39.55	+0.03
	4.545	39.48		-0.07
p-Bromobenzoic acid BrC_6H_4COOH	3.085	39.81	39.72	+0.09
	3.305	39.54		-0.18
2,3,3-Trimethylpentene-1-bromo-5 $C_8H_{15}Br$	2.660	41.88	41.91	-0.03
	2.840	42.00		+0.09
5,7-Dibromo-8-hydroxyquinoline $Br_2C_9H_4(OH)N$	2.895	52.80	52.75	+0.05
	3.610	52.74		-0.01
p-Dibromobenzene $C_6H_4Br_2$	5.605	67.45	67.74	-0.29
	5.110	67.63		-0.11

TABLE 4

Test material	Sample weight mg.	Iodine content % found	Iodine content % theoretical	Deviation %
Sample No. 281 $C_{22}H_{27}ON_2SI$	4.775	25.75	25.67	+0.08
	4.050	25.46		-0.21
Sample No. 366 $C_{21}H_{29}ON_2SI$	3.745	26.22	26.24	-0.02
	3.015	26.43		+0.19
Sample No. 207 $C_{20}H_{25}ON_2SI$	2.190	27.25	27.13	+0.12
	3.315	26.89		-0.24
α-(Nitrofurylpropyl) methylquinoline iodide $C_{17}H_{17}NNO_2OI$	4.380	29.95	29.92	+0.03
	4.140	29.90		-0.02
Iodoalkylate-1-methyl 2 (p-tolyl) benzimidazol $C_{16}H_{17}N_2I$	3.780	34.61	34.85	-0.24
	3.690	34.61		-0.24
m-Iodobenzoic acid $I\ C_6H_4COOH$	3.460	52.27	51.39	-0.12
	4.090	51.24		-0.15
β-Iodopropionic acid ICH_2CH_2COOH	9.250	63.40	63.47	-0.07

TABLE 2

Test material	Sample weight, mg.	Cl%, found: in a flask filled with O_2	Cl%, found: in O_2 stream	Cl%, found: by fusion with potassium
Copolymer of isoprene and chloro-styrene, Sample IVS* No. 13	5.290	13.42	—	—
	4.440	—	13.44	—
The same No. 25	2.910	24.55	—	—
	34.590	—	—	24.21
The same No. 32	5.670	11.05	—	
	7.860	—	10.98	
The same No. 34	3.515	11.12	—	—
	3.020	11.10	—	—
	5.460	—	10.96	—
The same No. 35	5.425	17.51	—	—
	5.060	17.82	—	—
	6.285	—	17.82	—
The same No. 39	5.490	22.82	—	—
	5.295	22.93	—	—
	3.245	—	22.63	—
	4.455	—	22.92	—
Three-dimensional polymer Mark KFUKh, Sample* No. 25	4.400	9.45	—	—
	4.520	9.57	—	—
	27.845	—	—	9.25
The same No. 1	2.685	10.61	—	—
	26.670	—	—	10.61
The same No. 2	4.140	10.80	—	—
	3.115	10.93	—	—
	36.940	—	—	11.19
The same No. 4	4.225	10.50	—	—
	5.060	—	10.73	—
Polyvinyl chloride	4.450	56.35	—	—
	3.470	56.43	—	—
	3.115	—	56.42	—
	3.305	—	56.23	—
Chloride residue in treated chitin	39.960	0.23	—	—
	51.310	0.18	—	—
	35.435	—	—	0.28
	28.290	—	—	0.22
The same No. 32	40.010	0.33	—	—
	44.730	0.34	—	—

* A sample prepared in the Institute of High Polymer Compounds.

TABLE 3

Test material	Sample weight mg.	Br% Found	Br% Theoretical	Deviation %
Methylbromophthalimide	5.000	33.24	33.21	+0.03
$BrCH_3C_8H_2O_2NH$	4.435	33.03	33.21	−0.18
Dimethylbromostyrene	2.745	37.81	37.91	−0.10
$(CH_3)_2BrC_6H_2CH=CH_2$	2.550	37.62	37.91	−0.29
p-Bromonitrobenzene	3.500	39.58	39.55	+0.03
$BrC_6H_4NO_2$	4.545	39.48	39.55	−0.07
p-Bromobenzoic acid	3.085	39.81	39.72	+0.09
BrC_6H_4COOH	3.305	39.54	39.72	−0.18
2,3,3-Trimethyl-5-bromo-1-pentene	2.660	41.88	41.91	−0.03
$C_8H_{15}Br$	2.840	42.00	41.91	+0.09
5,7-Dibromo-8-hydroxyquinoline	2.895	52.80	52.75	+0.05
$Br_2C_9H_4(OH)N$	3.610	52.74	52.75	−0.01
p-Dibromobenzene	5.605	67.45	67.74	−0.29
$C_6H_4Br_2$	5.110	67.63	67.74	−0.11

TABLE 4

Test material	Sample weight mg.	I% Found	I% Theoretical	Deviation %
Sample No. 281	4.775	25.75	25.67	+0.08
$C_{22}H_{27}ON_2SI$	4.050	25.46	25.67	−0.21
Sample No. 366	3.745	26.22	26.24	−0.02
$C_{21}H_{29}ON_2SI$	3.015	26.43	26.24	+0.19
Sample No. 207	2.190	27.25	27.13	+0.12
$C_{20}H_{25}ON_2SI$	3.315	26.89	27.13	−0.24
α-(Nitrofurylpropyl) methylquinoline iodide $C_{17}H_{17}NNO_2OI$	4.380	29.95	29.92	+0.03
	4.140	29.90	29.92	−0.02
Iodoalkylate-1-methyl 2 (*p*-tolyl) benzimidazole $C_{16}H_{17}N_2I$	3.780	34.61	34.85	−0.24
	3.690	34.61	34.85	−0.24
m-Iodobenzoic acid	3.460	52.27	51.39	−0.12
$I\ C_6H_4COOH$	4.090	51.24	51.39	−0.15
β-Iodopropionic acid ICH_2CH_2COOH	9.250	63.40	63.47	−0.07

the combustion products are absorbed with an alkali solution, and the halide is then determined titrimetrically in the same flask. The method can be used for both low molecular weight compounds and high molecular weight compounds which are difficult to burn. The presence of sulfur, of nitrogen in any form whatever, and of lead does not interfere with the determination. The method is distinguished by the simplicity of the apparatus used, and by the speed with which it can be carried out; it can be recommended for serial gravimetric analysis.

LITERATURE CITED

[1] F. Pregl and J. Grant, *Quantitative Organic Microanalysis*, Phila., 1946, p. 87.

[2] *Ibid.* p. 96.

[3] M. O. Korshun and M. N. Chumachenko, *Doklady Akad. Nauk SSSR* 99, 769 (1954).

[4] M. O. Korshun and N. S. Sheveleva, *Zhur. Anal. Khim.* **11**, 376 (1956).

[5] W. Schöniger, *Microchim. Acta* **1**, 123 (1955).

[6] W. Hempel, *Z. angew. Chemie* **13**, 393 (1892).

[7] E. Graefe, *Z. angew. Chemie* **26**, 616 (1904).

[8] J. Marcusson and H. Doscher, *Chemiker-Ztg.* **47**, 417 (1910).

[9] E. Votocek, *Chem. listy* **16**, 248 (1922).

[10] E. Votocek and L. Kasparek, *Chem. Abstr.* **17**, 1769 (1923).

[11] E. Votocek, *Chemiker-Ztg.* **42**, 257, 271, 317 (1918).

[12] O. Mikl and J. Pech, *Chem. listy* **66**, 382 (1952).

[13] O. Mikl and J. Pech, *Chem. listy* **47**, 904 (1953).

[14] F. Viebock, *Ber.* **65**, 493 (1932).

[15] A. Friedrich, *Die Praxis quantitativen organischen Mikroanalyse*, *Wien u. Liepsig*, 1939, p. 126.

[16] W. Pugh, *S. African Ind. Chemist* **4**, 182 (1950).

CHAPTER 38

Simultaneous Microdetermination of Carbon, Hydrogen, and Nitrogen in Nitro Compounds*

V. A. Klimova and K. S. Zabrodina

The methods described in the literature for the simultaneous micro determination of carbon, hydrogen and nitrogen in organic compounds are based on combustion of the substance to carbon dioxide, water and elemental nitrogen. Berraz [1] and Schöniger [2] determined the amounts of the resulting CO_2, H_2O and N_2 by measuring their pressure; Kirsten [3] proposed the measurement of the volumes of the combustion products. These methods require some elaborate apparatus and have not been developed sufficiently for securing dependable results.

Nitrogen oxides may form along with elemental nitrogen during the combustion of nitrogenous organic compounds in an oxygen stream; the amount of oxides depends, firstly, on what atoms were bound to the nitrogen [4-7]; thus, for example, considerably more nitrogen oxides are obtained during the combustion of nitro compounds than during the combustion of amines; secondly, the amount of nitrogen oxides formed depends upon the technique of combustion of the substance. Ingram [8] found that platinum aids the formation of nitrogen dioxide; however, in a combustion in a rapid oxygen stream he failed to oxidize nitrogen quantitatively to nitrogen dioxide even in nitro compounds.

Izvest. Akad. Nauk SSSR, Otdel. Khim. Nauk* **1959, 582-5.

TABLE 1
Analysis of *p*-Nitrobenzoic acid

Rate of O_2 in ml/min	Amount of N in %		
	Found	Calculated	Difference
6	8.34	8.38	-0.04
8	8.26	8.38	-0.12
13	7.92	8.38	-0.46
15	7.83	8.38	-0.55
28	7.55	8.38	-0.83

The method for simultaneous microdetermination of C, H and N in nitro compounds, proposed by us here, consists of the following: the substance under analysis is gradually distilled and burned in an oxygen stream over platinum. The resulting carbon dioxide, water and nitrogen dioxide are absorbed quantitatively by appropriate absorbents and the percentage content of C, H and N is calculated from the weight gain of the absorbents.

The method of combustion in the procedure suggested by us has decisive significance. Combustion of a substance with preliminary pyrolysis in a container, as is done in the microdetermination of C and H in the rapid method [6, 7] is inapplicable in the present method. As shown previously [7], the yield of nitrogen dioxide decreases during the pyrolysis and a considerable amount of elemental nitrogen is formed. For this reason we have attempted to run the combustion so that the vapors of the substance, mixed with oxygen, will reach the heated platinum prior to the beginning of the process of decomposition. This was achieved by distilling the substance at a relatively low temperature. The distillation must be run slowly, avoiding flashes which might occur on heated platinum with a high concentration of the vapors, so that pyrolysis, in place of the oxidation, is avoided.

The rate of oxygen flow also plays a very great role. As shown by our studies, the rate of 5-8 ml per min is the optimum value for the oxygen stream; with its increase to 15-30 ml per min we obtained incorrect results for the nitrogen determination (Table 1).

Oxides of nitrogen are quantitatively absorbed both by manganese dioxide [8] and by silica gel saturated with 0.02M $K_2Cr_2O_7$ solution in sulfuric acid of sp. gr. 1.84 [9], but the advantage of the latter substance lies in the fact that with the same layer thickness it absorbs a larger amount of nitrogen oxides. We found that a certain amount of nitrogen oxides may be retained by the water which condenses at the beginning of the absorption apparatus filled with Anhydrone; this leads to high results for hydrogen and low ones for nitrogen. In order to avoid this difficulty, the apparatus with Anhydrone was heated to 75-85° by means of an electric heater. The temperature of the apparatus with Anhydrone should not reach 100° [10].

A roll of silver mesh is inserted into the combustion tube, in place of the platinum, for analysis of nitro compounds which contain halogens.

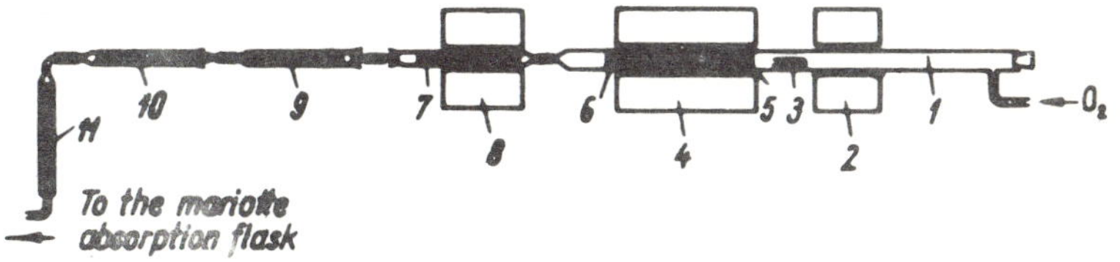

Diagram of the Apparatus. 1) combustion tube; 2) removable electric furnace, length 6 cm (200-350°); 3) quartz vessel; 4) electric furnace, length 14 cm (800-835°); 5) roll of platinum mesh, length 10 cm; 6) roll of silver mesh, length 5 cm; 7) water absorber; 8) removable electric furnace (75-85°); 9) nitrogen dioxide absorber; 10) carbon dioxide absorber; 11) guard tube.

The diagram of the installation for the simultaneous microdetermination of C, H and N in nitro compounds containing C, H, N, Cl and Br is shown in the figure.

Carbon dioxide is absorbed by Ascarite, while water is absorbed by Anhydrone. Glass wool is used for interlayers in filling the absorption apparatus with Anhydrone. A layer of silica gel 6-7 cm long, saturated with a solution of $K_2Cr_2O_7$ in sulfuric acid, is put in a Pregl absorption tube for absorption of nitrogen dioxide; on both sides of this layer are located layers of Anhydrone 1.5-2 cm long; these layers are separated by glass wool. Only the roll of platinum mesh 15 cm long may be placed into the zone of the longer furnace for the combustion of nitro compounds which do not contain halogen.

EXPERIMENTAL

In the analysis of low-boiling compounds or compounds which do not form tars on being heated, a sample of 2-8 mg is placed in a quartz capsule or quartz boat. The previously weighed absorption apparatus is attached to the combustion tube, after which the sample is pushed into the tube to a distance of 2-5 cm from the longer furnace, heated to 800-835° and oxygen introduced at a flow rate of 5-8 ml per min. The short, removable electric furnace, which may be heated to from 200° to 350°, depending on the boiling point of the substance, is placed at the quartz boat or at the closed end of the capsule with the sample, and the substance being analyzed is distilled into the zone in which the platinum is located by gradually moving the furnace over the sample in the direction of the longer electric furnace. The distillation is performed slowly in order to avoid flashing of the substance. The course of combustion is watched through the formation of yellow vapors of nitrogen dioxide at the end of the combustion tube. After the sample has evaporated, the short electric furnace is removed and the tube is heated by means of a gas burner throughout its length up to the longer electric furnace.

Depending upon the size of the sample and the rate of the oxygen flow, usually 100-300 ml of O_2 is required for combustion of the sample, while the displacement of the combustion products from the tube into the absorption apparatus requires 150-200 ml of O_2, i.e. the total time required is about one hour. The absorption tube is removed after completion of the displacement and weighed: for carbon dioxide at the 15th minute, for water at the 20th minute, and for nitrogen dioxide at the 25th minute. If the substance being analyzed has a high boiling point and yields tar on being heated, it is recommended that the sample be placed in a platinum boat. The short electric furnace heated to 200-250° is placed directly over the sample. After the distillation of the main part of the substance, the longer electric furnace is moved over the platinum boat for 5-7 minutes, after which the procedure is as indicated above.

The calculation for N is made by the following formula:

$$\% N = \frac{K_{NO_2} \cdot a \cdot 100}{b}$$

where $K_{NO_2} = 0.30447$, a = weight increase from NO_2 in milligrams, b = sample weight in milligrams.

SUMMARY

A gravimetric method for simultaneous microdetermination of C, H and N is proposed for nitro compounds containing C, H, N, O, Cl and Br, by combustion in an oxygen stream. The accuracy of the determination of C and H is ±0.2% and that for N ±0.3%.

TABLE 2

Substance	Amount of C in %			Amount of H in %			Amount of N in %		
	Found	Calc.	Diff.	Found	Calc.	Diff.	Found	Calc.	Diff.
Nitromethane	19.67	19.67	0.00	5.03	4.95	+0.08	22.76	22.95	-0.19
	19.70	19.67	+0.03	5.13	4.95	+0.18	22.81	22.95	-0.14
1,1-Dinitroethane	19.97	20.01	-0.04	3.44	3.36	+0.08	23.04	23.33	-0.29
	20.15	20.02	+0.14	3.40	3.36	+0.04	23.21	23.33	-0.12
4-Nitro-2,2,4-trimethylpentane	60.43	60.34	+0.09	10.56	10.76	-0.20	8.61	8.80	-0.19
	60.34	60.34	0.00	10.69	10.76	-0.07	8.89	8.80	+0.09
Nitrocyclohexane	55.64	55.79	-0.15	8.52	8.58	-0.06	10.55	10.85	-0.30
	55.93	55.69	+0.14	8.45	8.58	-0.13	10.60	10.85	-0.25
m-Dinitrobenzene	42.88	42.86	+0.02	2.39	2.40	-0.01	16.52	16.67	-0.15
	42.79	42.86	-0.07	2.42	2.40	+0.02	16.50	16.67	-0.17
p-Nitrotoluene	61.12	61.31	-0.19	5.13	5.15	-0.02	10.18	10.22	-0.04
	61.28	61.31	-0.03	5.17	5.16	+0.02	10.16	10.22	-0.06
o-Nitrophenol	51.71	51.80	-0.09	3.59	3.62	-0.03	9.96	10.07	-0.11
	51.70	51.80	-0.10	3.62	3.62	0.00	9.91	10.07	-0.16

m-Nitrobenzaldehyde	55.46	55.63	-0.17	3.28	3.34	-0.06	9.02	9.27	-0.25
	55.62	55.63	-0.01	3.31	3.34	-0.03	9.05	9.27	-0.22
p-Nitrobenzoic acid	50.45	50.31	+0.14	2.95	3.02	-0.07	8.28	8.38	-0.10
	50.21	50.31	-0.10	2.97	3.02	-0.05	8.34	9.38	-0.04
3,5-Dinitrobenzoic acid[a]	39.69	39.63	+0.06	1.93	1.90	+0.03	12.97	13.21	-0.24
	39.56	39.63	-0.07	1.95	1.90	+0.05	13.07	13.21	-0.14
Picric acid[a]	31.48	31.45	+0.03	1.48	1.32	+0.16	18.38	18.34	+0.04
	31.37	31.45	-0.08	1.39	1.32	+0.07	18.27	18.34	-0.07
ω-Nitrocamphene	66.25	66.27	-0.02	8.27	8.34	-0.07	7.72	7.73	-0.01
	66.16	66.27	-0.11	8.29	8.34	-0.05	7.70	7.73	-0.03
4-Nitro-5-methoxy-3-carbethoxy-2-methylbenzofuran	55.93	55.91	+0.02	4.83	4.69	+0.14	5.14	5.02	+0.12
	55.99	55.91	+0.08	4.77	5.69	+0.08	5.14	5.02	+0.12
Chlorotrinitromethane	6.47	6.47	0.00	0.00	0.00	0.00	22.54	22.66	-0.08
	6.63	6.47	+0.16	0.00	0.00	0.00	22.46	22.66	-0.20
Bromonitromethane	5.39	5.34	+0.05	0.00	0.00	0.00	18.48	18.70	-0.22
	5.32	5.34	-0.02	0.00	0.00	0.00	18.42	18.70	-0.28
5-Chloro-1,3-dinitrobenzene[a]	35.50	35.57	-0.07	1.32	1.49	-0.17	13.80	13.83	-0.03
	35.44	35.57	-0.13	1.45	1.49	-0.05	13.64	13.83	-0.19

[a]The sample was placed in a platinum boat.

LITERATURE CITED

[1] G. Berraz, *Anales inst. invest. cient. y tecnol. (Univ. nac. litoral, Santa Fe, Arg.)* **7**, 70 (1937).

[2] W. Schöniger, *Helv. Chem. Acta* **39**, 650 (1956).

[3] W. Kirsten, *Anal. Chem.* **26**, 1097 (1954).

[4] A. Friedrich, *Mikrochemie* **23**, 129 (1937).

[5] A. Heron, *Analyst* **72**, 142 (1947).

[6] M. O. Korshun and V. A. Klimova, *Zhur. Anal. Khim.* **3**,176 (1948).

[7] M. O. Korshun and V. A. Klimova, *ibid.* **4**, 292 (1949).

[8] G. Ingram, *Mikrochim. Acta* **1953**, 71.

[9] V. A. Klimova and M. O. Korshun, *Zhur. Anal. Khim.* **6**, 230 (1951).

[10] E.Abramson and A. Brochet,*Bull.soc. chim. France* **1954**, 367.

CHAPTER 39

Simultaneous Determination of Carbon, Hydrogen, and Halogen*

V. A. Klimova and E. N. Merkulova

The simultaneous determination of carbon, hydrogen, and halogen is usually carried out by the pyrolytic combustion of the organic compound [1] in a rapid stream of oxygen by the method of Korshun and Klimova [2, 3]. Carbon and hydrogen are determined as usual, and the halogen is absorbed by silver foil or gauze in a quartz apparatus attached to the combustion tube by a ground joint, at 550-600°. In the determination of halogen an error of up to ±0.6% is tolerated. Such a large deviation is explained by the authors as due to errors in weighing that may occur when the heavy quartz apparatus is used, and so much the more so since in this determination there is no conversion factor involved. But from our point of view the error in weighing is not the only reason for the lack of accuracy in the halogen determination. It has been observed that as a result of the reaction of silver, and especially of silver halide, with the quartz upon heating above the melting point of the silver halide (500-600°), the quartz is strongly attacked and in a short time ceases to maintain a constant weight. Thus, both the large weight of the quartz absorption apparatus and the heating of the silver above the melting point of the silver halides affect the accuracy of the halogen determination. It would be possible to eliminate contact of the silver halides with the quartz and simultaneously decrease the weight of the absorption apparatus by carrying out the absorption in a metal boat.

Izvest. Akad. Nauk SSSR, Otdel. Khim. Nauk* **1959, 781-6.

TABLE 1

	Name of compound	C content in %			H content in %			Hal content in %		
		Calc.	Found	Diff.	Calc.	Found	Diff.	Calc.	Found	Diff.
1	3-Chloro-4-(*p*-chlorophenyl)-	60.32	60.35	+0.03	4.05	4.21	+0.16	35.62	35.81	+0.19
	butyne	60.32	60.11	+0.03	4.05	4.11	+0.06	35.62	35.69	+0.07
2	β-Chloro-α-acetoxyethyl-	60.44	60.31	-0.13	5.59	5.66	+0.07	17.86	17.57	-0.29
	benzene	60.44	60.34	-0.10	5.59	5.59	0	17.86	17.86	-0.11
3	1,4-Dichloro-2,3-dibromo-2-butene	16.99	16.91	-0.08	1.43	1.39	-0.04	81.58	81.30	-0.28
		16.99	17.08	+0.09	1.43	1.49	+0.06	81.58	81.62	+0.04
4	Diethylaniline hydrochloride	64.67	64.53	-0.14	8.69	8.68	-0.01	19.10	19.27	+0.17
		64.67	64.46	-0.21	8.69	8.66	-0.03	19.10	19.13	+0.03
5	*p*-Nitrobenzoate of 1-(1,2-dibromo-	41.61	41.68	+0.07	3.49	3.63	+0.14	36.90	36.91	+0.01
	vinyl)-1-cyclohexanol	41.61	41.77	+0.16	3.49	3.62	+0.13	36.90	37.04	+0.14
6	Diethylethanolamine ethiodide	35.17	35.11	-0.06	7.37	7.64	+0.27	46.48	46.56	+0.08
		35.17	35.24	+0.07	7.37	7.59	+0.22	46.48	46.63	+0.15
7	Diethylvinyloxyethylamine	37.90	37.83	-0.07	7.07	7.01	-0.01	44.50	44.33	-0.17

TABLE 2

1[a]	Dichloroanthracene	68.02	68.20	+0.18	3.24	3.29	+0.05	28.74	28.96	+0.22
		68.02	68.17	+0.15	3.24	3.21	-0.03	28.74	28.78	+0.04
2[a]	Bromobenzoic acid	41.83	41.92	+0.09	2.50	2.60	+0.10	39.76	39.59	-0.17
		41.83	41.83	0.00	2.50	2.33	-0.17	39.76	39.61	-0.15
3[a]	3,4-Dibromo-2,5-dimethyl-2,5-hexanediol	31.81	32.11	+0.30	4.64	4.65	+0.01	52.93	53.08	+0.15
		31.81	32.01	+0.20	4.64	4.56	-0.08	52.93	52.81	-0.12
4[a]	β-Bromo-*p*-di(*tert*. butyl)-benzene	62.44	62.51	+0.07	7.86	7.94	+0.08	29.69	29.36	-0.33
		62.44	62.68	+0.24	7.86	7.98	+0.12	26.69	29.56	-0.13
5[a]	Hexachloroethane	10.14	10.10	-0.04	—	—	—	89.86	90.10	+0.24
		10.14	10.13	-0.01	—	—	—	89.86	89.76	-0.10
6[b]	Dichloroanthracene	68.02	67.96	-0.06	3.24	3.33	+0.09	28.74	28.77	+0.03
		68.02	67.96	-0.06	3.24	3.31	+0.07	28.74	28.74	0.00
7[b]	1-Chloro-2-methyl-2-*p*-cumylpropane	74.09	74.33	+0.24	9.09	9.16	+0.07	16.82	16.66	-0.16
8[b]	1,2,3,4-Tetrabromo-γ-methylindane	26.80	26.87	+0.07	1.80	1.72	-0.08	71.40	71.42	+0.02
		26.80	26.78	-0.02	1.80	1.80	0	71.40	71.37	-0.03
9[b]	Hexachloro-1,3-butadiene	18.42	18.67	+0.25	—	—	—	81.58	81.38	-0.20
		18.42	18.78	+0.36	—	—	—	81.58	81.50	-0.08
10[b]	*cis*-2-Methyl-1-(α, β-dibromo-ethynyl)-1-cyclohexanol	36.26	36.37	+0.11	4.71	4.82	+0.11	53.66	53.43	-0.23
		36.26	36.21	-0.05	4.71	4.80	+0.09	53.66	53.50	-0.16

[a] Combustion in a platinum boat; [b] Combustion in a quartz boat.

TABLE 3

	Name of compound	C content in %			H content in %		
		Calc.	Found	Diff.	Calc.	Found	Diff.
1	Ethyl (β-phenylethyl)dichlorosilane	50.61	50.81	+0.20	7.65	7.78	+0.13
		50.61	50.63	+0.02	7.65	7.85	+0.20
2	γ-Chloropropyldiethylchlorosilane	42.21	42.10	-0.11	8.10	8.18	+0.08
3	1-Trimethylsilyl-3-(dichloromethylsilyl)propane	36.68	36.41	-0.27	7.91	7.91	0
		36.68	36.46	-0.22	7.91	7.82	-0.09
4	β,β-Dichlorovinyltrimethylsilane	35.51	35.63	+0.12	5.96	6.04	+0.08
		35.51	35.74	+0.23	5.96	6.10	+0.14
5	γ-(*p*-thymyl)propyltrichlorosilane	44.87	45.13	+0.26	4.89	5.15	+0.16
		44.87	45.10	+0.23	4.89	5.13	+0.14
6	Phenyl-*p*-tolylborochloride	72.78	72.91	+0.13	5.64	5.57	-0.07
		72.78	72.79	+0.01	5.64	5.50	-0.14
7	Mixed anhydride of *p*-chlorophenylpyroboric and acetic acids	50.73	50.78	+0.05	3.72	3.63	-0.09
		50.73	50.88	+0.15	3.72	3.65	-0.07

TABLE [illegible] (cont'd)

	Name of compound	X content in %			B or Si content in %		
		Calc.	Found	Diff.	Calc.	Found	Diff.
1	Ethyl(β-phenylethyl)dichlorosilane	29.90	29.96	+0.06	11.84	11.81	+0.03
		29.90	—	—	11.84	11.86	+0.02
2	γ-Chloropropyldiethylchlorosilane	35.60	35.50	-0.10	14.10	14.23	+0.13
3	1-Trimethylsilyl-3-(dichloromethylsilyl)propane	30.48	30.84	-0.09	24.48	24.79	+0.31
		30.48	31.23	+0.30	24.48	24.72	+0.24
4	β,β-Dichlorovinyltrimethylsilane	41.93	41.73	-0.22	16.60	16.28	-0.32
		41.93	41.71	-0.22	16.60	16.46	-0.14
5	γ-(p-thymyl)propyltrichlorosilane	39.75	39.48	-0.27	10.49	10.47	-0.02
		39.75	39.53	-0.23	10.49	10.17	-0.32
6	Phenyl-p-tolylborochloride	16.53	16.32	-0.21	5.05	4.95	-0.10
		16.53	16.32	-0.21	5.05	4.97	-0.08
7	Mixed anhydride of p-chlorophenylpyroboric and acetic acids	18.72	18.87	+0.15	5.71	5.94	+0.23
		18.72	18.79	+0.07	5.71	5.77	+0.06

The absorption of acid gases in boats with various fillings has already been successfully employed in elementary analysis by Dennstedt [4]. Other authors also have used boats for this purpose. Thus, for example, in 1955 Kautsky and coworkers [5] absorbed halogen in a boat filled with silver at 500° in the simultaneous semimicro determination of carbon, hydrogen, silicon and halogen. Unfortunately they give the analytical figures for only two pure halohydrocarbons and one technical halosilane.

Attempts by us to use boats of stainless steel or palladium did not meet with success at all, since a considerable erosion of the metal occurred in the presence of halogen. Good results could be obtained by working at 500-550° in a platinum boat 70-80 mm long, filled with silver (1-1.5 g) in the form of foil or gauze. However, in this case it was necessary to decrease the rate at which the oxygen was passed through from 30-50 to 15-20 ml/min and to increase the duration of the combustion to 15-20 min. It is possible that lengthening the layer of silver would permit working at faster rates. In contrast to the previously used quartz apparatus, the boat satisfactorily maintained a constant weight, quickly reached the temperature of the balance room, and could be weighed in ten minutes, i.e., before weighing the absorption apparatus for the water and carbon dioxide.

Simultaneous determination of carbon, hydrogen, and halogen where the halogen is absorbed in silver foil or gauze in a platinum boat permits determination of all three elements with an error of not more than ±0.3% (Table 1). Nevertheless, working with foil or gauze has substantial drawbacks. Actually, the silver halides, which melt at 500-550° depending upon the amount accumulated, may run out of the boat, and it is therefore necessary to wash the boat after 4-5 determinations, i.e., at the end of each work day. Furthermore, even at such a high temperature only an insignificant part of the mass of silver is utilized. Consequently, we decided to replace the foil or gauze with a more suitable form.

A detailed investigation of the absorbing capacity of various preparations was carried out [6] in 1956. The authors proposed a method for the determination of halogen by combustion of the materials in a current of oxygen and catching the halogen in electrolytically deposited silver. In this work comparative data were obtained on the absorption of halogen at various temperatures by silver gauze and electrolytically deposited silver. It appeared that 50% of the electrolytically deposited silver was utilized, but only 2% of the silver gauze. Although Korshun and Sheveleva had shown that when the halogen was absorbed by silver foil or gauze in a layer 60-70 mm long at 550-600°, i.e., above the melting point of the silver halides, the metal was somewhat better utilized (by 5-10%); even this amount is not satisfactory. Therefore we decided to change over to catching the halogen in electrolytically deposited silver,† carrying out the absorption in a boat, and at a lower temperature (410-400°). We were able to utilize 50% and more of the silver and to obtain good analytical results not only for the halogen, but also for carbon and hydrogen (Table 2). This method was also useful for the simultaneous determination of the four elements in organosilicon and organo-

† The electrolytically deposited silver, according to data from X-ray phase analysis, had a crystalline structure with a face-centered cubic lattice: a = 4.08 Å. Crystals were large, with lower limit not less than 10^{-3} cm. This silver had a surface approximately 200 times greater than foil 0.1 mm thick. Measurement of the magnitude of the specific surface [7] of the electrolytically deposited silver showed that S = 0.6 ±0.2 m^2/g, while for the foil S = 0.03 m^2/g.

boron compounds (Table 3.). It should be noted that in the work with electrolytically deposited silver, one filling (0.4-0.7 g) was sufficient for 30-50 determinations, and thus the necessity for the tedious daily operation of washing the silver was eliminated.

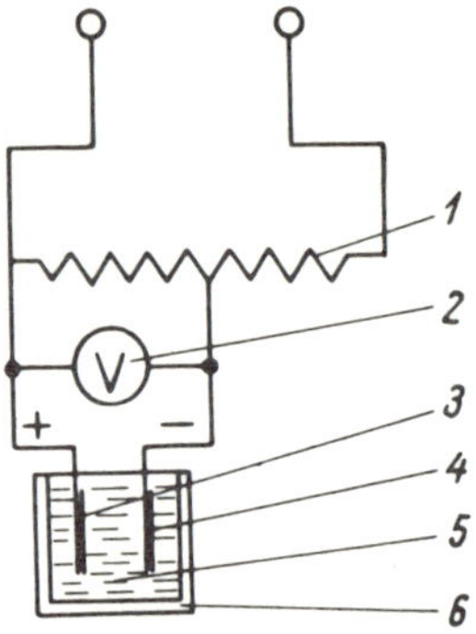

Fig. 1. Diagram of the preparation of electrolytically deposited silver; 1) rheostat; 2) voltmeter; 3) silver plate; 4) silver wire; 5) electrolyte; 0.5 N HNO_3 with 0.7 g $AgNO_3$; 6) 100-200 ml vessel.

Inasmuch as the temperature of absorption in the proposed method was lower than the melting point of the silver halides, it seemed possible to work also in a quartz boat. Corrosion of the quartz under these conditions was extremely slight. The halogen could be absorbed either after the combustion products had passed through the oxidizing zone of heating, with the boat placed in a ground-joint apparatus, or before this zone with the boat placed directly in the combustion tube. In the first case, as was shown in the work by Korshun and Sheveleva, it is necessary to have a platinum contact to accomplish quantitative oxidation of carbon monoxide, since the halogen passivates the process of oxidation. In the second case, the halogen is absorbed before the zone of oxidation and it therefore becomes unnecessary to have a platinum contact. Each of the methods has its advantages: placing the boat in the apparatus is more convenient than in the tube, but working in the tube eliminates the necessity not only for the platinum, but also for the ground-joint apparatus. Hence, the analyst himself can decide which method suits him better.

EXPERIMENTAL

For the electrolytic deposition of the silver, a silver plate was used as the anode and a silver wire served as the cathode, while the electrolyte was a solution of 0.7 g of silver nitrate in 0.5% nitric acid (Fig. 1). The silver was deposited in the form of threadlike crystals on the cathode by a current at a voltage of 1.3-2.5 V. At the conclusion of the electrolysis the silver was filtered off, washed with water, and dried in air. The boat filled with the silver that was obtained was heated for an hour in a current of oxygen at 420-440°.

The analysis was carried out in the apparatus used for the rapid method of determining carbon and hydrogen [1] (Fig. 2) or for the simultaneous determination of carbon,

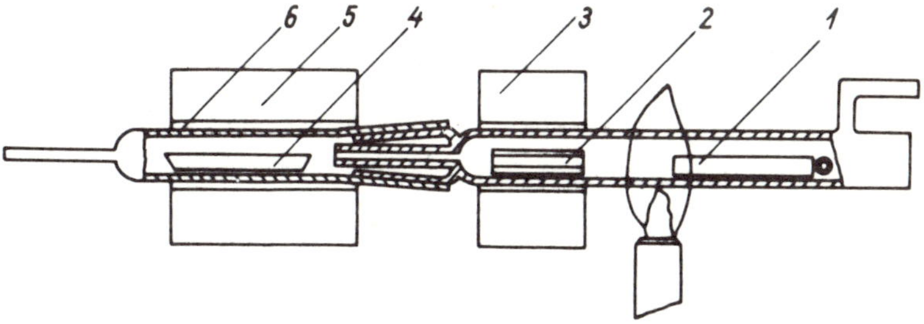

Fig. 2. Diagram of apparatus for simultaneous determination of halogens in the presence of a platinum contact: 1) boat with sample; 2) platinum contact; 3) electric furnace with temperature 950°; 4) boat with electrolytically deposited silver; 5) electric furnace with temperature 420-440°; 6) ground-joint apparatus.

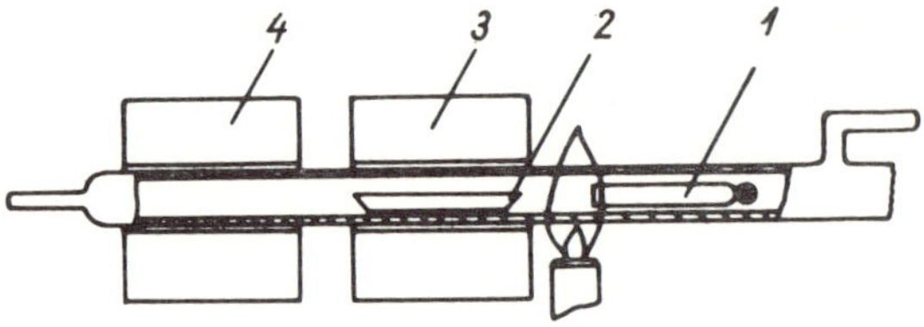

Fig. 3. Diagram of apparatus for determination of halides without platinum contact: 1) boat with sample; 2) boat with electrolytically deposited silver; 3) electric furnace with temperature 420-440°; 4) electric furnace with temperature 950°.

hydrogen, and sulfur or halogen [2, 3] (Fig. 3). Combustion of a 3-8 mg sample of compound was carried out in 15-20 min according to the rapid method with pyrolysis at an oxygen flow rate of 15-20 ml per min. A platinum or quartz boat 70-80 mm long, containing 0.4-0.7 g of electrolytically deposited silver, was placed either in the ground-joint apparatus (in this case there must be a platinum contact in the oxidation zone) or in the combustion tube ahead of a furnace at 900°, and was heated by a separate electric furnace at 410-440°, as shown in Fig. 2 and 3. Water and carbon dioxide were absorbed with Anhydrone and Ascarite, as usual. Upon completion of the combustion, the absorbers for water and carbon dioxide were disconnected, the furnace for heating the silver was moved away, and in 2 minutes the boat with the silver was taken out and placed on a metal block. The boat was weighed at the tenth minute, the water absorber at the fifteenth minute, and the carbon dioxide absorber at the twentieth minute.

In case it is necessary to work with foil or gauze instead of electrolytically deposited silver, only a platinum boat should be used and it should be heated to 500-550°. In this case, at the end of each work day, after 4-5 determinations, the boat with the spent silver is immersed overnight in an aqueous solution of ammonia. Before work is started, the silver is taken from the boat. The boat is washed with water, boiled for 5 minutes in nitric acid, again washed with water, and heated in an oxidizing flame of a gas burner. The silver is washed with water and reduced with methyl alcohol. The platinum boat can either be cast or folded from platinum foil. One filling of the boat (1-1.5 g) is sufficient for the absorption of 15 mg of chlorine.

The error in the determination by either method does not exceed ±0.3% for all three elements.

SUMMARY

1. Quantitative absorption of halogens by electrolytically deposited silver placed in a boat takes place at 410-440°. Use of such a means of absorption in the method for simultaneous determination of carbon, hydrogen, and halogen according to Korshun and Sheveleva permits determining all three elements with an error of not more than ±0.3%.

2. The proposed method is suitable for the analysis of organosilicon and organoboron compounds. The presence of nitrogen in a compound does not interfere with the determination.

LITERATURE CITED

[1] M. O. Korshun and V. A. Klimova, *Zhur. Anal. Khim.* **2**, 274 (1947).

[2] M. O. Korshun and N. S. Sheveleva, *Doklady Akad. Nauk SSSR* **60**, 63 (1948).

[3] M. O. Korshun and N. S. Sheveleva, *Zhur. Anal. Khim.* **11**, 376 (1956).

[4] D. M. Dennstedt, *Anleitung zur vereinfachten Elementaranalyse für wissenschaftliche und technische Zwecke*, Hamburg, 1906.

[5] H. Kautsky, G. Fritz, H. P. Siebel and D. Siebel, *anal. Chem.* **147**, 327 (1955).

[6] T. Mitsui and H. Sato, *Mikrochim. Acta* **1956**, 1603.

[7] A. M. Rubinshtein and V. A. Afanas'ev, *Izvest. Akad. Nauk SSSR, Otdel. Khim. Nauk* **1956**, 1294.

CHAPTER 40

Simultaneous Determination of Carbon, Hydrogen, Sulfur, and Halogens*

V. A. Klimova and G. K. Mukhina

Analysis of organic substances containing sulfur and halogen is carried out by the method of Korshun and Sheveleva [1], determining carbon, hydrogen, and the sum of halogens and sulfur. In this method, apart from the fact that sulfur and halogen are not determined separately, even the determination of the sum of the two elements is difficult. Taking advantage of the fact that cobaltic oxide, as we have found previously, at 400-500° absorbs only the oxides of sulfur and does not absorb halogens, we have developed a method for the simultaneous and separate determination of carbon, hydrogen, sulfur, and halogen. Phosphorus and silicon do not interfere with the determination of the other elements and can themselves be determined as ash. In the combustion of phosphorus-containing compounds the sample was covered with quartz [2] and in the case of silicon-containing compounds with asbestos [3].

The method here proposed consists of a pyrolytic decomposition of the substance in a rapid stream of oxygen [4]. The products of combustion are absorbed separately and determined by weighing, with the exception of sulfur, which is determined volumetrically [5]. Elements which on combustion give nonvolatile oxides are determined as ash from the increase in weight of the quartz capsule.

Izvest. Akad. Nauk SSSR, Otdel. Khim. Nauk* **1959, 2248-2250.

EXPERIMENTAL PART

The analysis was conducted in an apparatus adopted for the simultaneous determination of carbon, hydrogen, and sulfur [1] (Fig. 1).

A sample of the substance weighing 5-6 mg is placed in the quartz capsule *5*. The capsule with the sample is placed in the combustion tube *4*, which contains the platinum foil catalyst *13* and the quartz boat *12* with cobaltic oxide for absorption of oxides of sulfur. Halogen is retained at 420° by electrolytically precipitated silver [6, 7] placed in the quartz or platinum boat *11*. The boat with cobaltic oxide and the test tube with the sample are placed in the combustion tube as shown in Fig. 1. Before combustion the boat with silver is weighed and placed in the

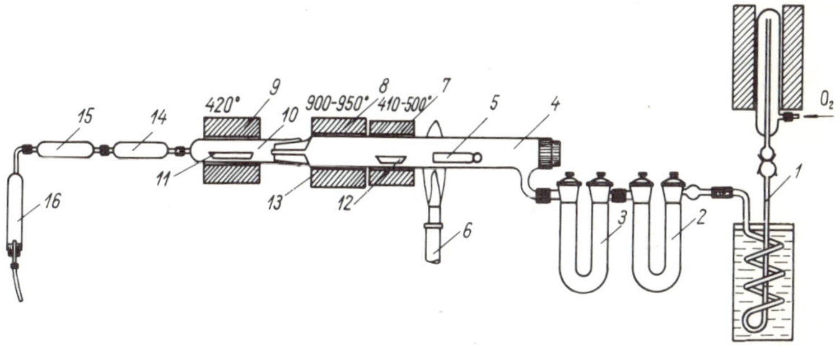

Fig. 1. Diagram of apparatus for simultaneous determination of carbon, hydrogen, sulfur, and halogen. 1) Gas-purification apparatus with electric furnace; 2) U-tube with Ascarite; 3) U-tube with anhydrone; 4) quartz combustion tube; 5) quartz test tube for sample; 6) burner; 7, 8, 9) electric furnaces; 10) quartz apparatus connected through ground joint to combustion tube; 11) quartz boat with electrolytically precipitated silver for absorption of halogen; 12) quartz boat with cobaltic oxide for absorption of oxides of sulfur; 13) platinum catalyst; 14) water absorption tube; 15) carbon dioxide absorption tube; 16) final tube with anhydrone.

ground-joint apparatus. The absorption tubes for water (14) and carbon dioxide (15) are then connected and the combustion is started. The rate of flow of oxygen during the entire analysis should be 15-20 ml per min.

Water and carbon dioxide are absorbed in the absorption tubes with Anhydrone and Ascarite in the usual manner. Oxides of sulfur are absorbed by cobaltic oxide and halogen by silver. The temperature to which the cobaltic oxide is heated for the absorption of oxides of sulfur in substances containing Cl and I is 470-500°, and 410-430° for those containing Br. At the end of the combustion, which requires 25-30 min, the furnaces heating the boats with the silver and cobaltic oxide are removed. The absorption tubes for water and carbon dioxide are then disconnected, the boats with the silver and cobaltic oxide removed, and after 15 min one weighs first the tube with the Anhydrone (H_2O), then the tube with the Ascarite (CO_2), and finally the boat with the silver halide (AgX).

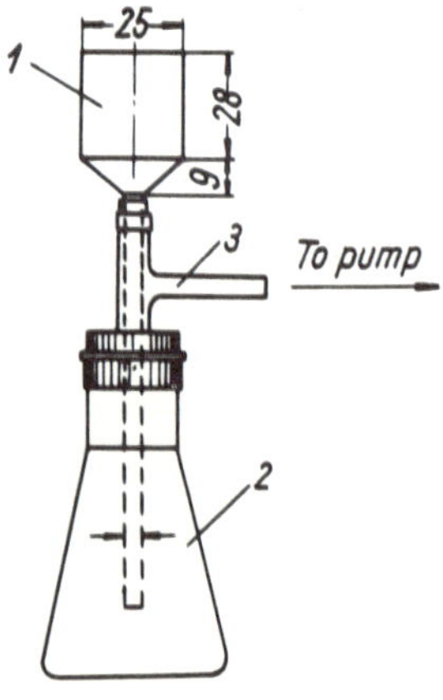

Fig. 2. Diagram of Filtration Apparatus.
1) micro funnel; 2) conical flask, 75-100 ml; 3) T-tube.
Legend: to pump.

TABLE 1

Substance	Temp. Co_2O_3, C°	% C Found	% C Calc.	% H Found	% H Calc.	% S Found	% S Calc.	% Halogen Found	% Halogen Calc.	% P Found	% P Calc.	% Si Found	% Si Calc.
2-(8-Chloronaphthyl) ethyl sulfone $C_{12}H_{11}ClO_2S$	470	56.45	56.57	4.24	4.35	12.73	12.58	13.86	13.92				
	485	56.14	56.57	4.20	4.35	12.50	12.58	14.17	13.92				
2-Iodo-5-ethylthiophene[a] C_6H_7IS	500	30.65	30.26	3.00	2.96	13.45	13'46	53.46	53.32				
	490	30.42	30.26	2.90	2.96	13.34	13.46	53.34	53.32				
2-(11-Bromoundecanoyl)thiophene[a] $C_{15}H_{23}BrOS$	430	54.85	54.37	7.04	7.00	9.98	9.68	24.35	24.12				
	420	54.84	54.37	7.01	7.00	9.83	9.68	24.52	24.12				
1-Ethylthio-3-oxychlorophosphino-4-chloro-1,3-butadiene [Translator's Note: sic] $C_6H_8Cl_3OPS$	480	27.16	27.14	2.96	3.04	12.03	12.07	40.26	40.07	11.70	11.67		
	490	27.19	27.14	2.93	3.04	11.82	12.07	40.02	40.07	11.86	11.67		
1-Propylthio-3-oxychlorophosphine-4-chloro-1,3-butadiene [Translator's Note: sic] $C_7H_{10}Cl_3OPS$	490	29.93	30.06	3.54	3.61	11.39	11.46	37.79	38.05	11.17	11.46		
	480	30.16	30.06	3.67	3.61	11.23	11.46	37.42	38.05	11.09	11.46		
Substance a[b]	480	47.06	47.01	5.42	5.41	18.18	18.05	22.70	22.50			16.58	16.50
	470	47.47	47.33	6.18	6.14	18.01	18.05	22.81	22.50			16.60	16.50

[a] Deviation from theoretical results was due to insufficient purity of the compounds analyzed.

[b] Substance "a" is a mixture of two compounds ($C_7H_{15}NSSi + C_6H_4ClNO_2$). The carbon and hydrogen percentages were calculated according to the amounts taken.

The contents of the boat with cobaltic oxide are transferred to a 50-ml beaker and 5 ml of distilled water is added. The beaker is covered with a watch glass and allowed to stand overnight. The solution is then filtered through a small glass filter directly into the titration flask (Fig. 2). The residue in the beaker is washed by decantation with hot water. The combined volume of the solution and washings should not exceed 20 ml. To the cooled solution are added 5 drops of 0.5 N NaOH, 1 ml of 25% acetic acid, and the mixture is heated for one minute to remove CO_2. The solution is then cooled, 20 ml of ethanol and 3 drops of 0.1% aqueous solution of the sodium salt of alizarinsulfonic acid are added, and the mixture is titrated with 0.02 M $Ba(NO_3)_2$ solution until the yellow color changes to pink. One ml of 0.02 M $Ba(NO_3)_2$ solution corresponds to 0.64 mg S. The 0.02 M $Ba(NO_3)_2$ solution is standardized with sodium sulfate under the same conditions.

To prepare cobaltic oxide, cobalt carbonate in the form of finely ground powder is placed in a porcelain crucible and heated in a muffle furnace at 800° until decomposition is complete. Before use the cobaltic oxide is ignited in a stream of oxygen at 410-500° for 15-20 min. The amount of the preparation used for each combustion is 0.3-0.4 g.

Examples of determinations are listed in Table 1.

SUMMARY

1. A method has been developed for the simultaneous determination of carbon, hydrogen, sulfur, and halogen.

2. The presence in the molecule of elements giving non-volatile oxides during combustion does not interfere with the determination. These elements can be determined at the same time as ash from the increase in weight of the quartz capsule.

LITERATURE CITED

[1] M. O. Korshun and N. S. Sheveleva, *Zhur. Anal. Khim.* **7**, 104 (1952).

[2] M. O. Korshun and E. A. Terent'eva, *Doklady Akad. Nauk SSSR* **104**, 707 (1955).

[3] V. A. Klimova and E. G. Bereznitskaya, *Zhur. Anal. Khim.* **12**, 424 (1957).

[4] M. O. Korshun and V. A. Klimova, *ibid.* **2**, 274 (1947).

[5] R. Geyer, *Z. anal. Chem.* **146**, 147 (1955).

[6] V. A. Klimova and E. N. Merkulova, *Izvest. Akad. Nauk SSSR, Otdel. Khim. Nauk* **1959**, 781.

[7] Tetsuo Mitsui and Hiromu Sato, *Mikrochim. Acta* **1956**, 1603.

CHAPTER 41

A New Method for the Simultaneous Determination of Silicon and Halogens in Organosilicon Compounds*

M. N. Chumachenko, M. O. Korshun, V. P. Burlaka, and V. N. Simonova

It was shown in earlier papers [1, 2, 3, 4] that the reductive method of decomposition of organic compounds, by heating with metallic K in a hermetically sealed steel microbomb, is effective and rapid. The effectiveness of this method is due to the fact that, at 850-900°, the metallic K, being in the vapor state, ensures complete reduction of all the elements in the organic molecule. We used this reductive decomposition for the determination of Si in organosilicon compounds. It is known that in the fusion of inorganic Si compounds with metallic K, the Si is reduced to the element [5], which dissolves readily even in weak alkali solutions, forming a soluble silicate.

In the present work we show that the above mentioned properties of Si can be used for the analysis of organosilicon compounds. By fusing organosilicon compounds with metallic K, and then dissolving the melt obtained in water, potassium silicate is formed, which can be determined by volumetric, colorimetric, or amperometric methods of analysis [6, 7, 8].

Another important advantage of this method is that, in the fusion of organic compounds containing Si and F, there

Doklady Akad. Nauk SSSR **133**, 138-40 (1960).

is no formation of volatile Si-F compounds, which is characteristic of oxidative decomposition methods. Because of this, the simultaneous determination of Si and F is possible.

A weighed 5-15 mg sample is placed in the steel test tube of the microbomb, together with 25-100 mg of metallic K. The bomb is hermetically sealed and heated in a furnace at 850-900° for 5 minutes. It is then cooled, the melt dissolved in water, the solution quantitatively transferred to a polythene beaker and placed for 10-15 minutes on a boiling water bath. After cooling, the solution is transferred to an alkali-resistant-glass measuring flask, the carbon filtered off, and aliquots are taken for determination of Si.

Colorimetric determination of Si. 10-25 ml of solution is placed into a 50 ml flask and made up to 35 ml with water; 1 ml $5NH_2SO_4$ and 3 ml of a 5% solution of ammonium molybdate are then added. The solution is stirred and left to stand for 10 minutes. 4 ml of 8N H_2SO_4 is then added in small portions to the flask. The mixture is diluted with water to 45 ml, and a drop of 30% $SnCl_2$ solution is added. The solution is made up to the mark with water, stirred, and the intensity of the color is measured after 5 minutes. The amount of Si (in mg) is determined graphically.

Amperometric titration. 5-25 ml of solution is placed in a cell and made up to 50 ml with 0.1N KNO_3. Measurements are carried out under a continuous stream of nitrogen. $Pb(NO_3)_2$ solution is gradually added from a burette, and galvanometer readings are taken after 1.5-2 minutes. Voltage is 1.5 V. The equivalence point is determined graphically (Fig. 1). Analytical results on pure substances, obtained by colorimetric and amperometric methods, are given in Table 1.

Acidimetric determination of Si. This is based on the reaction of soluble silicates with fluorides in the presence of a specific amount of HCl, with formation of K_2SiF_6 (6, 9, 10). The results of analyses are shown in Table 2.

TABLE 1

Compound	Calc. % Si	Colorimetric detn. % Si Found	Diff.	Amperometric titration. %Si Found	Diff.
Diethylsilanediol	23.35	23.37	+0.02	23.29	-0.06
$C_4H_{12}OSi$	23.35	23.07	-0.28	23.11	-0.24
Ethyltriphenylsilane	8.67	8.50	-0.17	8.60	-0.07
$C_{20}H_{20}Si$	8.67	8.87	+0.20	8.52	-0.15
Butyltrimethylsilane	21.73	21.50	-0.23	21.67	-0.06
$C_7H_{18}Si$	21.73	21.32	-0.41	21.68	-0.05
Tetraethylsilane	19.44	19.57	+0.13	19.33	-0.11
$C_8H_{20}Si$	19.44	19.23	-0.21	19.35	-0.09
1,4-Bis(trimethylsilyl)benzene	25.22	24.50	+0.28	25.20	-0.02
$C_{12}H_{22}Si_2$	25.22	25.01	-0.21		

TABLE 2

Compound	mg Sample	Calc.	%Si Found	Diff.
Octamethyltetracyclosiloxane	11.120	37.80	37.58	-0.22
$C_8H_{24}O_4Si_4$	8.964	37.80	37.62	-0.18
Hexamethyldiphenyltetracyclo-	18.615	26.66	26.82	+0.16
siloxane $C_{18}H_{22}O_4Si_4$	13.521	26.66	26.61	-0.05
Tetrakis(diethylphenylsiloxy)-	7.670	14.68	14.54	-0.14
titanium $C_{40}H_{60}O_4Si_4Ti$	8.722	14.68	14.71	+0.03
Hexyltrimethylsilane	14.520	17.73	17.70	-0.03
$C_9H_{22}Si$	8.870	17.73	17.62	-0.11

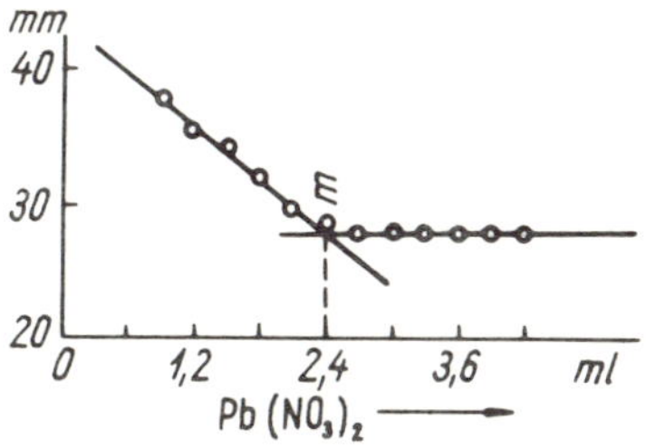

Fig. 1

Simultaneous determination of Si and halogens. A solution of the melt was made up to a determined volume and subjected to analysis in the form of aliquot samples. This permits the determination of several elements on one sample of the compound. In the present case, we determined Si and halogens simultaneously.

One aliquot sample was used for the determination of Si by any of the above methods, and another for the determination of Cl, Br, I, and F. Cl and Br were determined mercurometrically, I iodometrically [1, 2], and F thoriumetrically [3]. The experimental results are shown in Table 3.

Thus, a new method for determining Si has been developed, based on the reductive decomposition of organosilicon compounds with metallic K. It was shown that the Si contained in organic compounds, after decomposition with metallic K in a hermetically sealed bomb at 850-900° and solution of the melt in water, is quantitatively converted to the soluble silicate K_2SiO_3. Amperometric titration was used to determine Si in organic compounds. The simultaneous determination of Si and halogens, including F, was shown to be possible, with an accuracy for Si of $\pm 0.2\%$ by the amperometric method, $\pm 0.5\%$ by the colo-

TABLE 3

Simultaneous determination of Si and Halogen

Compound	%Si			%X		
	Calc.	Found	Diff.	Calc.	Found	Diff.
Ethylmethyldichlorosilane	19.58	19.60	+0.02	49.48	49.53	+0.05
$C_3H_8Cl_2Si$	19.58	19.57	-0.01	49.48	49.40	-0.08
1,5-Hexamethyl-2,3,4-trichlorophenyl-2,3,4-	20.82	20.85	+0.03	15.76	15.70	-0.06
trimethylpentasiloxane, $C_{27}H_{39}O_4Cl_3Si_5$	20.82	20.76	-0.06	15.76	15.60	-0.16
Bis(iodomethyl)tetramethyldisiloxane	13.55	13.65	+0.10	61.29	61.45	+0.14
$C_6H_{16}OI_2Si$	13.55	13.60	+0.05	61.29	61.55	+0.26
Diethylsilanediol + bromobenzoic acid	23.35	23.40	+0.05	39.76	39.70	-0.06
	23.35	23.45	+0.10	39.76	39.65	-0.11
Diethylsilanediol + Teflon	23.35	23.35	0.00	76.00	76.20	+0.20
	23.35	23.44	+0.09		75.90	-0.10

We also analyzed γ-(α,α,β-trifluoro-β-chloroethoxy)propylmethyldichlorosilane. The Si, F, and Cl percentages were determined on one sample.

Found: %Si = 9.73; 9.73 %F = 19.75; 19.99 %Cl = 36.73; 36.72
Calc.: %Si = 9.69 %F = 19.68 %Cl = 36.74

rimetric method and ±0.3% by the volumetric method, and an accuracy of ±0.3% for halogen.

LITERATURE CITED

[1] M. N. Chumachenko, Thesis, Inst. of Heteroorganic Compounds of the Academy of Sciences of the U.S.S.R. (1954).

[2] M. O. Korshun and M. N. Chumachenko, *Doklady Akad. Nauk SSSR* **99**, 769 (1954).

[3] M. O. Korshun, V. A. Klimova, and M. N. Chumachenko, *Zhur. Anal. Khim.* **10**, 358 (1955).

[4] N. P. Fedorovskaya, I. M. Khaskina, and M. N. Chumachenko, *Trudy Inst. Goryuchikh Iskopaemykh Akad. Nauk SSSR* **8**, 213 (1959).

[5] G. I. Klyukovskii and L. A. Manuilo: *The Physical Chemistry and Chemistry of Silicon*, p. 7 (Moscow, 1957).

[6] A. D. Vorob'eva, *Zavodskaya Lab.* **5**, 166 (1936).

[7] A. P. Kreshkov: *The Analysis of Organosilicon Compounds*, p. 126 (Moscow, 1953).

[8] M. T. Berkovich, *Zavodskaya Lab.* **16**, 558 (1950).

[9] I. M. Korenman: *Quantitative Microchemical Analysis*, p. 183 (Moscow, 1949).

[10] Z. Šír and R. Komers, *Chem. listy* **50**, 88 (1956).

CHAPTER 42

Microdetermination of Carbon, Hydrogen, Mercury, and Halogen from One Sample of an Organic Material*

M. O. Korshun, N. S. Sheveleva, and N. E. Gel'man

Methods for the gravimetric determination of carbon, hydrogen, and mercury based on a combination of elemental analysis with weighing of the mercury in the form of gold amalgam [1, 2] are inadequate and therefore are not used.

The absorption of mercury by gold foil or wire when organic compounds are analyzed has been employed for its quantitative microdetermination by Boëtius [3], and Korshun and Lavrovskaya [4]. We have made use of a rapid method for the microdetermination of carbon and hydrogen described by Korshun and Klimova [5-8] in conjunction with catching the mercury with gold for the simultaneous determination of these elements, employing for this purpose a previously developed technique of combustion in an insert tube [cartridge] [9].

The essence of this method consists in subjecting a weighed sample of the compound to pyrolytic combustion, retaining the mercury with gold in a quartz insert tube which is placed within the combustion tube, and determining it from the change in weight of the insert tube. The carbon and hydrogen were determined as usual [5-8]. The same portion of gold can be used for a long time, since it is easily freed of the mercury, as it accumulates, by simple heating in a current of air. In blank runs and

Zhur. Anal. Khim.* **15, 99-103 (1960).

when compounds that do not contain mercury are burned, the insert tube with the gold remains constant in weight within the limits of ± 0.010-0.020 mg (Table 1), which guarantees the determination of mercury with an accuracy of up to 0.6-0.7% absolute.

TABLE 1
Weight of Insert Tube with Gold, mg[a]

Expt. No.	Before expt.	After passage of 600 ml O_2	Change in weight
1	4.840	4.828	-0.012
2	3.478	3.492	+0.014
3	3.800	3.790	-0.010
4	1.030	1.026	-0.004
5	8.750	8.730	-0.020

[a]Reading from rider and microscale is shown without accounting for tare.

If halogen is present in the compound, it can also be determined simultaneously with C, H, and Hg. For this purpose the products of pyrolytic combustion after leaving the capsule pass first over heated silver to absorb the halogen, and then over cooled gold, which retains the mercury.

In a similar way it is also possible to determine sulfur, if the compound does not happen to contain halogen [10, 11]. The results of the simultaneous determination of C, H, and Hg, and also of C, H, Hg, and halogen or sulfur are given in Table 2.

This does not exhaust the possibilities of simultaneous determination of mercury and other elements in organic compounds. For example, it may be anticipated that various metals that are weighed in the form of the oxides in the residue, after combustion of the same sample in which carbon and hydrogen are run, can be determined simultaneously with mercury, and sometimes also with halogen.

TABLE[a] 2

Compound	%Hg		%C		%H		%X or %S	
	Found	Diff.	Found	Diff.	Found	Diff.	Found	Diff.
$C_{11}H_{12}HgO$								
2-Methyl-4-phenylmercuributyn-3-ol-2	55.89	+0.29	36.86	+0.25	3.37	+0.02		
C-36.61%; H-3.35%; Hg-55.60%	55.63	+0.06	36.65	+0.04	3.34	-0.01		
$C_{18}H_{18}Hg$								
Bis(bicyclo[2,2,1]hepten-3-ylethynyl)mercury	45.77	-0.36	49.56	-0.15	4.14	-0.03		
C-49.71%; H-4.17%; Hg-46.13%	46.65	+0.48	49.63	-0.08	4.28	+0 11		
$C_3H_3Cl_3Hg$								
α,β-Dichloroallylmercury chloride	58.02	+0.05	10.47	+0.06	1.00	+0.13	30.74	0.00
C-10.41%; H-0.87%; Cl-30.74%; Hg-57.97%	57.67	-0.30	10.54	+0.13	0.96	+0.09	30.84	+0.10
$C_{14}H_{14}HgI_4$								
Tropylium mercuritetraiodide	22.60	+0.07	18.68	-0.20	1.55	-0.03	56.97	-0.04
C-18.88%; H-1.58%; I-57.01%; Hg-22.53%	22.34	-0.19	18.80	-0.08	1.56	-0.02	57.38	+0.37
$C_{20}H_{18}Br_2HgN$								
Dimethylaminophenylmercuri-4,4-dibromodiazoaminobenzene	29.73	0.00	35.71	+0.12	2.68	-0.01	23.80	+0.12
C-35.59%; H-2.69%; Hg-29.73%; Br-23.68%	30.03	+0.33	35.58	-0.01	2.72	+0.03	23.61	-0.07

$C_{18}H_{16}HgN_2O_2$ Dimethylaminophenylmercuri-4-nitrosonaphthol-1	40.46	−0.24	44.09	+0.24	3.27	0.00		
C-43.85%; H-3.27%; Hg-40.70%	40.83	+0.13	43.97	+0.12	3.31	+0.04		
$C_{22}H_{20}HgN_2O_2$ Dimethylaminophenylmercuri-α-benzylmonoxime	36.40	−0.41	48.39	−0.09	3.84	+0.14		
C-48.48%; H-3.70%; Hg-36.81%	36.36	−0.45	48.53	+0.05	3.83	+0.13		
$C_{20}H_{18}BrHgN_5O_2$ Dimethylaminophenylmercuri-4-bromo-4-nitrodiazoaminobenzene	31.72	+0.42	37.78	+0.30	2.99	+0.16		
C-37.48%; H-2.83%; Hg-31.30%; Br-12.47%	31.52	+0.22	37.61	+0.13	2.82	−0.01		
$C_{22}H_{23}Hg_2N_3O_2S$ Bis(dimethylaminophenylmercuri) benzenesulfonamide	50.30	0.00	33.43	+0.23	3.11	−0.03	4.12	+0.09
C-33.20%; H-3.14%; Hg-50.30%; S-4.03%	50.10	−0.20	33.33	+0.13	3.03	−0.11	4.09	+0.06
$C_{14}H_6Br_8Fe_2Hg_5O_4$ Mercury derivative of cyclopentadienyl iron[b]	50.32	−0.02	8.44	0.00	0.59	+0.27	31.81	−0.29
C-8.44%; H-0.32%; Br-32.10%; Fe-5.61%; Hg-50.34%	50.14	−0.20	8.33	−0.11	0.58	+0.26	32.22	+0.12

[a]All analyses given in Table 2 were carried out with samples of 5-8 mg.

[b]In these same samples 5.46 and 5.40% Fe were found in simultaneous determination.

As an example we can cite the analysis of the mercury derivative of cyclopentadienyl iron, in which we determined, in one sample, carbon, hydrogen, mercury, iron, and halogen (Table 2).

The analysis of compounds containing nitrogen deserves particular attention. According to the observations of Boëtius [3], when compounds containing mercury and nitrogen are burned in oxygen, high results are obtained for mercury because of the reaction between the oxides of nitrogen and the mercury. Boëtius therefore recommended carrying out the combustion of nitrogen-containing compounds in a current of inert gas and reducing the oxides of nitrogen with metallic copper placed in the combustion tube. The method of Boëtius, like that of Korshun and Lavrovskaya [4], was developed on the basis of the principles of Pregl's classical microanalysis, with slow vaporization of the sample and quantitative oxidation of the vapors of the compound by an oxidizing filling in the combustion tube. It is known that under these conditions the presence of nitrogen in the compound is responsible for the formation of considerable amounts of oxides of nitrogen [12, 13], the fixation of which is absolutely necessary in order to obtain correct results for the elemental analysis. It is a different matter with the rapid pyrolytic combustion, where the output of oxides of nitrogen, even in the analysis of nitro compounds, does not exceed 12-14% of the total amount of nitrogen [7, 2]. In order to establish whether mercuric nitrate is produced under the conditions of rapid pyrolytic combustion of nitrogen-containing compounds, we carried out a series of combustions of picric acid in a combustion tube containing an insert tube with gold, in which mercury was present in the form of the amalgam and a deposit of metallic mercury on the sides of the insert tube. The change in weight of the insert tube after the experiment did not exceed the usual error in weighing (Table 3). After these tests the insert tube was repeatedly washed with water and the wash water was tested for nitrate ion with diphenylamine. No nitrate ion was detected. Thus, we came to the conclusion that under our conditions mercuric nitrate is not produced. Determinations that were made of carbon, hydrogen, mercury, and halogen in compounds that contained nitrogen in oxidized form confirmed this conclusion (Table 2).

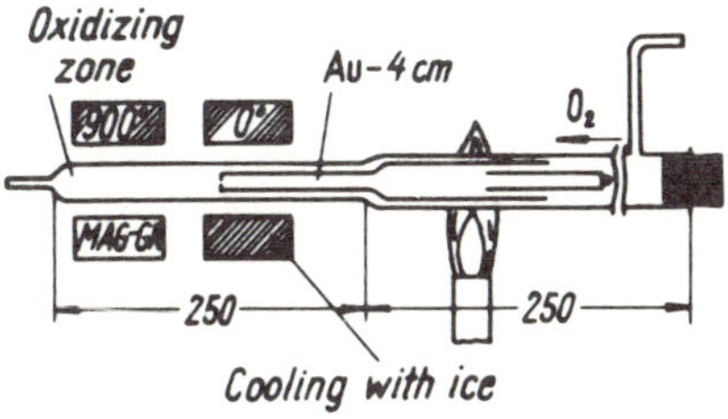

Fig. 1.

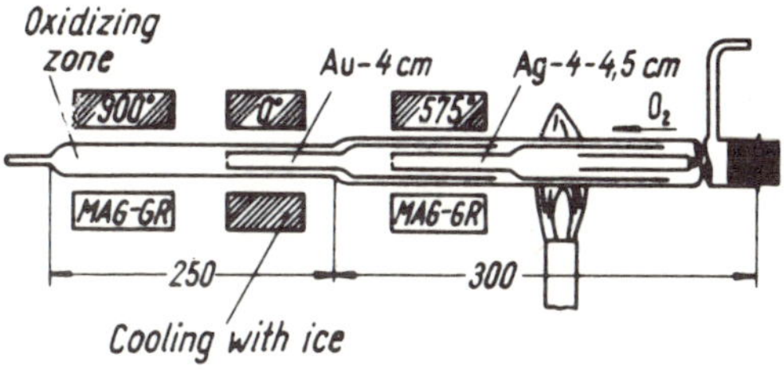

Fig. 2.

Apparatus. The same apparatus is used as for the simultaneous determination of C, H, and halogen in an insert tube [9]. The arrangement of the insert tube with the gold, the sample, and the oxidizing zone are shown in Fig. 1. If the compound contains halogen, then an insert tube with silver is also placed in the combustion tube, as shown in Fig. 2. In this case the length of the wide portion of the combustion tube should be not less than 30 cm.

TABLE 3

Weight of Insert Tube with Gold, mg

Expt. No.	Before combustion	After combustion	Diff.	Remarks
1	7.660	7.672	+0.012	5-6 mg samples of picric
2	1.991	1.972	-0.019	acid were burned. 7-11
3	1.951	1.963	+0.012	mg of mercury was in
4	1.963	1.959	-0.004	the insert tube.

Reading from rider and microscale is shown without accounting for tare.

For rapid cooling of the absorption apparatus and the insert tube, we used in our laboratory metallic blocks with channels corresponding to the dimensions of the Pregl apparatus and insert tube (Figs. 3 and 4).

Reagents. Thin gold wire or foil is used to retain the mercury. The halogen is absorbed by metallic silver in the form of gauze, foil, or wire weighing about 3 g. It is impossible to work with silvered pumice in this case, because it retains part of the mercury and the results of the analysis for mercury and halogen are disturbed. The rest of the reagents used are the same as for the determination of carbon and hydrogen [5-11].

Analytical procedure. We shall discuss only the procedures connected with the determination of mercury and the placing of the insert tube and the sample tube in the combustion tube, since the part of method [9] relating to the determination of carbon, hydrogen, and halogen is unchanged.

Before the insert tube is used, it is boiled in 1:1 HCl for 10 min, washed, dried, and heated in the combustion tube. In the narrow part of the insert tube which has been heated is placed a layer of gold 4 cm long. The gold should lie loosely, so that it does not offer any resistance to the gas flow. The insert tube with the gold is tared, using a tare flask with pieces of glass. Before work is started, a blank test is run, at which time the insert tube is placed in the combustion tube, as is shown in Fig. 1. If the insert tube already contains mercury absorbed in previous experiments, then for the blank run, as in a combustion, the layer of gold must be cooled with ice in order not to carry very small particles of the mercury beyond the insert tube in the stream of hot oxygen issuing from the heated combustion tube. When organomercury compounds are burned, the insert tube, with the sample tube (5-8 mg sample) placed in it, is moved into the tube to a distance of not less than 5 cm from the oxidizing zone,

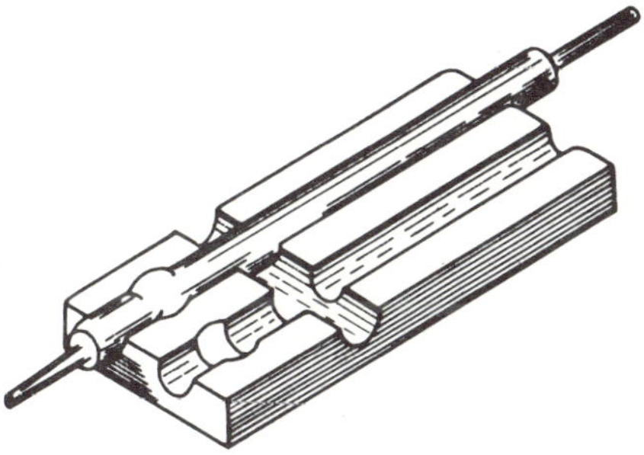

Fig. 3.

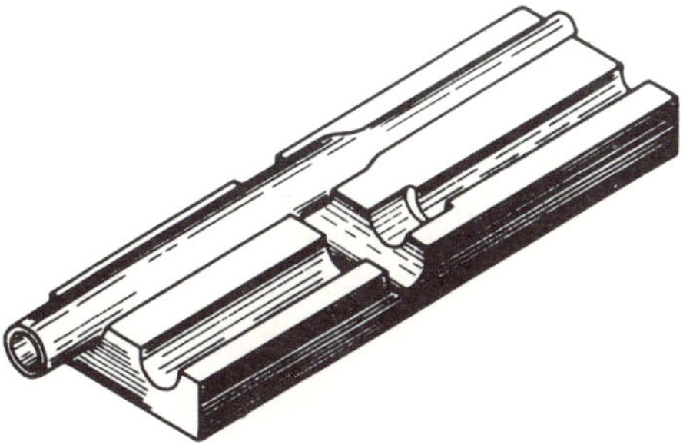

Fig. 4.

the rate of flow of the oxygen is set at 30-35 ml per min, and combustion of the sample is started with the gas burner. At the end of the combustion, the wide part of the insert tube is heated with the burner. The narrow part should not be heated, to avoid vaporization of the mercury accumulating there. It should be noted that the mercury only partially falls onto the gold. Part of it deposits on the cold sides of the narrow portion of the insert tube in the form of a very thin coating. When the combustion is ended, the burner is extinguished, cooling is discontinued, and oxygen is passed through the combustion tube for 10 minutes longer. During this time the sample for the next determination is weighed. Then the absorption apparatus is disconnected and the tube is immediately closed with a stopper. The insert tube is transferred to the weighing room, placed on a metal block standing near the weights (Fig. 4), and is weighed 25 minutes after removal from the combustion tube [9].

If halogen is present in the compound to be analyzed, then it is necessary to place in the combustion tube another quartz insert tube, filled in its narrow portion with metallic silver (see Fig. 2). If the wide part of the combustion tube is 30 cm long, the two insert tubes are freely arranged and the combustion takes place far enough from the rubber stopper. It is necessary to pay careful attention, so that the whole layer of silver will be heated to 575° and the mercury will not be retained in the colder ends of the layer as silver amalgam. The combustion is carried out as described above. At the end of the combustion, the burner is extinguished, the heater is taken away from the silver, and oxygen is passed for 10 minutes longer: for the first five minutes without discontinuing the cooling, and then for another five minutes after removing the ice. The insert tubes with the silver and the gold are weighed 25 and 30 minutes, respectively, after removing them from the combustion tube.

At the end of each working day it is recommended that the gold be regenerated. For this purpose the insert tube is placed in a pure quartz tube with a drawn-out end, the tube is connected, through a small wash bottle filled with 1:1 nitric acid, with a Mariotte flask or a water pump, and air is sucked through the tube while the gold is heated with a burner or furnace. When all the mercury has been expelled from the surface of the gold and consequently from the insert tube, the insert tube is taken out and the quartz tube is washed with nitric acid.

SUMMARY

A method has been developed for the simultaneous microdetermination of carbon, hydrogen, mercury, and halogen in organic compounds, based on pyrolytic combustion of the organic compound and gravimetric determination of the four elements. The carbon, hydrogen, and halogen are determined by a previously described method, and the mercury is retained in the combustion tube in a special insert tube filled with gold and is determined from the increase in weight of the insert tube. The deviation of the values obtained from the calculated ones for mercury does not exceed 0.6-0.7% absolute.

The possibility of determining mercury in compounds containing nitrogen has been established and it has been demonstrated that under the conditions of pyrolytic combustion mercuric nitrate is not produced.

It has been shown that when the compound contains a heavy metal that does not yield volatile salts with halogen, it can be determined simultaneously with carbon, hydrogen, mercury, and halogen.

LITERATURE CITED

[1] E. Frankland and B. F. Duppa, *Ann.* **130**, 107 (1864).

[2] A. Abelmann, *Ber.* **47**, 2935 (1914).

[3] M. Boëtius, *J. prakt.Chem.* **151**, 279 (1938).

[4] M. O. Korshun and E. V. Lavrovskaya, *Zhur. Anal. Khim.* **3**, 322 (1948).

[5] M. O. Korshun and V. A. Klimova, *ibid.* **2**, 274 (1947).

[6] M. O. Korshun and V. A. Klimova, *ibid.* **3**, 176 (1948).

[7] M. O. Korshun and V. A. Klimova, *ibid.* **4**, 292 (1949).

[8] M. O. Korshun and N. E. Gel'man, *Novye metody elementarnogo mikroanaliza.* [New Methods of Elemental Microanalysis], Goskhimizdat [State Scientific and Technical Publishing House of Chemical Literature], Moscow-Leningrad, 1949.

[9] M. O. Korshun, N. E. Gel'man and N. S. Sheveleva, *Zhur. Anal. Khim.* **13**, 695 (1958).

[10] M. O. Korshun, *ibid.* **7**, 101 (1952).

[11] M. O. Korshun and N. S. Sheveleva, *ibid.* **7**, 104 (1952)

[12] A. Friedrich, *Mikrochemie* **23**, 129 (1937).

[13] A. E. Heron, *Analyst* **72**, 142 (1947).

CHAPTER 43

Analysis of Organic Fluorocompounds—The Use of Pyrohydrolysis for the Simultaneous Microdetermination of Fluorine, Carbon, and Hydrogen*

N. E. Gel'man, M. O. Korshun, and K. I. Novozhilova

The essence of the method which we should like to suggest, and which was developed on the basis of work published by us earlier on the analysis of organic fluorocompounds [1-3], consists in determining carbon and hydrogen by pyrolytic combustion of the sample in a layer of magnesium oxide, absorption of fluorine at the moment of decomposition in the form of magnesium fluoride, hydrolytic decomposition of the latter by superheated steam at a high temperature, and titration of the fluoride ion obtained.

The capacity of halides to hydrolyze has been used several times in inorganic chemistry, for technological purposes [4-8], and thermodynamic studies [9-16]. The effect of superheated steam in an oxidizing atmosphere is also regarded as a most efficient method for the quantitative decomposition of thermostable fluorides in the elemental analysis of organic fluorocompounds, and has found wide application [17-21]. It should be pointed out, however, that the introduction of water into the combustion zone of the material restricts the analyst to the determination of carbon alone, and special methods are necessary to establish the amount of hydrogen [22-25]. Attempts have been made in organic analysis to determine fluorine directly by

Zhur. Anal. Khim. **15**, 222-26 (1960).

decomposing the organic fluorides in a stream of superheated steam and liberating the fluorine as HF [26]. Such a technique has, however, been found to be applicable only to solid, nonvolatile fluorides, and, in view of numerous limitations and low accuracy [27], has not been developed further. Recently, high temperature hydrolysis or so-called "pyrohydrolysis" has been used in inorganic analysis for the determination of fluorine in heavy metals and their fluorides [28-33].

There are no clear indications in the literature as to the possibility of quantitative pyrohydrolysis of magnesium fluoride, which in its properties occupies an intermediate position between the heavy metal fluorides, which are readily broken down by steam, and the fluorides of the alkali and alkaline earth metals, which are difficult to hydrolyze and which demand the use of special catalysts to make the process quantitative. Domange [11] did not get reproducible results during the pyrohydrolysis of MgF_2, and was of the opinion that the reason for this was the sintering of MgO particles and the difficult accessibility of the material to the steam. Warf [28], on the basis of single experiments, classes magnesium fluoride as a readily hydrolyzable material, although his results clearly show that pyrohydrolysis of MgF_2 may be incomplete, since the fluorine deficit corresponds to an enhanced magnesium content, as determined by Warf on the basis of the MgO residue. The fact that pyrohydrolysis of MgF_2 is incomplete was confirmed by experiments which we carried out under conditions differing from those described by both Domange and Warf. We subjected to pyrohydrolysis in a platinum tube, magnesium fluoride dispersed in granules of magnesium oxide, in a layer of which combustion of an organic fluoro compound had been carried out beforehand. During these experiments the same portion of magnesium oxide was used several times, and was placed in turn during the analysis in the combustion tube where the fluorine was absorbed, and in the tube used for pyrohydrolysis where it was liberated as HF. The fluorine content was about 0.25% of the weight of magnesium oxide taken. The first experiments with a new fluorine-free magnesium oxide preparation always gave low results for fluorine in the condensate after pyrohydrolysis, but, after four to five cycles of combustion and pyrohydrolytic treatment, carried out with the same preparation of MgO, we started to get correct results for fluorine (Table 1), and subsequently, the preparation could be used with good results for an indefinitely long time. Thus, it was established that it should be possible to pyrohydrolyze magnesium fluoride quantitatively in a layer of magnesium oxide after appropriate treatment of the latter. In order to establish the correctness of Domange's suggestion, we carried out a combustion of organic fluoro compound with a new portion of magnesium oxide, pyrohydrolyzed the MgO granules after using it once, and then ground it up in an agate mortar and treated it repeatedly with steam under the same conditions. The amount of fluorine which was additionally liberated from the ground magnesium oxide amounted to about 0.5% of the fluorine, i.e., approximately

TABLE 1

Preparation	Number of combustions	%F found	Loss of F	
			%	mg
Mg O, analytical grade,	1—3	37.09[a]	4.95	0.283
GOST 4526-48	4	41.83	0.21	0.012
Mg O, analytical grade,	1—4	40.68[a]	1.36	0.094
Kahlbaum	5	41.86	0.18	0.013

[a]Mean value.

one tenth of the total loss of fluorine. On pyrohydrolyzing ground-up preparations which had been used earlier in granular form, negligible amounts of fluorine were again liberated. Obviously the sintering of MgO particles on the surface of the granules bears only an insignificant share of the blame for the incomplete pyrohydrolysis of magnesium fluoride when new portions of magnesium oxide are used.

During the analysis of organic compounds containing several elements, magnesium oxide, in addition to absorbing fluorine, also absorbs other elements whose behavior during pyrohydrolysis differs. Phosphorus, silicon, and boron do not give volatile hydrolysis products. Sulfur, which is quantitatively absorbed by magnesium oxide, only partially passes into the condensate as sulfate and does not interfere with fluorine determination. Thus, during the analysis of β, β'-difluorodiethylmethylthiophosphinate, of the 1.8 mg of sulfur present in the sample, only 0.658 mg was found in the hydrolyzate (by difference between the results of acidimetric and thoriumetric [34] titration). The extent to which chlorine, bromine, and iodine are absorbed by magnesium oxide decreases with increasing atomic weight of the halogen. We used the method of determining halogen in one case [35], in order to establish what are the amounts of the various halogens that pass into the silver after combustion of several materials in a magnesium oxide layer (Table 2). In view of the obvious absorption of appreciable amounts of chlorine and bromine by magnesium oxide, we checked on the possibility of determining carbon and hydrogen in halogen derivatives without using a platinum catalyst, and obtained satisfactory results (Table 3). Obviously, the presence of small amounts of halogen in the gaseous products of pyrolytic combustion issuing from the tube containing MgO does not inhibit an

TABLE 2

Test material	X	Calc.	Found on the silver after combustion in a MgO layer	Absorbed by the magnesium oxide	Length of MgO layer cm
Hexachloroethane	Cl	89.86	3.13	86.73	8
C_2Cl_6		89.86	3.79	86.07	8
			—	89.86	14
Bromobenzoic acid	Br	39.75	22.07	17.68	8
$C_7H_5BrO_2$		39.75	17.54	22.21	8
		39.75	2.64	37.11	14
		39.75	5.53	34.22	14
Bis(iodomethyl)-					
tetramethyl	I	61.29	55.48	5.81	8
disiloxane		61.29	57.79	3.50	8
$C_6H_{16}I_2Si_2O$					

All the experiments were carried out with 4-8.5 mg samples.

oxidation of CO to CO_2 [36] in the oxidizing section of the combustion tube.

At the same time it was shown that magnesium chloride can be quantitatively pyrohydrolyzed, and determined together with fluoride by titrating an aliquot of the hydrolyzate (Table 4).

Examples of simultaneous determination of F, C, and H in various organic fluoro compounds are given in Table 3.

In the method suggested it was found expedient to combine the usual methods of elemental analysis, i.e., combustion in a quartz tube, with the possibility of a further quantitative liberation of fluorine. For this purpose it was important to make a correct evaluation of the loss of fluorine arising from its interaction with silica at a high temperature. This interaction has not been examined under the conditions used for elemental analysis of fluorides, and there is no unanimous opinion on this question. While one analyst thinks that heat-

TABLE 3

Test Material	C % Found	C % Diff.	H % Found	H % Diff.	F % Found	F % Diff.
β,β′-Difluorodiethylmethylthiophosphinate	29.43	+0.01	5.41	-0.02	18.52	-0.10
$C_5H_{11}F_2O_2PS$	29.35	-0.07	5.41	-0.02	18.45	-0.17
Calculated: C 29.42%; H 5.43%; F 18.62%						
Tris(*m*-nitrophenyl)oxonium borofluoride	46.06	-0.02	2.56	-0.02	15.74	-0.46
$C_{18}H_{12}BF_2N_3O_7$	46.19	+0.11	2.66	+0.08	15.74	-0.46
Calculated: C 46.08%; H 2.58%; F 16.20%						
Anilide of α-hydroperfluoroisobutyric acid+diethylsilanediol						
$C_{10}H_7F_6NO+C_4H_{12}SiO_2$						
1) C 41.85%; H 6.80%; F 18.48%	41.92	+0.07	6.72	-0.08	18.32	+0.16
2) C 42.12%; H 6.32%; F 21.04%	42.19	+0.07	6.35	+0.03	21.26	+0.22
1,1-Dibromo-2,3,3-trifluorochloro-3-bromopropane	9.65	-0.18	0.25	-0.02	15.42	-0.02
$C_3HBr_3ClF_3$	9.75	-0.01	0.34	+0.07	15.22	-0.22
Calculated: C 9.76%; H 0.27%; F 15.44%						
The hydrate of the ethylamide of α,γ, γ-trifluoro-	30.59	—	3.80	-0.05	23.97	-0.22
γ-chloroacetoacetic acid $C_6H_9ClF_3NO_3$	30.46	-0.13	3.73	-0.12	24.11	-0.08
Calculated: C 30.59%; H 3.85%; F 24.19%						

Note: All the analyses were carried out with 4-9 mg samples; the length of the magnesium oxide layer was 8 cm.

TABLE 4

Test Material	C %		H %		Cl %		F %	
	Found	Diff.	Found	Diff.	Found	Diff.	Found	Diff.
Monohydromonochloroperfluoroethylene C_4HClF_7	20.33	+0.02	0.29	−0.14	14.94	−0.05	64.12	−0.13
Calculated: C 20.31%; H 0.43%; Cl 14.99%; F 64.27%	20.51	+0.20	0.46	+0.03	15.26	+0.27	63.94	−0.33
α,α,α-Difluorochloro-β,β,β-dichlorofluoro-ethylene $C_2Cl_3F_3$	12.94	+0.12	0.04	+0.04	56.78	+0.03	30.35	−0.07
Calculated: C 12.82%; Cl 56.75%; F 30.42%	12.73	−0.09	0.05	+0.05	56.50	−0.25	30.30	−0.12

Note: All the analyses were carried out with 5-7 mg; the length of the magnesium oxide layer was 8 cm.

TABLE 5

Fluorine Found by Pyrohydrolytic Treatment, %

Quartz tube fragment	Broken pieces of quartz	Broken quartz etched with alkali	Magnesium oxide from the sleeve	Magnesium oxide from the capsule	Total	Calculated	Diff.
0.65*	—	—	51.67	—	52.32	52.39	-0.07
0.41**	—	—	41.40	—	41.81	42.04	-0.23
2.16**	3.50	—	36.15	—	41.81	42.04	-0.23
2.00**	5.60	—	34.14	—	41.74	42.04	-0.30
1.89*	—	11.39	38.72	—	52.40	52.39	+0.01
0.59*	—	18.46	23.40	—	42.45	42.04	+0.41
—	—	—	—	52.63	52.63	52.39	+0.24
—	—	—	—	52.51	52.51	52.39	+0.12

*Combustion of the amide of α-hydroperfluoropropionic acid.

**Combustion of the anilide of α-hydroperfluoroisobutyric acid.

ing a fluoride in a quartz tube forms not only hydrogen fluoride and silicon tetrafluoride, but entrains a certain amount of the fluorine by silica in the form of nonvolatile compounds [37-39], others have not observed this phenomenon [26, 40]. Our earlier attempts at determining C, H, and F, by absorbing the latter inside the combustion tube, led us to the conclusion that it is impossible to liberate the fluorine completely from the quartz combustion apparatus in a stream of oxygen. We demonstrated the presence of fluorine in the apparatus by means of the following experiments. In an empty combustion tube was placed a quartz sleeve filled in its narrow part with magnesium oxide granules, a 9-cm capsule containing a fluoride sample was then fitted into the sleeve, so that the edge of the capsule touched the MgO in the sleeve. When combustion was complete the capsule was broken up and the pieces of quartz subjected to pyrohydrolysis. The magnesium oxide was pyrohydrolyzed separately. 2) The experiment was repeated with a layer 8 cm long of broken quartz 1-2 mm in size in the capsule. The broken bits of the capsule, the quartz pieces, and the MgO from the sleeve were pyrohydrolyzed separately. 3) The same experiments were repeated using broken quartz etched beforehand with alkali and then washed with acid and water. 4) The test material was ignited in a tube filled with magnesium oxide; after combustion the MgO was strewn into the tube used for pyrohydrolysis, while the capsule was broken up and the fragments pyrohydrolyzed separately. As is evident from the results given in Table 5, it was established experimentally in the course of these experiments that fluorine is absorbed by silica during pyrolytic combustion of organic fluoro compounds, and that there exists a relation between the amount of fluorine absorbed and the magnitude of the quartz surface. It was established at the same time that during combustion of a fluoride in a capsule filled with magniesium oxide, that fluorine is not absorbed on the quartz walls of the capsule, and that the traces of magnesium oxide remaining on the capsule walls after strewing the MgO do not contain appreciable amounts of fluorine; consequently, strewing the MgO granules, after combustion of the organic fluorocompound, from the capsule into the tube used for pyrohydrolysis, cannot be a source of error during fluorine determination.

The observations described above have been used for developing a method for the simultaneous determination of fluorine, carbon, and hydrogen.

SUMMARY

It has been established that fluorine absorbed by magnesium oxide during combustion of fluoroorganic compounds can be quantitatively liberated by pyrohydrolysis, and can be determined by titration.

It has been established that during combustion of organic fluoro compounds in the absence of magnesium oxide, the fluorine is absorbed by silica in amounts whch are proportonal to the surface area of the quartz, and can be liberated by pyrohydrolysis.

It has been shown that it is possible to carry out a simultaneous microdetermination of carbon, hydrogen, and fluorine in various heteroorganic compounds; nitrogen, phosphorus, silicon, boron, sulfur, chlorine, and bromine do not interfere with fluorine determination.

LITERATURE CITED

[1] N. E. Gel'man and M. O. Korshun, *Doklady Akad. Nauk SSSR* **89**, 685 (1953).

[2] N. E. Gel'man, M. O. Korshun, and N. S. Sheveleva, *Zhur. Anal. Khim.* **12**, 526 (1957).

[3] M. O. Korshun, N. E. Gel'man and K. I. Glazova, *Doklady Akad. Nauk SSSR* **111**, 1255 (1956).

[4] M. J. Pelouze, *Compt. rend.* **52**, 1267 (1861).

[5] W. Moldenhauer, *Z. anorg. Chem.* 51, 369 (1906).

[6] F. K. Pershke and G. I. Chufarov, *Zhur. Khim. Prom.* **7**, 332 (1930).

[7] A. G. Bergman and A. P. Obukhov, *Trudy Gos. Inst. Priklad. Khim.* **25** (Goskhimizdat, Leningrad) p. 1934.

[8] R. K. Iler and E. J. Tauch, *Trans. Am. Inst. Chem. Eng.* 37, 853 (1941).

[9] P. L. Robinson, H. C. Smith, and H. V. A. Briscoe, *J. Chem. Soc.* **1926**, 836.

[10] E. B. R. Prideaux and E. C. Roper, *J. Chem. Soc.* **1926**, 898.

[11] L. Domange, *Ann. Chim.* **7**, 225 (1937).

[12] E. Briner and N. Gagnaux, *Helv. Chim.* Acta **31**, 556 (1948).

[13] E. Briner and P. Roth, *Helv. Chim. Acta* **31**, 556 (1948).

[14] E. Briner, *Compt. rend.* **228**, 1591 (1949).

[15] C. W. Koch, A. Broido, and B. B. Cunningham, *J. Am. Chem. Soc.* **74**, 2349 (1952).

[16] C. W. Koch and B. B. Cunningham, *J. Am. Chem. Soc.* 75, 796 (1953).

[17] W. C. Schumb and K. I. Radimer, *Anal. Chem.* **20**, 871 (1948).

[18] O. J. Milner, *Anal. Chem.* **22**, 315 (1950).

[19] R. Belcher and R. Goulden, *Mikrochemie ver Mikrochimica Acta* **36-37**, 679 (1951).

[20] H. E. Freier, B. W. Nippolt, P. B. Olson, and D. G. Weiblen, *Anal. Chem.* **27**, 146 (1955).

[21] V. P. Mashovets and E. L. Zakharov, *Zhur. Priklad. Khim.* **29**, 1512 (1956).

[22] J. F. Miller, H. Hunt, H. B. Hess, and E. T. McBee, *Anal. Chem.* **19**, 146, (1947).

[23] F. E. McKenna, H. F. Priest, and E. Staple, Natl. Nuclear Energy Series. Div. VIII, Vol. I, Analyt. Chem. Manhattan District, 248 (1950).

[24] W. H. Pearlson, T. J. Brice, and J. H. Simons, *Ind. Eng. Chem. Anal. Ed.* **18**, 330 (1946).

[25] C. J. Rodden, Natl. Nuclear Energy Series Div. VIII, I, Anal. Chem. Manhattan District 271 (1950).

[26] R. R. Rickard, F. L. Ball, and W. W. Harris, *Anal. Chem.* 23, 919 (1951).

[27] J. Simons, *Fluorine and its Compounds* [Russian translation] (IL, Moscow, 1956), p. 271.

[28] J. C. Warf, W. D. Cline, and R. D. Tevebaugh, *Anal. Chem.* **26** 343 (1954); Natl. Nuclear Energy Series, Div. VIII, I, Anal. Chem. Manhattan District, 729.

[29] C. D. Susano, J. C. White, and J. E. Lee, Jr., *Anal. Chem.* **27**, 453 (1955).

[30] P. C. Davis, Atomic Energy Commission, J. S. C-83 (1950).

[31] J. E. Lee, J. H. Edgerton, and M. T. Kelley, *Anal. Chem.* **28**, 1441 (1956).

[32] A. R. Gahler and G. Porter, *Anal. Chem.* **29**, 296 (1957).

[33] J. O. Hibbits, *Anal. Chem.* **29**, 1760 (1957).

[34] M. O. Korshun, V. N. Klimova, and M. N. Chumachenko, *Zhur. Anal. Khim.* **10**, 358 (1955).

[35] N. E. Gel'man, M. O. Korshun, and N. S. Sheveleva, *Zhur. Anal. Khim.* **13**, 695 (1958).

[36] M. O. Korshun and N. S. Sheveleva, *Zhur. Anal. Khim.* **11**, 376 (1956).

[37] D. M. Hubbard and A. L. Henne, *J. Am. Chem. Soc.* **56**, 1078 (1934).

[38] J. D. Calfee, N. Fukuhara, et al., *J. Am. Chem. Soc.* **62**, 267 (1940).

[39] L. Mázor, *Microchimica Acta* **1956**, 1757.

[40] N. V. Sokolova and K. T. Loseva, *Zhur. Anal. Khim.* **13**, 349 (1958).

CHAPTER 44

Simultaneous Microdetermination of Fluorine, Carbon, and Hydrogen*

N. E. Gel'man, M. O. Korshun, and K. I. Novozhilova

The present paper gives a detailed description of a method of simultaneous determination of fluorine, carbon, and hydrogen, the principles of which were put forth in previous communications [1-3]. The method has already been in use in this Laboratory for three years, and its reliability has been demonstrated by the analysis of several hundred organic fluorine compounds. It has been used for the analysis of solid, liquid, and gaseous substances containing in addition to carbon, hydrogen, and fluorine also oxygen, nitrogen, sulfur, chlorine, bromine, boron, silicon, and phosphorus.

Principle of the method. The substance is subjected to pyrolytic combustion in a quartz capsule in a layer of magnesium oxide. Carbon and hydrogen are determined as usual from the weights of CO_2 and H_2O, while fluorine remains in the layer of magnesium oxide as magnesium fluoride. The magnesium oxide containing the MgF_2 formed is then swept with steam at 1000°, displacing all fluorine in the form of HF, which is absorbed in a receiver with distilled water. The fluoride ion is determined by thoriumetric titration. Examples of the determination of C, H, and F in a variety of substances are listed in the Table.

Apparatus. A standard apparatus for the determination of carbon and hydrogen is used, with some modifications and additions. The combustion tube has the usual length of 550 mm, with an outside diameter of 14 mm and an inside diameter of 12 mm. The quartz capsules for the

Zhur. Anal. Khim.* **15, 342-346 (1960).

TABLE 1

Simultaneous Determination of Fluorine, Carbon, and Hydrogen

Compound	% C		% H		% F	
	Found	Diff.	Found	Diff.	Found	Diff.
Hydroquinone di-β-hydroperfluoroisobutyl ether,	32.71	-0.25	0.90	-0.29	59.18	-0.41
$C_{14}H_6F_{14}O_2$	32.68	-0.28	0.90	-0.29	59.40	-0.19
Calcd.: C 32.96; H 1.19; F 59.59%						
Perfluoroallyl phenyl ether,	48.24	+0.01	2.41	+0.16	42.22	-0.17
$C_9H_9F_5O$	47.99	-0.24	2.25	0.00	42.15	-0.24
Calcd.: C 48.23; H 2.25; F 42.39%						
α,β-Difluorobromoethylene,	9.29	+0.05	—	—	28.98	-0.27
$C_2Br_2F_4$	9.20	-0.04	—	—	29.13	-0.12
Calcd.: C 9.24; F 29.25; Br 61.51%						
Difluorochloroacetamide,	18.26	-0.28	1.42	-0.14	29.57	+0.28
$C_2H_2C1F_2NO$	18.49	-0.05	1.48	-0.08	29.53	+0.18
Calcd.: C, 18.54; H 1.56; F 29.35; C1 27.39%						

N-Ethyl-α-hydroperfluoroisobutyramide	32.35	+0.05	3.17	+0.01	51.15	+0.06
$C_6H_7F_6NO$	32.54	+0.24	3.08	−0.08	50.97	−0.12
Calcd.: C 32.30; H 3.16; F 51.09%						
α-Hydroperfluorovinylacetopiperidine,	47.86	−0.14	4.87	−0.05	33.58	−0.17
$C_9H_{11}F_4NO$	48.16	+0.16	4.89	−0.03	34.00	+0.25
Calcd.: C 48.00; H 4.92; F 33.75%						
BF_3-acetic acid complex,	32.32	−0.16	5.58	+0.13	15.71	−0.09
$C_2H_4O_2.20.5BF_3$	32.35	−0.13	5.60	+0.15	15.53	−0.27
Calcd.: C 32.48; H 5.45; F 15.80%						
Bis(β-fluoroethyl) methylphosphonate	31.75	−0.17	5.87	−0.02	20.04	−0.16
$C_5H_{11}F_2O_3P$	32.03	+0.11	5.93	+0.04	20.18	−0.02
Calcd.: C 31.92; H 5.89; F 20.20%						
O,O-Diethyl S-β-fluoroethyl thiophosphate,	33.39	+0.04	6.60	+0.07	9.36	+0.57
$C_6H_{14}FO_3PS$	33.59	+0.24	6.56	+0.03	9.06	+0.27
Calc.: C 33.35; H 6.53; F 8.79%						

Note: All analyses were run with 4-9 mg samples.

samples are 90 mm long, with 8-8.5 mm outside and 6-6.5 inside diameter. Thick-walled quartz capillaries (Fig. 1), which are used repeatedly, are employed with liquid samples. The use of glass capillaries is not recommended since during combustion the magnesium oxide becomes contaminated with glass particles, which may interfere with the subsequent pyrohydrolysis and damage the apparatus. The manipulation of the capsule filled with the heated magnesium oxide is conveniently done with a pair of tweezers with soldered-on curved tips which securely grip the test tube. The apparatus requires two MASN-11 furnaces for heating the gas-purifying tube and the silver layer† in the combustion tube when substances are analyzed which contain other halogens in addition to fluorine, one hinged furnace type MA-14, and one MAG-6R electric burner or three electric heaters type MAG-6R for the combustion of the sample and ignition of magnesium oxide. A general view of the combustion tube is shown in Fig. 2.

The pyrohydrolysis apparatus (Fig. 3) consists of a steam generator, a steam superheater, a pyrohydrolysis tube, a receiver, and electric heaters. The steam generator is a wide-neck flask of 0.5-1 liter capacity. The flask is heated on an electric hot plate or a flask heater [heating mantle?] powered through an autotransformer. The steam superheater is made of clear fused quartz. The exit tube has a male ground joint for connection to the pyrohydrolysis tube. The steam superheater is heated by an MA-02-20 furnace to 1000-1100°. To reduce heat losses at the exit of the steam, the exit tube is wrapped with asbestos cord. The pyrohydrolysis tube should be made of a material that is resistant to prolonged action of HF mixed with steam at 1000-1100°. In addition to platinum, which satisfies these requirements best, it is also possible to use soft nickel tubing N-3 according to GOST [State All-Union Standard] 849‡; this can be

† The silver may also be heated with an MAG-6R electric heater.
‡ Tubing made by the Revdin plant was used in the present work.

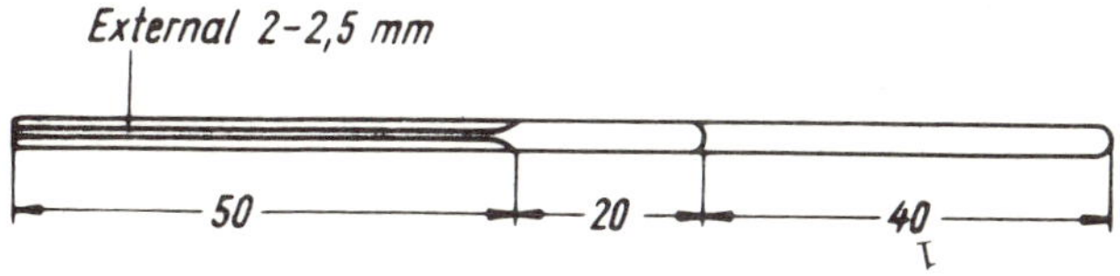

Fig. 1. Quartz capillary. Legend: Outside diameter 2-2.5 mm.

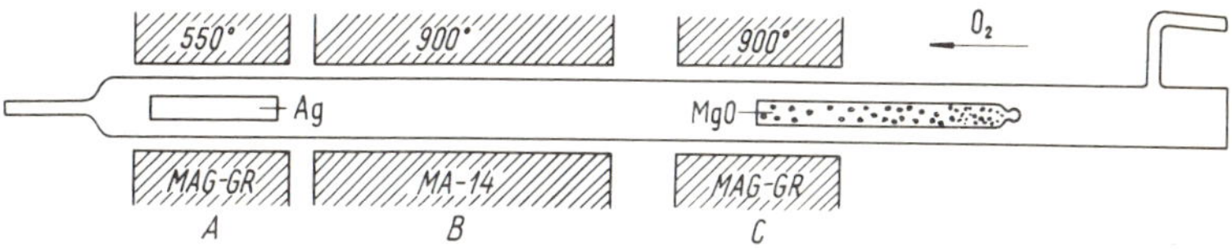

Fig. 2. Combustion tube. Legend: a) MAG-6R; b) MA-14; c) MAG-6R.

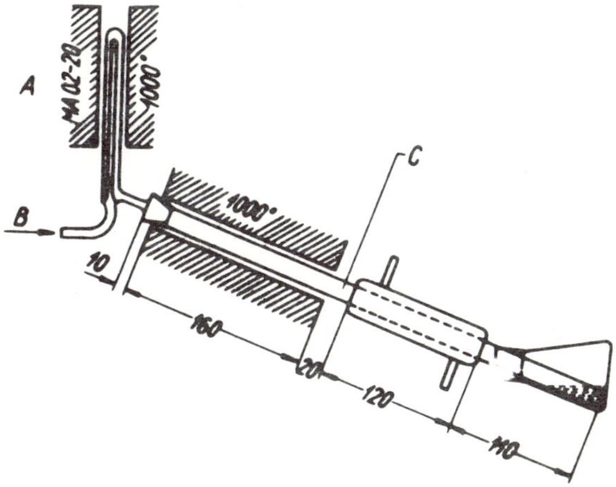

Fig. 3. Pyrohydrolysis apparatus. Legend: a) MA-02-20; b) Steam from steam generator; c) For platinum tube: outside diameter 11 mm, inside diameter 10 mm. For nickel tube: outside diameter 13 mm, inside diameter 11 mm.

used for more than one hundred determinations and thanks to its low cost and availability can compete with platinum. The tube is 400-425 mm long, with an inside diameter of 10-11 mm, and a wall thickness of 1 mm with nickel and 0.5 mm with platinum. The portion of the tube which is connected to the steam superheater carries a ground joint (it was found in practice that the ground joint can be dispensed with if the male joint fits the tube sufficiently firmly). Next to the ground joint, 1 cm from the end of the tube, the tube is provided with a winding of heat-resistant alloy IE-595 (wire thickness 0.5 mm, approximate length 7 m, 100-110 turns). It is also possible to use a removable electric furnace of type MA-02-20 at 1150° (in which case the pyrohydrolysis tube should be 4 cm longer), but a winding is far more convenient. At a distance of 2 cm from the furnace, a brass condenser 12-14 cm long is soldered onto the nickel tube; when a platinum tube is used, a quartz or brass jacket is installed with the aid of pieces of rubber tubing. The cold end of the tube at the exit of the condenser is 11-12 cm long. In order to hold the layer of magnesium oxide in the hottest portion of the tube during pyrohydrolysis, a partition of close-woven platinum gauze, double-folded, is pushed into the tube with a rod and positioned so that the magnesium oxide would lie in the center of the heated zone during the pyrohydrolysis. The receiver is a 100-ml wide-mouthed Erlenmeyer flask.

The titration of the fluoride ion is carried out with a 10-ml microburette having 0.02-ml divisions.

Preparation of magnesium oxide. Magnesium oxide powder is mixed to a paste, placed in a Petri dish, and dried at 110-120° until the surface becomes cracked; the mass is then crushed in a porcelain mortar and passed through a sieve. Material retained by sieves No. 1 and 2 is used. A portion of the screened material is placed in a column 17-18 cm long in an old quartz combustion tube and heated at 900° as in the determination of C and H in fluorides [4], connecting the tube to a working apparatus for determining C and H or to a special apparatus. The drying from the moment of turning on the furnaces is 1.5 hours. This treatment of the magnesium oxide results in a decrease of its volume by a factor of 1.5-2. The magnesium oxide is then transferred to a bottle closed with a rubber stopper, in which it is stored until the final conditioning in the pyrohydrolysis tube. Directly before use for an analysis, the required amount of MgO, measured with a quartz sample capsule, is poured into the pyrohydrolysis tube, previously heated to 1000-1100°, steam is passed through for 5-10 min, another half of the test tube of MgO is added, and the pyrohydrolysis is continued for a further 30 minutes. There are produced dense, dust-free granules which are transferred to the quartz test tube, placed in the combustion tube, and ignited at 900° for 20 min in the MA-14 furnace at an oxygen flow rate of 30 ml per min. The same procedure is used to prepare a second portion of magnesium oxide, and in carrying out analyses the two portions are used alternately. To obtain correct results for fluorine with a new magnesium oxide preparation, it is necessary to carry out four or five combustions of a pure fluoride, pyrohydrolyzing the preparation after each combustion [3] (correct values of C and H are already obtained in the first combustion) ; a determination of F, C, and H in these cases is not necessary.

Preparation of buffer solution [5]. One hundred milliliters of exactly 1 N solution of monochloroacetic acid is prepared, and neutralized to phenolphthalein with 1 N alkali. Another 100 ml of the monochloroacetic acid solution is added, and the volume is brought to 500 ml. The resulting solution should have a pH of 2.9-3.0 (checked with a potentiometer).

Preparation of thorium nitrate solution. To prepare a 0.025 N solution of thorium nitrate, 3.45 g of dry $Th(NO_3)_4$.

$4H_2O$ is dissolved in one liter of water. The solution is standardized by weighing an ignited dry residue.

When the pyrohydrolysis is conducted in a nickel rather than a platinum tube, a certain amount of fluorine may be lost in the apparatus. Our observations have shown this loss to be proportional to the amount of fluorine in the sample. In this case it is thus recommended to standardize the solution according to the results of combustion of two or three samples of the pure substance.

Analysis. The determination of C, H, and F consists of two operations: the combustion of the sample and the pyrohydrolysis of magnesium fluoride. For the sake of simplicity, we shall describe these operations consecutively although in actual fact the analyst performs simultaneously the combustion of a sample and the pyrohydrolysis of magnesium oxide from the preceding combustion.

Determination of carbon and hydrogen. A blank experiment is carried out at the beginning of the working day. At the time of weighing the absorption apparatus, a 9-cm quartz test tube with the last-pyrohydrolyzed MgO is placed in the combustion tube, positioning it in the part of the tube that is heated by the MA-14 furnace to 900°. The test tube with MgO is ignited for 20 min. The weighed absorption tubes are connected to the combustion tube, and with the aspirator stopcock closed, the test tube with MgO is withdrawn from the furnace into the cool part of the tube for 10 min. The weighed sample of the substance (4-8 mg) is now prepared in a clean quartz capsule, covered with the magnesium oxide that was allowed to cool in the tube, the capsule is placed in the combustion tube, the aspirator stopcock is opened, and the combustion is started. High-boiling nonvolatile liquids are placed in a thick-walled quartz capillary (see Fig. 1), which is pushed into the capsule with the open end towards the bottom, and covered with magnesium oxide as with a solid sample. The combustion of the sample is carried out with the aid of an electric heater heated to 900° or with a gas burner at an oxygen flow rate of 30-35 ml per min. At the beginning of the combustion the electric burner should cover 3-4 cm of the capsule from the open end of the latter. The electric heater and behind it the MA-14 furnace are moved toward the closed end of the capsule, observing separation of carbon on the magnesium oxide, which can readily be used as a guide in controlling the rate of combustion. After 10-15 min the combustion is concluded, and the capsule with magnesium oxide is held in the heated zone for 5-7 min to decompose the magnesium carbonate formed. The electric heater is then completely removed from the tube, the long furnace is moved away, and the absorption tubes are detached from the apparatus and weighed as usual. The capsule with the magnesium oxide in which fluorine was absorbed is taken out of the combustion tube and transferred to the apparatus for pyrohydrolytic decomposition. The other portion of magnesium oxide that has just been pyrohydrolyzed is poured into a quartz capsule and inserted in the combustion tube as described above. It must be kept in mind during all the operations with magnesium oxide that it should be exposed to air only during the time required for pouring it from one capsule into the other, i.e., for a few seconds. The speed with which the magnesium oxide is manipulated determines to a considerable degree the magnitude and reproducibility of the corrections introduced to compensate for the absorption of CO_2 and H_2O from air by the oxide. Magnesium oxide which is to be used in the combustion or for the ignition before combustion must not be left in air.

The performance of the apparatus is confirmed by combustion of a pure fluorine-containing substance, and appropriate corrections for the weights of CO_2 and H_2O are introduced. The magnitude of these corrections depends mainly on the temperature conditions during the conditioning of the magnesium oxide, and also on the time during which the conditioned oxide was exposed to air in transferring it from one capsule into the other; it is 175-200μg for water and 100-150μg for carbon dioxide. Despite the magnitude of these corrections, they are quite reproducible and show sufficient stability to give reliable results in the determination of carbon and hydrogen.

When the compound contains other halogens in addition to fluorine, the end of the combustion tube is packed over a length of 5-8 cm with metallic silver or silver precipitated on granulated pumice, heated to 575 or 425° respectively, [6] with an electric heater or an MA-SN-11 electric furnace. When the substance contains chlorine or bromine, the platinum catalyst in the oxidation zone may be dispensed with [3].

When analyzing compounds containing sulfur, phosphorus, silicon, or boron, the capsule with magnesium oxide is placed in an empty combustion tube.

Determination of fluorine. At the start of work, the electric heating of the pyrohydrolysis tube, the superheater furnace, and the flask heater are turned on. When the water in the steam generator starts to boil and the necessary steam supply rate (0.5 ml per min as condensate) has been reached, the steam superheater is connected to the pyrohydrolysis tube heated to 1000-1100°, which contains the magnesium oxide pyrohydrolyzed on the preceding day, and the latter is swept with steam for 15 min. The magnesium oxide is then poured into a quartz capsule and the latter is placed in the combustion tube (see above), while the now empty pyrohydrolysis tube is charged with magnesium oxide which remained unpyrohydrolyzed after a combustion carried out on the preceding day; the end of the pyrohydrolysis tube is dipped in water (10-15 ml) in a 100-ml Erlenmeyer flask and the tube is connected to the steam superheater. After 20 min the flask is lowered so that the end of the tube clears the surface of the liquid, the tube is rinsed externally with a wash bottle, and after another 10 min the receiver is removed. Passing steam through the pyrohydrolysis tube is continued until the beginning of the next determination. The hydrolyzate is transferred to a 100-ml volumetric flask, filled to the mark with distilled water, and three 25-ml aliquots are measured out into 100-ml Erlenmeyer flasks. To each aliquot is added 0.4 ml of a 0.1% solution of sodium alizarinsulfonate, and the solution is neutralized with 0.02 N NaOH to violet color and titrated with thorium nitrate [5].

The lower concentration limit that can be determined is about 1 mg of fluorine in the aliquot of the solution, which corresponds to a consumption of about 2 ml of the thorium nitrate solution. At higher concentrations of the fluoride ion the change of the solution to pink occurs before the equivalence point and the results are too low. It is therefore recommended, in analyzing fluorine-containing substances with a high fluorine content, to dilute the hydrolyzate to 250 ml.

One ml of 0.025 N thorium nitrate corresponds to 0.475 mg F.

In a six-hour working day an analyst can carry out four determinations.

SUMMARY

A method is described for the simultaneous microdetermination of fluorine, carbon, and hydrogen in solid and liquid organic substances containing various elements.

LITERATURE CITED

[1] N. E. Gel'man and M. O. Korshun, *Doklady Akad. Nauk SSSR* **89**, 685 (1953).

[2] M. O. Korshun, N. E. Gel'man and K. I. Glazova, *ibid.* **111**, 1225 (1956).

[3] N. E. Gel'man, M. O. Korshun and K. I. Novozhilova, *Zhur. Anal. Khim.* **15**, 222 (1960).

[4] N. E. Gel'man, M. O. Korshun and N. S. Sheveleva, *ibid.* **12**, 526 (1957).

[5] M. O. Korshun, V. A. Klimova and M. N. Chumachenko, *ibid.* **10**, 358 (1955).

]6] M. O. Korshun, N. E. Gel'man and N. S. Sheveleva, *ibid.* **13**, 695 (1958).

CHAPTER 45

Conductometric Microdetermination of Carbon and Hydrogen in Organic Compounds*

N. E. Gel'man and Ven-Yun Van

The so-called "manual" methods are predominant to a considerable extent in elemental organic analysis. When using these methods the results of the determinations depend to a certain extent on the analyst's individual characteristics. We undertook the problem of developing an objective instrumental completion of the carbon and hydrogen determination. A conductometric method was chosen to serve as a means of measuring the amounts of final combustion products—carbon dioxide and water. The existing modifications of the conductometric determination of small amounts of carbon [1-7] were not found to be sufficiently accurate for organic analysis, in which it is necessary to determine the high carbon content of the sample with a relative error of 0.2-1%. The necessary accuracy was obtained by constructing a cell for electrical conductance measurements according to the specific requirements of elemental microanalysis. As a basis for conductometric hydrogen determination, the reaction of water with carbon at elevated temperatures in an inert gas atmosphere, which proceeds according to the following scheme, $C + H_2O = CO + H_2$, was considered. This reaction has been used for the direct oxygen determination [8-11], for iodometric hydrogen determination in organic compounds [12], and also for determining small amounts of water in gases [13]. We used it for conductometric hydro-

Zhur. Anal. Khim. **15**, 487-494 (1960).

gen determination in organic substances together with carbon determination on a single sample [14].

The determination is performed in two stages. First, a weighed portion of the organic substance is subjected to pyrolytic combustion in a clean tube in a rapidly flowing stream of oxygen [15]. The resulting water is frozen with a dry ice and acetone mixture, and the carbon dioxide is introduced into the cell containing dilute alkali for electrical conductance measurements. Afterwards, the condensed water is evaporated in an argon or nitrogen stream onto a catalyst layer of incandescent platinized carbon black [11], where the oxygen of the water reacts with carbon to form CO. Next the CO is oxidized over heated cupric oxide [11] to CO_2, and the latter is also introduced into the cell for electrical conductance measurements. Thus both carbon and hydrogen are determined by the change in electrical conductance of the absorbing solution.

Conductometric carbon and hydrogen determination is advantageously distinguished from the common gravimetric determination by better reproducibility of the results, which is attributable to the greater objectivity of measurements and elimination of the effect on the accuracy of the results of such causes as, for example, changes in atmospheric conditions, electric charges on the surface of the equipment that is weighed, and errors in large and small weights. Furthermore, this method of measurement permits us to follow the combustion process of the sample with time, which undoubtedly is of interest. Finally, the electrical conductance measurement can most certainly be made automatic.

Simultaneously with developing the instrumental execution of the analysis, it is proposed in this paper to burn the sample without the analyst's attention. The procedure for carbon and hydrogen determination consists of three steps: weighing the sample, combustion of the substance, and measurement of the final combustion products. The last two steps are the most difficult ones. Attempts at automation of the elemental microanalysis were made repeatedly, but all of them were concerned only with the second step, namely —with the combustion of the sample. For this purpose several types of furnaces were proposed, which moved at a given speed and stopped or shut off with increase in pressure within the tube[16-19]. The most expedient way of automating this step is the establishment of optimum combustion conditions at high temperatures and with a large excess of oxygen, when it is not necessary to regulate the combustion process. As an example, the combustion of organic halogen derivatives by Wickbold [20] in an oxygen-hydrogen flame may be cited, where both gases are fed directly from the cylinders at a high velocity, or the combustion of metals by Malissa [7] at a temperature of 1000° with an oxygen feed rate of hundreds of milliliters per minute. However, even with considerably smaller oxygen rates and at not as high temperatures we succeeded in effecting a rapid quantitative combustion. By increasing the ratio of oxygen to the sample five-fold by decreasing the sample to 1 mg, it was possible to carry out the combustion of the sample in 3-5 minutes without the analyst's supervision.

The carbon and hydrogen content of 65 organic compounds of different compositions and characteristics [30] was determined by the proposed method. Examples are given in Table 1.

Apparatus. The apparatus for C and H determination consists of the combustion unit, equipment for conversion of water and instruments for electrical conductance measurements.

Equipment for combustion (Fig. 1). The usual equipment for rapid C and H determination is used [15, 21], but instead of the absorbing apparatus and the Mariotte flask, a trap for freezing the water and a cell for CO_2 determination by changes in the electrical conductance are attached to the combustion tube.

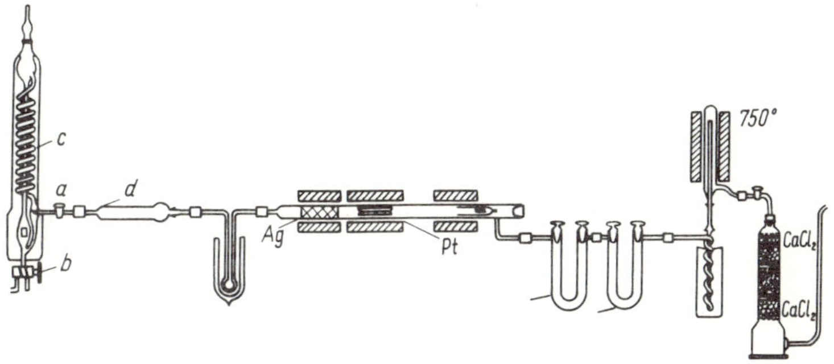

Fig. 1. Apparatus for combustion and carbon determination. Legend (from left to right): Anhydrone; Ascarite.

The trap consists of a heavy-walled, U-shaped, quartz capillary tube with a capillary diameter of 1 mm, the length of each arm of the U-part being 70 mm and the length of each side-tube being 50 mm.

Equipment for conversion of water (Fig. 2) consists of a system for removing traces of oxygen from the inert gas (argon or nitrogen) (see [11]), a trap with frozen-out

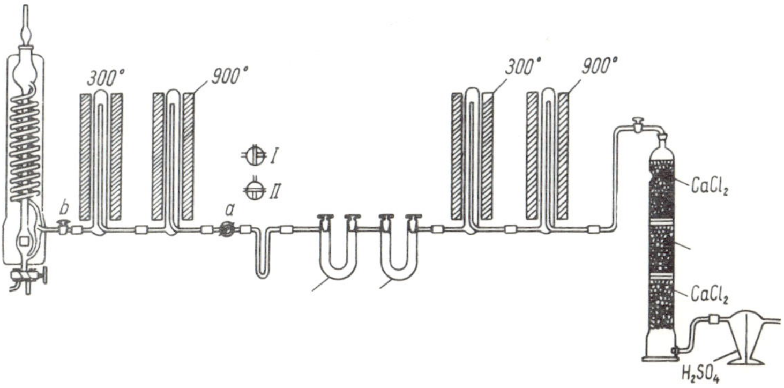

Fig. 2. Apparatus for water conversion. Legend (from left to right): CuO; Pt-black; Anhydrone; Ascarite; CuO; Pt-black; $CaCl_2$; Ascarite; $CaCl_2$; conc. H_2SO_4; Ascarite.

TABLE 1
Microdetermination of Carbon and Hydrogen

Substance	mg Sample	C % Found	C % Diff.	H % Found	H % Diff.
CH_3COCH_3	1.470	61.96	−0.07	10.47	+0.05
62.03% C; 10.42% H	0.906	61.80	−0.23	10.56	+0.14
$C_2H_5OC_2H_5$	1.853	65.03	+0.17	13.69	+0.18
64.86% C; 13.51% H	0.716	64.86	0.00	13.56	+0.05
C_6H_5COOH*	1.229	69.02	+0.16	5.16	+0.18
68.86% C; 4.98% H	1.020	68.82	−0.04	5.04	+0.06
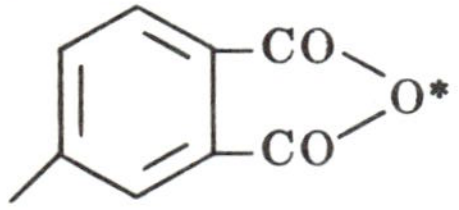 $(CH_3)_2C{:}CH—(CH_2)_2$	1.052	72.05	+0.20	7.74	+0.04
71.85% C; 7.70% H	1.172	71.76	−0.09	7.66	−0.04
$C_{10}H_8$	0.988	93.56	−0.08	6.47	+0.15
93.68% C; 6.32% H	0.710	93.38	−0.30	6.60	+0.28
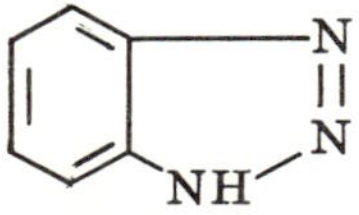	1.006	60.77	+0.27	4.33	+0.10

60.50% C; 4.23% H	0.934	60.47	−0.03	4.24	+0.01
$C_6H_2(OH)(NO_2)_3$	2.257	31.50	+0.05	1.35	+0.03
31.45% C; 1.32% H	1.652	31.67	+0.22	1.30	−0.02
$CCl_3CH_2CHBrCH_2OC(:O)CCl_3$	2.248	17.90	−0.04	1.19	−0.07
17.94%; C1.26% H	2.974	18.05	+0.11	1.16	−0.10
$CCl_2{=}CClCH_2SO_2C_6H_5$	1.112	37.82	−0.02	2.46	+0.01
37.84% C; 2.45% H	1.448	37.98	+0.14	2.63	+0.17
$(CF_2CF_2CF_2CF_2CONH_2)_2$	1.738	24.35	−0.24	0.77	−0.05
24.59% C; 0.82% H	1.689	24.44	−0.15	0.77	−0.05
$(CH_3)_3SiCH\ (CH_3)\ CN$	1.755	56.59	−0.04	10.08	−0.22
58.63% C; 10.30% H	1.487	56.82	+0.19	10.21	−0.09
$(HO)_2$ B 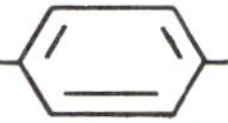 B $(OH)_2$	1.077	43.34	−0.17	5.02	+0.15
43.51% C; 4.87% H	0.911	43.47	−0.04	5.12	+0.25
$(CH_3)_2NC_6H_4HgNHC_6H_4NO$*	1.980	38.33	+0.23	3.51	+0.11
38.10% C; 3.40% H	1.743	38.32	+0.22	3.44	+0.04
$(\eta C_3H_7)_2SnO$*	1.320	32.80	+0.19	6.44	+0.06
32.61% C; 6.38% H	1.374	32.68	+0.07	6.36	−0.02

*Combustion carried out with an automatically moving furnace.

water, a Dewar flask, a three-way stopcock, a catalyst section with platinized carbon black heated to 900° and an oxidizing section with cupric oxide to which the cell is directly attached. In charging the catalyst unit with platinized carbon black, it is necessary to place at the bottom a sufficient amount of crushed quartz so that the platinized carbon black is heated to the required temperature throughout its entire length.

The substances to be analyzed are weighed on a Bunge microbalance.

The apparatus for electrical conductance measurements was assembled along the lines of a Wheatstone bridge and consists of an oscillator (ZG-10) as a source of alternating current, a direct current bridge (MTV), three component resistances forming the three arms of the Wheatstone bridge, a capacitor (ME-3) for compensation and self-induction, an electronic oscillograph (EO-7) as a null instrument and two working cells for the electrical conductance measurements.

Cell for the electrical conductance measurements. By burning 1 mg of the organic substance, 3.5 mg of carbon dioxide is obtained, which enters into the cell together with a stream of oxygen which is passed through the apparatus at the rate of 30 ml per min, so that a large part of the CO_2 is introduced into the cell in the course of 2-3 minutes. It was necessary, under these conditions, to attain complete CO_2 absorption and equalization of concentration of the absorbing solution in the shortest possible time. Cells, constructed as described earlier [3, 6, 22] and intended for determining small amounts of carbon, were found to be unsuitable for measuring such large amounts of CO_2. The cell, as modified by the authors and meeting the established requirements, is described below (Fig. 3). The gas containing CO_2 enters the cell through the side stopcock *a* and the capillary tip *b* and bubbles along the 2-meter long coil *c* through the absorbing solution, producing simultaneously a circulation of liquid throughout the whole cell. The carbon dioxide is absorbed

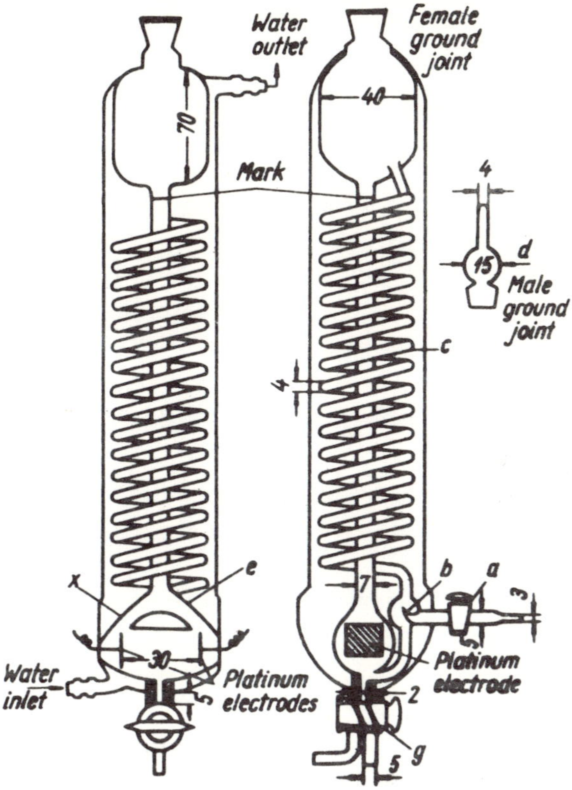

Fig. 3. Cell for measuring the electrical conductance.
Legend: mark.

by the alkali, and the carrier gas escapes to the atmosphere through the protecting tube *d* containing Ascarite. The electrode chamber has the form of a horizontally placed cylinder. As a result of the inlet tube bifurcation, the solution reaches the electrodes directly through the tubes *e* and *f*, the dead space between the electrode and the chamber [cell] wall is eliminated and the thorough mixing of the solution is considerably accelerated. Two platinum foils, 13 x 13 mm in size, serve as the electrodes. The distance between them is 30 mm. Smooth surface and the vertical positioning of the electrodes make it possible to reduce to a minimum the errors which arise as a result of the settling of the heavy $BaCO_3$ particles (when $Ba(OH)_2$ solution is used as the absorbent) on the platinum surface. The cell is filled (up to the mark) and the solution is changed by means of the lower three-way stopcock *g*. The cell has a volume of 55 ml. It has a water jacket and is thermostated at 25 ±0.1°. Experiments conducted with two cells connected in series indicated that with a gas rate of 30 ml per min the CO_2 is absorbed completely. By increasing the velocity 1.5-fold, that is, to 45 ml per min, a small amount of CO_2 is detected in the second cell (Table 2). A similarly constructed cell is also used for the determination of hydrogen in organic substances [14].

Reagents. 1) Cupric oxide wire. Carbon black, granulated, containing 50% platinum, prepared as described by Bondarevskaya [23]. The residue after evaporation is ignited without additional drying by gradually raising the furnace temperature.

2) Anhydrone. 3) Ascarite. 4) Nitrogen or argon of the highest purity.

5) 0.01 N NaOH or $Ba(OH)_2$ solution, prepared by dissolving, respectively, 2 g of NaOH or 8 g of $Ba(OH)_2 \cdot 8H_2O$ (analytical grade) in 5 liters of thoroughly boiled water which does not contain any CO_2. After standing for 24 hours the clear portion of the solution is decanted off.

Analytical procedure. Sample weighing. The analysis is performed with about 1 mg sample, that is, one fifth of that used in microanalysis. In order to eliminate errors connected with the weighing of relatively large quartz capsules in which combustion takes place, the sample is weighed in a small platinum boat, which is placed, prior to combustion, in the empty quartz capsule always placed inside the combustion tube. This method of introducing the sample decreases the magnitude of the blank run, as a minimal amount

TABLE 2
Verification of Completeness of CO_2 Absorption in the Cell

Substance	mg Sample	Gas flow rate ml/min	%C found (relative) In the 1st cell	In the 2nd cell	Total
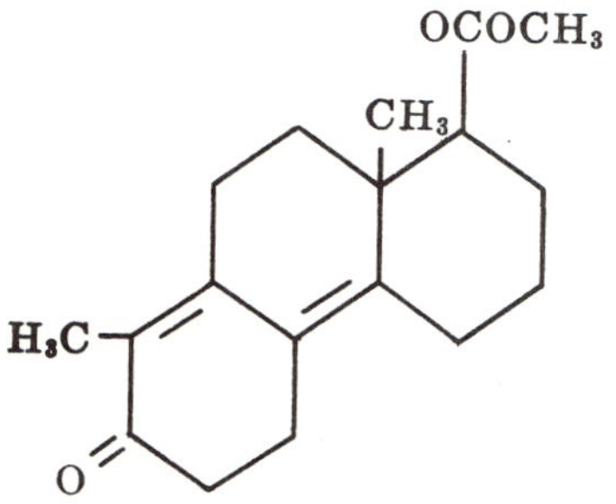	1.015	30	100.2	0	100.2
	0.987	30	99.9	0	99.9
	0.994	45	99.4	0.6	100.0
% C = 75.00	1.002	45	96.6	3.0	99.9

of the contaminated laboratory air is introduced into the apparatus, and it decreases the equipment purging time with oxygen prior to combustion.

Analytical procedure using 0.01 N $Ba(OH)_2$ solution as the absorbent. The apparatus is assembled as shown in Figs.1 and 2. The cell for electrical conductance measurements (Fig. 3) is rinsed twice with 0.01 N $Ba(OH)_2$ solution. For filling and emptying the cell the stopcock *b* is used. The side stopcock *a* (Fig. 1) must be closed. The cell is then filled up to the mark with the same solution, and by gently turning the stopcock *a* oxygen is started on its way through the coil *c*. The position of the side stopcock is so regulated that the solution level in the cell rises up to the line which is previously drawn with paint on the outer surface of the glass jacket surrounding the coil. This line represents a gas flow rate of 30 ml per minute.

While the apparatus is being purged with oxygen (5 min), approximately 1 mg of the sample is weighed,.and then, with stopcock *a* (Fig. 1) closed, the quartz capsule is removed from the combustion tube, the boat with the sample is placed in it, and the capsule is inserted back into the tube. The stopcock *a* is opened once again and within 2-4 min the resistance readings become sufficiently stable. (The resistance value is about 0.2 ohm per minute with a total resistance of 330 ohms). The measurement is made with an AC frequency of 1000 cps, a potential of 4 volts, and a resistance of 100 ohms for each of the two arms and maximum amplification of the vertical and horizontal inputs of the oscillograph. When the bridge is not balanced, Lissajous' figure is observed on the oscillograph screen. Depending on the sign of the inclination angle and the form of the figure, the resistance of the third arm is raised or lowered as well as the capacitance, until the ellipse is transformed into a horizontal straight line, which indicates the balancing of the bridge. After taking the instrument reading, the water trap is immersed in the acetone mixture contained in the beaker flask, which has a temperature of about −70°, and the MAG-6R electric furnace is so placed on the combustion tube that it covers the open end of the quartz capsule and comes to the edge of the boat. Combustion requires 3-5 minutes, whether the furnace is moved manually (3-4 times) or automatically over the sample capsule. It is not necessary to watch the combustion process. During the combustion, potential is not applied to the cell for electrical conductance measurements in order to avoid the occurrence of the Joule effect.

Ten minutes after the first reading, the instrument reading is noted, the stopcock *a* is closed, the trap with frozen water is disconnected and, without removing it from the Dewar flask, is attached to the apparatus for conversion of the water (Fig. 2). Usually during the gas flow, water freezes in the first arm of the U tube which, upon changing over must be turned toward the argon gas purification system; otherwise, upon preliminary purging of the apparatus with argon, some loss of water might occur. The cell of the water conversion apparatus is rinsed and filled with barium hydroxide solution during the sample combustion. After connecting the water trap, the three-way stopcock *a* (Fig. 2) is placed in position I for 3-4 minutes, and the oxygen is expelled with argon flowing at the rate of 40-50 ml/min. Afterwards the stopcock *a* is moved to position II (Fig. 2); and by slowly opening the side stopcock *b* argon is allowed to flow through the cell at a rate of 25-30 ml/min. When the change in instrument readings does not exceed 0.1-0.3 ohm per minute, the initial reading is taken, the Dewar flask, is removed, and the vertical tubular furnace heated to 300° is placed over the U-part of the trap. The water evaporates and is carried by the gas stream over the platinized carbon black layer heated to 900°. The resulting carbon monoxide, with the argon stream, enters the oxidizing apparatus filled with CuO at 300°, where it is oxidized to

CO_2, and the latter enters the cell for electrical conductance measurements and is absorbed by the alkali, while the argon escapes to the atmosphere. After 10 minutes the instrument reading is taken. In the case of a series of determinations, it is possible to burn the next sample in the first apparatus while water is being determined in the second.

Substances containing C, H, O and N are analyzed in the empty tube. If sulfur is present, some silver is placed in the tube, and in the presence of a halogen, platinum catalyst is also included [24]. Silicon-, phosphorus- and fluorine-containing organic compounds are analyzed under conditions described earlier [25-28]. Mercury does not interfere, as it is retained in the water trap. Samples of low-boiling liquids are weighed in open quartz capillaries [29]. Nitrogen oxides are absorbed outside the combustion tube [21] (Fig. 1 d).

Analytical procedure when using 0.01 N caustic soda solution as the absorbent. In this case the necessity for rinsing the cell between the runs is eliminated; it is sufficient to pour off the used NaOH solution and replace with fresh absorbent. The change in resistance (and consequently in electrical conductance) of the solution at the end of the experiment, as a result of carbonate ion formation, is almost one half of that when $Ba(OH)_2$ is used and when $CO_3^=$ and HCO_3^- ions are practically absent in the solution.

In practice it is possible to work with both absorbents, provided sufficient accuracy in resistance measurements is attainable; however, preference should be given to the caustic soda.

Calculation. The C and H percentages are obtained from calibration curves [14] which are established from the results of carbon and hydrogen determinations by the described method on a series of different sized samples of a pure test material (for example, sucrose). The value of carbon or hydrogen is plotted on the abscissa axis in micrograms; and on the ordinate axis the decrease in the electrical conductance Δa in $ohms^{-1}$. The latter is equal to $1/R_1 - 1/R_2$, where R_1 and R_2 are the initial and final resistances in ohms of the absorbing solution. Final calculation for carbon and hydrogen is made as follows:

$$\%C = \frac{a \cdot 100}{e} \qquad \%H = \frac{b \cdot 100}{e}$$

where a is the weight of carbon in micrograms found from the calibration curve, b is the same for hydrogen, and e is the sample weight in micrograms.

Sample calculation in the analysis of benzoic acid.

$$e = 1229\ \mu g$$
$$R_1^{tot} = 346.2\ \text{ohms}, \qquad R_2^{tot} = 406.3\ \text{ohms}$$
$$\Delta d_{tot} = 1/346.2 - 1/406.3 = 427 \times 10^{-6}\ \text{ohm}^{-1},$$

Hence from the calibration curve, the decrease corresponds to

$$a = 848\ \mu g \text{ and } 848 \times 100/1229 = 69.02\%\ C$$
$$\text{Also } R_2^{init} = 373.9\ \text{ohms},\ R_2^{init} = 403.0\ \text{ohms},$$
$$\Delta d_{init} = 1/373.9 - 1/403.0 = 193 \times 10^{-6}\ \text{ohm}^{-1},$$
$$\text{hence } b = 63.4\mu g, \qquad \frac{63.4 \times 100}{1229} = 5.16\%\ H$$

The authors express their sincere thanks to N. A. Balashova for her valuable suggestions and directions and also for the interest shown in this work; to Yu. S. Solomatin for assembling the electrical conductance apparatus and for the construction of the automatic movable furnace.

SUMMARY

A conductometric method is developed for the microdetermination of carbon and hydrogen in organic compounds. The method is based on the pyrolytic combustion of the substance, freezing out of water, and separate determination of both carbon and hydrogen by the amount of carbon dioxide which is absorbed by the alkali for measurement in an electrical conductivity cell.

For the hydrogen determination conversion of water in an inert gas stream by the incandescent platinized carbon black is used, whereby oxygen of the water is quantitatively converted into carbon monoxide with further oxidation to carbon dioxide and final conductometric measurement.

The method opens the possibility of automating all of the analytical steps except weighing the sample, and is suitable for the analysis of solid and liquid substances containing nitrogen, halogens (including fluorine), sulfur, boron, silicon, phosphorus and metals.

LITERATURE CITED

[1] W. Koch and H. Malissa, *Arch. Eisenhuttenw.* **27**, 695 (1956).

[2] R. Boulin, *Chim. anal.* 40, 72 (1958).

[3] P. A. Kryukov and N. E. Shults, *Gidrokhim. Mater.* 23, 110 (1955).

[4] P. A. Kryukov and E. V. Rengarten, *Zhur. Anal. Khim.* **10**, 51 (1955).

[5] B. Bruns, G. B. Braude and D. Tsiklis, *Ibid.* 2, 294 (1947).

[6] W. Schmidt and D. Baasch, *Forschungsberichte des Wirtschafts und Verkehrsministeriums Nord-Rhein—Westfahlen* [Research Reports of the Ministry of Industry and Commerce North Rhine—Westfalia], No. 67, 1954.

[7] H. Malissa, *Mikrochim. Acta* 1957, 553.

[8] M. Schutze, *Z. anal. Chem.* **118**, 244, 248 (1939).

[9] W. Zimmermann, *Ibid.* 118, 258 (1939).

[10] M. O. Korshun, *Zavodskaya Lab.* 10, 241 (1941).

[11] M. O. Korshun and E. A. Bondarevskaya, *Trudy Nauch.-Issledovatel. Inst.*, Goskhimizdat [Transactions of the Scientific Research Institute, State Chemical Publishing House], p. 127, Moscow, 1958.

[12] J. Unterzaucher, *Chem.-Ing.-Tech.* 1950, 39; *Mikrochim. Acta* 1957, 448.

[13] N. Shurmovskaya and L. Kupriyanova, *Zhur. Anal. Khim.* **7**. 41 (1948).

[14] N. E. Gel'man, M. O. Korshun, Ven-Yun Van and N. A. Balashova, *Doklady Akad. Nauk SSSR* **129**, 1046 (1959).

[15] M. O. Korshun and V. A. Klimova, *Zhur. Anal. Khim.* **2**, 274 (1947); M. O. Korshun, *Ibid.* **7**, 96 (1952).

[16] H. Reihlen, *Chem. Fabrik* 1934, 63.

[17] F. O. Fischer, *Anal. Chem.* **21**, 827 (1949).

[18] B. Bobransky, *Mikrochim.* Acta **1956**, 1735.

[19] T. Kono, *Ibid.* **1958**, 461.

[20] R. Wickbold, *Angew. Chem.* **64**, 133 (1952); **66**, 173 (1954).

[21] M. O. Korshun and V. A. Klimova, *Zhur. Anal. Khim.* **3**, 176 (1948); **4**, 292 (1949); V. A. Klimova and M. O. Korshun, *Ibid.* **6**, 230 (1951); M. O. Korshun and N. E. Gel'man, *Novye metody elementarnogo mikroanaliza* [New Methods of Elemental Microanalysis], Goskhimizdat [State Chemical Publishing House], Moscow, 1949.

[22] A. S. Smith, *Ind. Eng. Chem., Anal. Ed.* **6**, 293 (1934).

[23] E. A. Bondarevskaya, "Izuchenie sovremennogo metoda pryamogo opredeleniya kisloroda v organicheskikh soedineniyakh i razrabotka metoda odnovremennogo opredeleniya kisloroda i galoida" [A Study of the Modern Method of Direct Oxygen Determination in Organic Compounds and Development of a Method for the Simultaneous Determination of Oxygen and Halogen], Candidate Dissertation, Inst. Elementoorganicheskikh Soedinenii [Inst. of Heteroörganic Compounds], *Akad. Nauk SSSR*, Moscow, 1958.

[24] M. O. Korshun and N. S. Sheveleva, *Zhur. Anal. Khim.* **7**, 104 (1952); *Ibid.* 11, 376 (1956).

[25] V. A. Klimova, M. O. Korshun and E. G. Bereznitskaya, *Doklady Akad. Nauk SSSR* 84, 1175 (1952); *Zhur. Anal. Khim.* **11**, 223 (1956).

[26] M. O. Korshun, E. A. Terent'eva and V. A. Klimova, *Ibid.* **9**, 275 (1954).

[27] M. O. Korshun and E. A. Terent'eva, *Doklady Akad. Nauk SSSR* 100, 707 (1955).

[28] N. E. Gel'man, M. O. Korshun and N. S. Sheveleva, *Zhur. Anal. Khim.* 12, 526 (1957).

[29] N. E. Gel'man, M. O. Korshun and K. I. Novozhilova, *Ibid.*

[30] Ven-Yun Van "Issledovanie v oblasti primeneniya konduktometricheskogo izmereniya dvuokisi ugleroda k mikroopredeleniyu ugleroda, vodoroda i kisloroda v organicheskikh soedineniyakh" [Study of the Application of Conductometric Carbon Dioxide Measurement in the Microanalysis of Carbon, Hydrogen and Oxygen in Organic Compounds], Candidate Dissertation, Inst. Elementoorganicheskikh Soedinenii, Akad. Nauk SSSR, Moscow, 1960.

CHAPTER 46

Simultaneous Microdetermination of Fluorine, Carbon, and Hydrogen in Low-Boiling and Gaseous Compounds*

N. E. Gel'man, M. O. Korshun, and K. I. Novozhilova

The present article is devoted to a description of the application of a method for the simultaneous determination of fluorine, carbon, and hydrogen [1-8] to the analysis of low-boiling and gaseous fluoro organic compounds. The basic difficulties in the analysis of such materials are connected with taking samples and the introduction of samples into the combustion tube. Several methods have been described in the literature for the analysis of volatile liquids. In most cases a capillary containing the sample is sealed up and it is then opened either in the combustion tube itself by means of one of several devices, or it is opened immediately before being introduced into the tube [4-13]. Vessels of special design have also been suggested which are connected to the combustion tube and serve for breaking the ampoule containing the sample, and for mixing the vapors of the test material with oxygen [14-17]. As we have mentioned previously [3], the use of glass capillaries for holding the sample is not desirable during the combustion of fluoro organic liquids in a magnesium oxide layer, since contamination of the magnesium oxide by lumps of fused glass can lead to damage of the apparatus and to incorrect experimental results. Accordingly, for the analysis of high-boiling liquids we have used a quartz capillary, the end of which was left open. The weight of the capil-

Zhur. Anal. Khim. **15**, 628-634 (1960).

lary plus sample hardly changes even when the liquid material has a boiling point below 100°.

Since combustion of a liquid is no more complicated than combustion of a solid if the sample is taken in an open capillary, it was of interest to check the possibility of also using such a capillary in the analysis of low boiling liquids. As long ago as 1932, Pirsch [18] found that an aliquot of diethyl ether does not lose weight as long as the liquid layer in the capillary is separated from the surrounding medium by a few cubic millimeters of air found in the narrow end of the capillary. We established that the weight of various volatile liquids remains constant in a series of quartz capillaries when the diameter of the open end is from 0.2-0.3 mm. As can be seen from the results given in Table 1, thanks to the small diameter of the open end of the capillary and the small evaporation surface, the losses resulting from volatilization are comparatively small, are readily reproducible, and are constant with time.

In those cases where loss of sample is appreciable, as for example, during the analysis of pentafluoropropane, knowing the change in weight of capillary plus sample per unit time, and the time which elapses between reading the balance and introducing the capillary plus test material into the combustion tube, it is possible to calculate how many micrograms must be subtracted from the weight taken.

Table 2 contains analytical results for liquid materials, samples of which were taken in open capillaries. It should be noted, however, that considerable skill is required on the part of the analyst when dealing with materials boiling at 20°, while for materials which are gaseous at room temperature, the method described above for taking samples generally cannot be used.

TABLE 1
Weighing of Volatile Materials in Open Capillaries

B. p. of material °C	First reading mg	Readings at fixed time intervals after first reading, mg minutes						Change in weight in one minute[a] mg
		1	2	3	4	5	10	
78.4	3.195	3.195	3.195	3.195	3.195	3.195	3.192	0.000
56	1.225	1.220	1.218	1.212	1.210	1.205	1.188	-0.004
	9.210	9.208	9.200	9.192	9.188	9.182	9.155	-0.006
34.6	7.125	7.120	7.112	7.109	7.103	7.100	7.080	-0.004
	1.570	1.566	1.558	1.552	1.547	1.541	1.516	-0.005
27	3.430	3.420	3.412	3.401	3.990	3.980	3.970	-0.010
	5.160	5.151	5.139	5.130	5.120	5.110	5.101	-0.010
20	3.150	3.128	3.102	3.080	—	—	—	-0.023
	4.160	4.135	4.111	4.088	—	—	—	-0.024

[a]Balance room temperature about 20°.

TABLE 2
Determination of Fluorine, Carbon, and Hydrogen in Volatile Materials, Samples of Which Were Taken in Open Capillaries

Test material	B.p.	C % Found	C % Diff.	H % Found	H % Diff.	F % Found	F % Diff.
1-Ethylperfluoroisobutylene $C_6H_5F_7$	75°	34.36	+0.07	2.58	+0.18	63.70	+0.40
Calculated: C = 34.29%;		34.41	+0.12	2.66	+0.26	63.40	+0.10
H = 2.40%; F = 63.30%							
1-Bromo-2-hydroperfluoroisobutane	56°	17.23	+0.13	0.39	+0.04	54.15	+0.06
C_4HBrF_8		17.19	+0.09	0.38	+0.03	53.82	+0.27
Calculated: C = 17.10%; H = 0.35%; F = 54.09%							
3,3,3,1-Tetrafluoro-2-trifluoroisobutane	40°	26.10	+0.02	1.66	+0.04	72.29	+0.01
$C_4H_3F_7$		25.98	-0.10	1.53	-0.09	72.08	-0.20
Calculated: C = 26.08%; H = 1.62%; F = 72.28%							
1,2-Dihydroperfluoroisobutane	35°	23.88	+0.12	1.15	+0.16	74.91	-0.33
$C_4H_2F_8$		23.91	+0.15	1.06	+0.07	74.74	-0.50
Calculated: C = 23.76%; H = 0.99%; F = 75.24%							
3,3,3,2,1-Pentafluoropropane	20°	26.96	+0.12	2.45	+0.22	70.41	-0.47
$C_4H_3F_5$		26.96	+0.12	2.50	+0.27	70.57	-0.31
Calculated: C = 26.84%; H = 2.23%; F = 70.88%							

Note: All the analyses were carried out on samples of 4-8 mg.

Samples of gaseous compounds can be taken either by volume — by means of a gas buret — or by weight — in a sealed glass capillary. The latter technique has the advantage that the percentage content of the elements in the sample can be calculated, without having to know the specific gravity of the material; however, it is suitable only for readily condensed gases which boil within the temperature range +20° to —45°. For reasons outlined before [3], a capillary containing a sample cannot be opened in the magnesium oxide layer, and, consequently, in this case it is necessary to break it inside the combustion tube.

Of the known methods used for this purpose we chose the "stopcock opener" suggested by Yu. N. Dagaevaya and K. I. Novozhilova. The end of the sealed capillary is placed in a hole in the plug of a stopcock inserted between the combustion tube and the apparatus used for preliminary purification of the oxygen (Fig. 1). Upon turning the stopcock the capillary is broken and the liquid is swept by the oxygen stream into the combustion tube. This apparatus had the drawback that upon breaking the capillary tube the liquid sample fell on the stopcock plug and thereby led to possible losses of material. We have improved the design of this stopcock by placing the stopcock in the

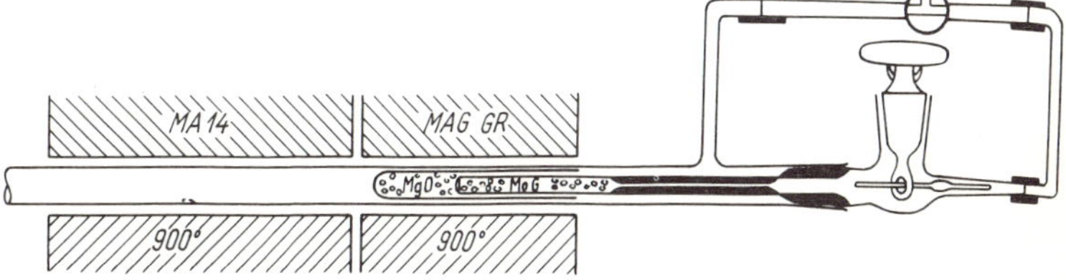

Fig. 1

TABLE 3
Determination of Fluorine, Carbon, and Hydrogen in Gases

Test material	B. p.	C % Found	C % Diff.	H % Found	H % Diff.	F % Found	F % Diff.
Monohydroperfluoroisobutylene C_4HF_7	+18°	26.06[b]	-0.31	0.64	+0.10	73.13	+0.11
Calculated: C = 26.37% H = 0.54%; F = 73.02%		26.08[b]	-0.29	0.67	+0.13	72.89	+0.13
Monohydroperfluoroisobutane C_4HF_9	+13°	21.62[b]	-0.21	0.53	+0.07	77.76	+0.05
Calculated: C = 21.83%; H = 0.46%; F = 77.71%		21.70[b]	-0.13	0.42	+0.04	77.60	-0.11
3,3,3,2,1,1-Hexafluoropropane $C_3H_2F_6$	+5°	23.33[b]	-0.34	1.28	-0.03	74.53	-0.47
Calculated: C = 23.67%; H = 1.31%; F = 75.00%		23.30[b]	-0.37	1.23	-0.08	74.93	-0.07
Monohydroperfluoropropane C_3HF_7	-17°	21.13[b]	-0.04	0.46	-0.13	77.92	-0.31
Calculated: C = 21.17%; H = 0.59%; F = 78.23%		20.95[b]	-0.22	0.57	-0.02	78.52	+0.29
3,3,3,2,1-Pentafluoropropene-1 C_3HF_5	-20°	26.98[b]	-0.29	0.90	+0.15	72.04	+0.07
Calculated: C = 27.27%; H = 0.75%; F = 71.97%		27.21[b]	-0.06	0.79	+0.04	71.71	-0.26

Hexafluoroazoxymethane $C_2F_6N_2O$ Calculated: C = 13.19%; F = 62.63%	+8°	13.16[a]	−0.03	0.00	0.00	62.34	−0.29
		13.31[b]	+0.12	0.15	+0.15	62.36	−0.27
Trifluoronitromethane CF_3NO_2 Calculated: C = 10.44%; F = 49.56%	−31.2°	10.43[a]	−0.01	0.07	+0.07	49.60	+0.14
		10.60[b]	+0.16	0.09	+0.09	49.39	−0.17
Difluorochloromethane $CHClF_2$ Calculated: C = 13.89%; H = 1.17%; F = 43.95%	−40°	13.78[a]	−0.11	1.10	−0.07	44.15	+0.20
		13.69[b]	−0.20	1.20	+0.03	43.95	0.00
3,3,3-Trifluoropropyne-1 C_3HF_3 Calculated: C = 38.31%; H = 1.07%; F = 60.61%	−46°	38.24[a]	−0.07	1.17	+0.10	60.52	−0.09
		38.46[b]	+0.15	1.16	+0.09	60.24	−0.37
1,1,2-Trifluoroethene C_2HF_3 Calculated: C = 29.28%; H = 1.22%; F = 69.48%	−51°	29.45[a]	+0.17	1.17	−0.05	69.72	+0.24
		29.40[a]	+0.12	1.12	−0.10	69.30	−0.18

Note: All analyses were carried out on 4-9 mg samples.

[a]The analysis was carried out by means of a gas buret.

[b]The analysis was carried out by means of a capillary opener.

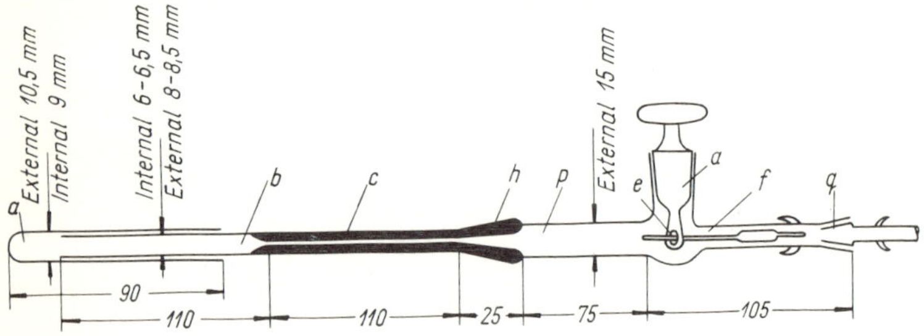
External 10,5 mm
Internal 9 mm
Internal 6–6,5 mm
External 8–8,5 mm
External 15 mm
a
b
c
h
p
e
a
f
q
90
110
110
25
75
105

Fig. 2

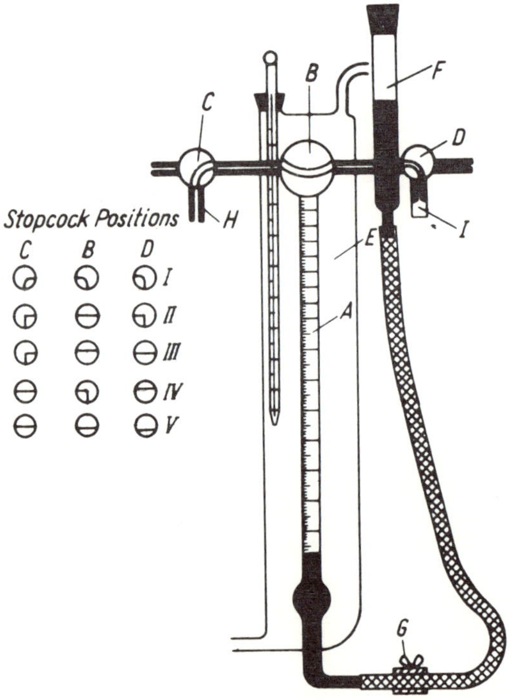
Stopcock Positions
C
B
D
I
II
III
IV
V
C
B
F
D
H
I
E
A
G

Fig. 3

top part of the apparatus, thereby eliminating the drawback indicated above. An added advantage of this design for the stopcock (Fig. 2) is that the capacity of the widened part of the capillary which is located inside the ground joint connection of the stopcock can be increased when it is necessary that the samples be copiously diluted with oxygen for safe combustion.

Materials which boil at temperatures below +20° can also be analyzed by means of a gas buret (Fig. 3). Table 3 contains the results obtained during the analysis of gaseous materials, when various methods of taking samples were used.

Below is given a description of those stages in the analysis of volatile and gaseous compounds which differ from the corresponding stages in the analysis of solid samples described earlier [3].

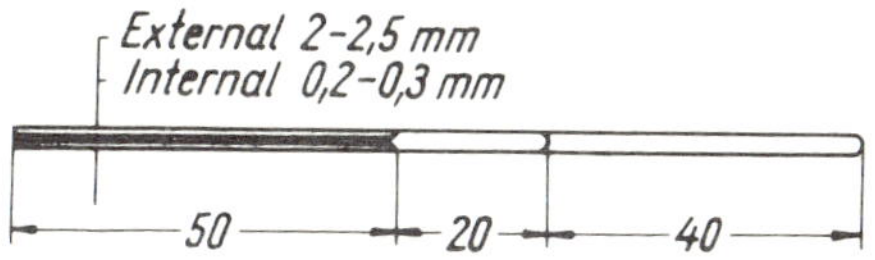

Fig. 4

Analysis of volatile liquids boiling above 20°. Samples are taken in a quartz capillary (Fig. 4). In order to suck the liquid into the widened part of the capillary and to create an air stopper (blanket) in the narrow end, the capillary reservoir is cooled by means of special forceps which have cold compartments (Fig. 5).

The forceps with cold compartments can be constructed by soldering on to ordinary forceps hollow, brass, semicylindrical compartments with small longitudinal channels on the clamping surfaces, which enable the capillary to be held reliably. Lumps of dry ice are placed in the cooling compartments. When a sample is located in the bottom of the widened part of the capillary, no movement

Fig. 5

of the aliquot along the capillary is observed during weighing.

Analytical procedure. Before introducing the sample, that part of the combustion tube where the calcined magnesium oxide to be used for successive determinations is located is cooled with dry ice. The quartz capillary plus sample is placed in a tube, covered with magnesium oxide, and introduced into the combustion tube so that the closed end of the tube is in the cooled zone. At the beginning of combustion, the cooled part of the tube is insulated from heat by means of an asbestos screen. Then, as the electric oven is moved forward, cooling is removed and the analysis is finished as usual by gradually moving the oven onto the whole layer of magnesium oxide. A combustion lasts about 20 min.

Analysis of materials boiling in the range +20 to —45°. Into a small Dewar vessel fixed in an inclined position, and containing a cooling mixture whose temperature is a few degrees lower than the boiling point of the sample, is introduced a sealed ampule containing the sample, the ampule itself being fitted into a two-hole rubber stopper. After the ampule has been immersed for 3-5 min, the drawn-out end which sticks out through the rubber stopper is cut and broken. The widened part of a thick-walled glass capillary which has been weighed beforehand is gently warmed over a flame and the drawn-out part of the capillary submerged in the liquid sample; the widened part of the capillary is held by means of the forceps with the cooled compartments (Fig. 6). When enough liquid has collected in the capillary, its drawn-out end is withdrawn from the ampule, and is sealed, while the widened part is still held in the cooled compartments of the tongs. Using the simple equipment described it is possible to take a sample in such a way that it is all found in the widened part of the capillary.

The sample is introduced into the combustion tube by means of the capillary breaker (Fig. 2) which is a quartz stopcock *d* with a stopper terminating in a eyelet *e*. On one side of the stopcock is fused a tube *f* through which the sealed capillary containing the sample is introduced, while on the other side is fused a widened tube *p*, which is connected via the capillary *c* with the tube *b* filled with magnesium oxide. Combustion of the sample occurs in tube *b*. The capillary breaker is connected to the combustion tube (Fig. 1) through a joint *h* which is not lubricated. It could be connected by means of a rubber stopper,

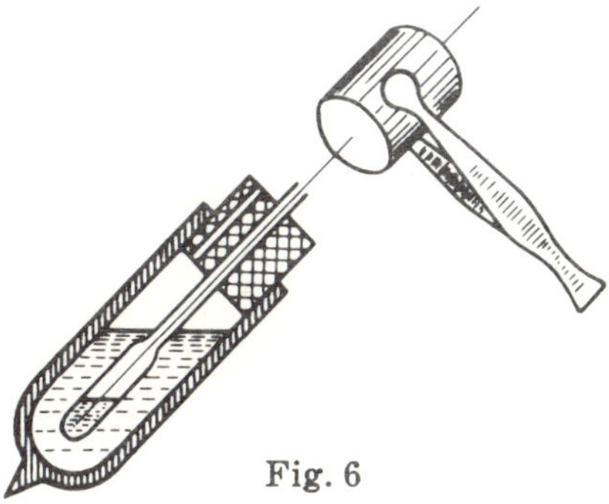

Fig. 6

but this is not so convenient. The ground joint of the tap is lubricated with silicone grease.

In the case of these low-boiling liquids, approximately one and a half times more magnesium oxide is used than for the analysis of solid and high-boiling fluorides [3]. A blank run is made at the beginning of a working day. Magnesium oxide which has been pyrohydrolyzed beforehand is then strewn into tube *a* (Fig. 2), and from tube *a* into tube *b*. Tube *a* in which some magnesium oxide is left is fitted on to tube *b* so that a continuous MgO layer 9-10 cm long is obtained; the capillary breaker is connected by means of the joint *h* to the combustion tube and one of the capillaries containing a sample is placed in *f* so that the drawn-out end fits into the eyelet *e* of stopcock *d*. The joint *g* is connected to an oxygen supply, and, while the absorption apparatus is being weighed [3], the magnesium oxide is calcined, by drawing the furnace over the tube and passing oxygen through the capillary breaker only (Fig. 1). The absorption apparatus is now connected to the combustion tube and the furnace moved slightly in the direction of the oxygen stream, so that tube *b* containing the magnesium oxide sticks out from the furnace by a distance of 2 cm, while the rest of the magnesium oxide is in the calcined zone. In order to prevent heating of capillary *c*, an asbestos screen which presses tightly against the electric heater is fitted on the combustion tube; the O_2 supply stopcock is opened and that part of the capillary breaker in which the widened part of the capillary is located is cooled with dry ice, tube *p* is also cooled. After 2-3 min the capillary tip is broken by rotating the stopcock *d*, and the sample allowed to issue slowly from the capillary, the rate at which it issues being controlled by alternately applying and removing the cooling. This process takes about 15 min. When all the sample has come out of the capillary, cooling is stopped and all the magnesium oxide is calcined for 7-10 min. Next, by rotating the three-way stopcock, the oxygen stream is switched into the side entrance of the combustion tube, after a further 5-7 min the absorption apparatus is disconnected, the capillary breaker is switched off, all the magnesium oxide is strewn into the tube *a*, and the latter is transferred into the pyrohydrolysis tube as described previously [3].

Combustion as a whole takes about 25-30 min when the oxygen throughput is 20-25 ml/min.

Analysis of materials boiling below —45°. Gases boiling below —45° are sampled by means of the usual type of gas buret A (Fig. 3) having a volume of 2 ml. One filling of the buret is sufficient for 1-2 determinations. The buret is permanently connected into the setup used for carbon determination, between the capillary breaker and the gas-purifying system (Fig. 1). When the analyst is not working with a capillary breaker, it is necessary to have a similar connecting quartz tube without a stopcock. The buret is provided with a crescent stopcock B by means of which it is possible either to introduce the gas to be analized or oxygen into the combustion tube, and with two additional stopcocks C and D which allow the buret to be filled with gas without disconnecting it from the apparatus. Water from a thermostat is introduced into the water-jacket E of the buret. The zero reading of the buret is located at the stopcock B. The buret is filled with mercury or gas by means of leveling bulb F and the screw clamp G.

When sampling, the gas passes into the buret either directly from a cylinder, or from a gas pipet. The leveling bulb F containing mercury is fixed in a position above stopcock B, the stopcock is turned to position I (Fig. 3), and, by opening clamp G, the graduated part of the buret gradually filled with mercury; stopcock B is turned to position II and the clamp closed. The gas inlet tube of the cylinder or a gas pipet is connected to the outlet H and gas passed through stopcocks C, B and D for ½ minute. During this operation the small drop of mercury which has entered into the opening in stopcock B during the operation of filling the buret is blown out by the gas stream into outlet I and is removed. The bulb F is then lowered, the clamp G is opened slightly, and, by careful rotation of stopcock B into position I the gas is slowly let into the buret, after which stopcock B is turned back again into position II, and oxygen passed through all three stopcocks (position III). During this operation the volume of the gas is read off at atmospheric pressure, and after setting the stopcocks in position IV, combustion is started.

At the beginning of a working day a blank run is carried out, oxygen being passed through the side inlet of the combustion tube. While the buret is being filled with gas sample, and after the blank run has been carried out, oxygen is passed through stopcocks B, D, and C. The magnesium oxide is then prepared as indicated above; the weighed absorption apparatus is connected to the combustion tube and combustion of the gas is started. For this purpose, bulb F containing mercury is fitted in a position above the levels of stopcocks B, C, and D which are in position IV, and the screw clamp G is opened so that the gas issues from the buret into the combustion tube at the rate of about 0.2 ml/min. During this operation the oxygen is passed through the side inlet. When the requisite amount of gas has been passed into the apparatus, the stopcocks are turned to position V for 10 min, and the gas supply line swept with oxygen. The oxygen is passed for a further 10 min into the side inlet of the combustion tube, after which combustion is complete. The total time taken for a combustion is 25-30 min, for an oxygen flow rate of about 20-25 ml/min.

SUMMARY

A method is described for the simultaneous determination of fluorine, carbon, and hydrogen in volatile and gaseous compounds. It has been established that it is possible to take 4-9 mg samples of volatile liquids boiling at 20° and higher by means of quartz capillaries with open ends. A device is described which enables the capillaries containing samples to be opened outside the combustion tube, and a technique has been developed for taking samples of materials boiling in the range of +20° to −45° in sealed capillaries.

LITERATURE CITED

[1] M. O. Korshun, N. E. Gel'man, and K. I. Glazova, *Doklady Akad. Nauk SSSR* **111**, 1225 (1956).

[2] N. E. Gel'man, M. O. Korshun, and K. I. Novozhilova, *Zhur. Anal. Khim.* **15**, 222 (1960).

[3] N. E. Gel'man, M. O. Korshun, and K. I. Novozhilova, *Zhur. Anal. Khim.* **15**, 342 (1960).

[4[F. Pregl, *Quantitative Organic Microanalysis* [Russian translation] (Moscow, 1934) p. 66.

[5] L. J. Clark, *J. Am. Chem. Soc.* 34, 746 (1912).

[6] S. Avery and D. Hayman, *Ind. Eng. Chem., Anal. Ed.* **2**, 236 (1930).

[7] H. Levin and K. Uhrig, *ibid.* **9**, 326 (1937).

[8] V. A. Aluise, *ibid.* **10**, 56 (1938).

[9] M. O. Korshun and N. E. Gel'man, *New Methods of Elemental Microanalysis* [in Russian] (Goskhimizdat, Moscow-Leningrad, 1949) p. 45.

[10] F. O. Fischer, *Anal. Chem.* **21**, 827 (1949).

[11] W. Schenk, *Anal. Chem.* **26**, 788 (1954).

[12] J. M. Corliss, *Anal. Chem.* **29**, 1902 (1957).

[13] R. N. McCoy and E. L. Bastin, *Anal. Chem.* **28**, 1776 (1956).

[14] E. E. Reid, *J. Am. Chem. Soc.* **34**, 1037 (1912).

[15] J. R. Baily, *Ind. Eng. Chem., Anal. Ed.* **5**, 171 (1933).

[16] J. H. Brunn and W. B. M. Faulconer, *Ind. Eng. Chem., Anal. Ed.* **8**, 315 (1936).

[17] W. C. Schumb and K. J. Radimer, *Anal. Chem.* **20**, 871 (1948).

[18] J. Pirsch, *Ber.* **65**, 865 (1932).

CHAPTER 47

The Quantitative Determination of Carbon and Hydrogen-containing Organic Substances in the Presence of a Surface-combustion Catalyst*

P. N. Fedoseev and M. M. Pavlenko

In our first report [1] we showed that by using flameless, surface-combustion catalysts in quantitative elementary organic analysis it is possible to simplify the method of determining carbon and hydrogen and at the same time to reduce the duration of the analysis to 10-15 minutes.

This article is devoted to the development of a method for the quantitative determination of carbon and hydrogen in nitrogen-containing organic substances in the presence of a surface-combustion catalyst, chromium sesquioxide.

The method described by us is a rapid method.

In the rapid combustion of an organic substance containing nitrogen the danger of the passage of the nitrogen oxides formed during burning into the weighed absorption tubes is particularly great.

Since the complete absorption of nitrogen oxides during rapid combustion is difficult to accomplish inside the tube we decided to test a method of removing the nitrogen oxides outside the tube with the aid of the procedure described in a book by Korshun and Gel'man [2]. These authors used the following oxidizing absorbents to remove the nitrogen oxides: 1) permanganate and concentrated sulfuric acid, 2) bichromate and concentrated sulfuric acid.

**Zhur. Anal. Khim.* 6, 317-320 (1951).

A check of the activity of these oxidizing mixtures showed them to be completely suitable for the analysis. However, we discovered that chromic and permanganic anhydrides are very volatile. This inevitably led to higher figures for the percentage content of carbon. The passage of these substances is easily detected by using solid potassium hydroxide as the gas absorbent. This turns green with chromic acid and rose with manganese heptoxide.

After testing a series of preparations, the substance most suitable for the purpose of retaining volatile oxidants proved to be a saturated solution of ferrous sulfate in 25-30% sulfuric acid.

We established that volatile chromium and manganese oxides are hardly retained at all by concentrated sulfuric acid and anhydrous calcium chloride.

In our work we used an ordinary spiral potash bulb instead of the special foam-extinguisher apparatus described in the literature for the absorption of nitrogen oxides [2].

The bulb of the potash apparatus is removed from the combustion tube, and filled with glass wool to hold back the spatter and spray of the foam. The spiral potash tube differs to advantage from the foam-extinguisher tube described by M. O. Korshun in its larger active surface and easier availability.

The set-up of the absorption tubes is pictured in Fig. 1.

The method for the quantitative determination of carbon and hydrogen in organic, nitrogen-containing substances is described below.

In the determination of carbon and hydrogen in substances that do not contain nitrogen the nitrogen oxide absorption tubes are removed.

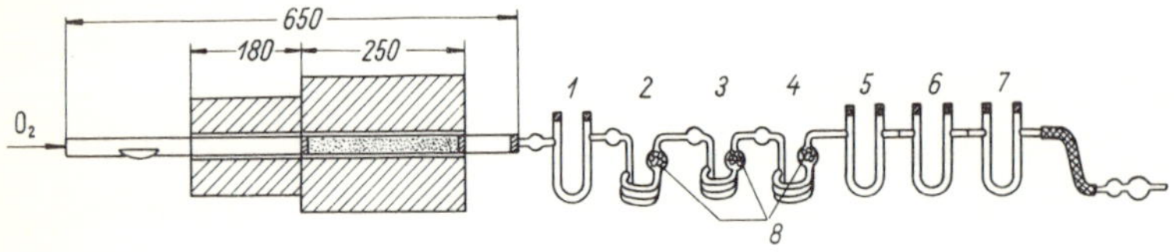

Fig. 1. Set-up of the equipment. 1 & 5 — anhydrous calcium chloride; 2 & 3 — oxidizing mixture; 4 — solution of ferrous sulfate in 25% sulfuric acid; 6 & 7 — pieces of potassium hydroxide; 8 — glass wool.

EXPERIMENTAL

Our experiments showed that the analysis may be conducted in ordinary high-melting glass tubes since it is not necessary to heat the combustion tube above 600-650°. Complete and rapid oxidation of the combustion products to CO_2 and water takes place at this temperature.

The apparatus is set up as shown in Fig. 1. Pressed chromium sesquioxide crushed to small pieces (18-25 cm) is placed inside the combustion tube and supported at both ends with small asbestos plugs.

We recommend using U-tubes and spiral potash bulbs as the absorbing equipment. Two spiral potash tubes are filled with an oxidizing mixture comprising concentrated sulfuric acid and potassium permanganate (0.1-0.2%), to no more than one bend. These tubes are intended to oxidize and absorb the nitrogen oxides formed as a result of the combustion.

One spiral tube is filled with an almost saturated solution of ferrous sulfate in 20-25% sulfuric acid to decompose and retain the volatile higher chromium and manganese oxides.

To retain the water vapors carried away from the spiral filled with the solution of iron salt, a U-tube filled with anhydrous calcium chloride is attached. The details can be seen in Fig. 1.

Course of the analysis. Oxygen is passed through the assembled apparatus and the catalyst tube is heated for 10-20 minutes to remove the water vapor and other impurities.

The absorption tubes are disconnected and carried to the balance room where they are weighed after 20 minutes.

The small electric furnace is switched off. The substance being analyzed is weighed in a porcelain boat, covered with dry fused oxide, which must be kept in stock, and is placed inside the tube. Weighed samples of liquids are taken in glass ampules, placed in the boat, and chromic oxide is added. The weighed absorption tubes are connected, the small electric furnace is turned on, and a stream of oxygen from a gasometer or oxygen cylinder is passed through. The large electric furnace should not be switched off during the entire course of the work. The substance being analyzed is ignited by moving the electric furnace to the edge of the boat. After the sample has been ignited, the electric furnace is returned to its former place and the combustion process is followed by the slow movement of the glowing zone in the catalyst. The stream of oxygen is so regulated that no carbon separates during the combustion process. If carbon does appear, the rate at which the oxygen is passed through must be increased. After the combustion is over — this lasts no more than 3-5 minutes — the entire length of the tube is brushed with a burner flame for 5-10 minutes; after this, the combustion may be considered complete.

When the amount of nitrogen is small, one spiral tube is sufficient; two or three tubes must be used, however, when there is a large amount of nitrogen present.

TABLE 1

Substance	Calculated %C	Calculated %H	Found %C	Found %H	Difference %C	Difference %H
Aniline	77.38	7.57	77.52	7.69	-0.05	-0.10
	77.38	7.57	77.48	7.63	-0.09	-0.00
	77.38	7.57	77.44	7.41	-0.16	-0.16
Acetanilide	71.08	6.71	71.16	6.66	+0.08	-0.05
	71.08	6.71	71.21	6.69	+0.13	-0.02
Diphenylamine	85.17	6.51	84.94	6.59	-0.23	+0.08
	85.17	6.51	85.30	6.46	+0.13	-0.05
	85.17	6.51	85.17	6.43	-0.01	-0.08
	85.17	6.51	85.26	6.53	+0.09	+0.02
Oxanilide	69.98	5.03	69.82	5.19	-0.16	+0.15
	69.98	5.03	70.24	5.15	+0.26	+0.12
	69.98	5.03	70.00	5.22	+0.02	+0.19
p-Nitrotoluene	61.30	5.14	61.29	5.05	-0.01	-0.09
	61.30	5.14	61.16	5.03	-0.14	-0.11
p-Nitroanisole	54.89	4.61	55.08	4.72	+0.19	+0.11
	54.89	4.61	55.11	4.68	+0.22	+0.07
	54.89	4.61	55.12	4.58	+0.23	-0.03
p-Nitroaniline	55.17	4.38	52.03	4.39	-0.06	+0.01
	55.17	4.38	51.94	4.21	-0.23	-0.17
	55.17	4.38	52.32	4.55	+0.15	+0.17
m-Nitroaniline	55.17	4.38	52.24	4.41	+0.07	+0.03
	55.17	4.38	52.05	4.54	-0.09	+0.16
	55.17	4.38	52.34	4.31	+0.17	-0.07
o-Nitroaniline	55.17	4.38	52.15	4.49	-0.02	+0.11
	55.17	4.38	52.23	4.29	+0.09	-0.12
o-Nitro-*p*-toluidine	55.25	5.30	55.37	5.22	+0.12	-0.08
	55.25	5.30	55.54	5.47	+0.29	+0.17
	55.25	5.30	55.22	5.16	-0.03	-0.14
m-Nitro-*p*-toluidine	55.25	5.30	55.11	5.56	-0.14	+0.26
	55.25	5.30	55.13	5.35	-0.11	+0.05

The oxidizing mixture is changed when the green color turns brown ($KMnO_4$) and the yellow turns green ($K_2Cr_2O_7$).

1.5-2 liters of oxygen is passed through during the analysis.

With tubes of the size indicated it is recommended that the weighed samples of the substance be no more than 0.1500 g.

When easily subliming substances are burned, it is preferable to use a long boat (10-12 cm) or to distribute the substance evenly in two boats. This guarantees uniform combustion of the substance.

From a difficult, cumbersome, protracted process, quantitative analysis by the method described by us has become a convenient, reliable, and rapid procedure.

The analysis does not demand great effort of the worker and is automatic to a considerable degree. In the course of a work day 5-6 analyses can be carried out.

The results of the analyses of organic nitrogen-containing substances are given in the table above.

CONCLUSIONS

1. A method of quantitative determination of carbon and hydrogen in nitrogen-containing organic substances has been described. This method consists in burning the substance in a stream of oxygen in the presence of a surface-combustion catalyst-chromium sesquioxide.

2. The method described permits use of ordinary high-melting tubes for the entire analysis. An analysis lasts 10-20 minutes. The accuracy of the analysis (in the determination of carbon) lies within the limits of $\pm 0.3\%$ C. and of $\pm 0.2\%$ H in the determination of hydrogen. The analysis demands no great effort of the worker and to a considerable extent is automatic.

3. Nitrogen oxides formed as a result of the combustion are absorbed outside the tube with a solution of potassium permanganate in concentrated sulfuric acid. An ordinary spiral potash tube is used for this purpose.

4. The data presented above give a basis for expecting wide acceptance of our method in many different types of laboratory.

LITERATURE CITED

[1] P. N. Tedoseev and M. M. Pavlenko, *Zhur. Anal. Khim.* **5**, 929 (1950).

[2] M. O. Korshun and N. E. Gel'man, *New Methods of Elementary Microanalysis*, State Chemical Publishing House 1949, pp. 29, 46-49.

CHAPTER 48

A New Method for the Quantitative Determination of Nitrogen in Organic Substances*

P. N. Fedoseev and N. P. Ivashova

Many methods are known for the quantitative determination of nitrogen in organic compounds [3, 4, 6-10]. The most widely used of these are the Dumas and Kjeldahl methods. The Dumas method is considered classical and is suitable for the analysis of all organic substances that contain nitrogen. Among its shortcomings, however, are its clumsiness, complicated apparatus, and the length of time required to carry out the analysis. The Kjeldahl method is used extensively in plant laboratories and some scientific research institutes because it affords the possibility of conducting several analyses simultaneously and does not require complicated equipment.

The drawback of the Kjeldahl method is its limited applicability.

As is known, nitro, nitroso, and azo compounds and nitrogenous heterocycles can be analyzed by the Kjeldahl method only with extreme difficulty.

This article describes the new method that we have developed for the quantitative determination of nitrogen in any kind of organic substance.

The principle of the quantitative determination of nitrogen in organic substances consists in heating the nitrogen-containing substance together with powdered mag-

**Zhur. Anal. Khim.* 7, 112-15 (1952)

nesium in a hydrogen atmosphere [1, 2, 5]. The nitrogen of the organic substance reacts quantitatively with the magnesium to form magnesium nitride, which gives ammonia when decomposed with water. The ammonia is driven off into a standard acid solution and determined quantitatively.

The reaction between magnesium and the nitrogen-containing organic substance is conducted in test tubes or glass tubes. Upon heating, the magnesium partly reacts with the glass, forming small quantities of magnesium silicide from which gaseous silanes highly combustible in air form upon subsequent treatment of the reaction mass with acid. When the reaction mass is treated with water, however, there is almost no combustion of the silanes. The small flashes appearing at the start of the treatment do not affect the accuracy of the analysis.

Under our experimental conditions powdered magnesium reacts easily not only with the nitrogen of the organic substances but also with the halogen and sulfur. This affords the possibility of quantitatively determining nitrogen, sulfur, and halogen simultaneously in one weighed

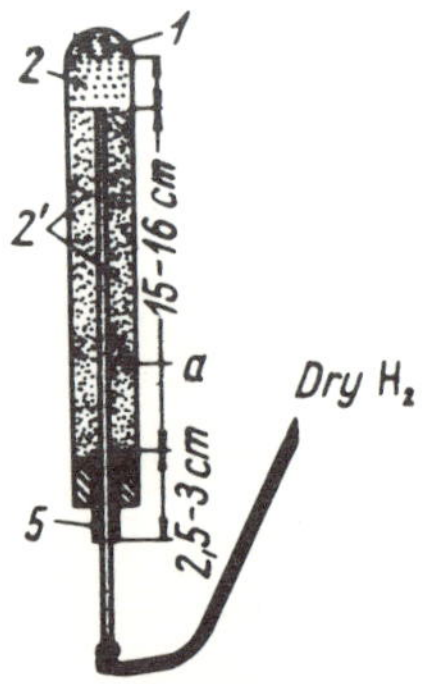

Fig. 1
Test Tube
1—Weighed sample; 2—first layer of magnesium; 2′—magnesium; a—glass inlet tube for the H_2; b—glass exit tube for the gases.

sample. We are working along these lines. From four to five analyses can be made in one day on a single apparatus. The analysis lasts 1½ hours.

The accuracy of the analysis lies within the limits of 0.2-0.3%.

EXPERIMENTAL

A weighed sample (0.1000-0.1500 g.) of the organic substance is placed in a test tube (Fig. 1) and covered with a layer (1.5-2 cm.) of powdered magnesium.

A thin glass tube *a* that does not reach the bottom of the test tube is inserted. Then magnesium is poured into the test tube almost up to the stopper. A short glass tube *b* somewhat larger in diameter than *a* is inserted in the hole in the stopper. The stoppered tube is turned upside down and tube *a*, which projects from the end, is connected to a Kipp aparatus.

In the analysis of liquids, the weighed sample is introduced in a glass ampule.

Air is expelled from the test tube by a slow stream of hydrogen over a period of 10-15 minutes. After 15 minutes, while hydrogen is still being passed through, tube *a* is carefully removed from the test tube, the Kipp apparatus is disconnected, and rubber tubing having clamps at both ends and filled with water is put over tube *b*. The tubing is filled with water so as to prevent atmospheric nitrogen from entering the reaction vessel. After this, the weighed sample of the subtance is mixed with the magnesium by careful shaking. The clamps on the rubber tubing are opened 1-2 minutes after the combustion furnace has been switched on and are closed when the reaction stops. The test tube with the weighed sample can also be heated with gas burners.

In our work we used a specially constructed electric furnace, which is diagrammed schematically in Fig. 2.

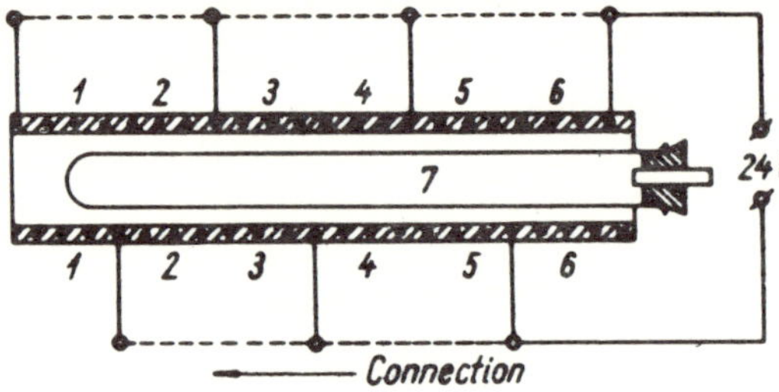

Fig. 2
Electric Furnace
1, 2, 3, 4, 5, and 6 — Sections of the furnace; 7—test tube.

The furnace consists of 4-6 sections which are switched on in stages. This furnace construction makes it possible to increase the heating zone and bring it closer to the weighed sample in uniform steps.

The test tube must not be heated to more than dull redness (600-650°). The organic substance usually decomposes almost completely when the first two sections of the furnace (5-6 cm.) are switched on and are at dull red heat.

When necessary (slow evolution of gas), the combustion rate is increased by switching on the rest of the sections of the furnace. It is easy to follow the burning process by noting the speed with which the bubbles of gas formed in the apparatus upon decomposition of the organic substance are evolved (Fig. 3). For this purpose the rubber exit tube is immersed in water under a cylinder or left loose in a crystallizing pan filled with water.

The combustion is considered over after the whole test tube has been at a dark red heat for a few minutes. The evolution of gas bubbles begins very slowly. The combustion process takes 25-30 minutes. The test tube is carefully cooled and, as quickly as possible, is broken in a mortar and transferred to a round-bottomed flask (0.5-0.8 liter). 20-25 ml. of 10% H_2SO_4 is run in to dissolve the

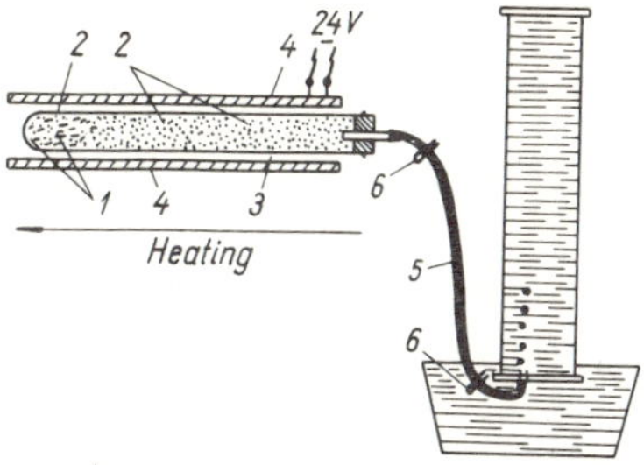

Fig. 3. 1—Weighed sample; 2—magnesium; 3—test tube; 4—furnace; 5—rubber tubing; 6—clamps.

magnesium nitride completely. The silane evolved ignites in the presence of air and may produce explosion of the detonating mixture (hydrogen and atmospheric oxygen). Consequently, it is necessary to expel the air with carbon dioxide before adding the acid. Then an excess of strong alkali is added and the ammonia is driven off into a standard solution of 0.1 N hydrochloric or sulfuric acid. The process of driving off the ammonia lasts 30-35 minutes. The excess acid is titrated with 0.1 N alkali in the presence of methyl orange and the percentage content of nitrogen is calculated.

The analytical data presented in the table indicate the satisfactory accuracy of the determination of nitrogen in organic substances.

CONCLUSIONS

1. A method was described for the quantitative determination of nitrogen in any type of organic substance. This method consists in heating the organic nitrogen-containing substance with magnesium in a hydrogen atmosphere.

TABLE 1

Exp. no.	Compound	Empirical formula	Calc. N%	Found N%	Difference
1	Nitrobenzene	$C_6H_5NO_2$	11.38	11.23	−0.15
			11.38	11.19	−0.19
2	*o*-Nitrotoluene	$C_7H_7NO_2$	10.21	10.03	−0.18
			10.21	10.14	−0.07
3	Nitrophenol	$C_6H_5NO_3$	10.07	10.01	−0.06
			10.07	9.77	−0.30
4	*o*-Nitroanisole	$C_7H_7NO_3$	11.56	11.37	−0.19
			11.56	11.36	−0.20
5	Aniline	C_6H_7N	15.04	14.85	−0.19
			15.04	15.32	+0.28
6	Toluidine	C_7H_9N	13.07	13.28	+0.21
			13.07	13.08	+0.01
7	Dimethylaniline	$C_8H_{11}N$	11.56	11.30	−0.26
			11.56	11.75	+0.19
8	Diphenylamine	$C_{12}H_{11}N$	8.28	8.25	−0.03
			8.28	8.31	+0.03
9	*m*-Nitroaniline	$C_6H_6N_2O_2$	20.28	20.05	−0.23
			20.28	19.90	−0.30
10	*p*-Nitroaniline	$C_6H_6N_2O_2$	20.28	20.40	+0.12
			20.28	20.38	+0.10
11	*o*-Nitro-*p*-toluidine	$C_7H_8N_2O_2$	18.41	18.15	−0.26
			18.41	18.40	−0.01
12	Ethylaniline	$C_8H_{11}N$	11.56	11.49	−0.07
			11.56	11.69	+0.09
13	Diphenylthiourea	$C_{13}H_{12}N_2S$	12.27	12.32	+0.05
			12.27	12.28	+0.01
14	Sulfanilic acid	$C_6H_7NO_3S.H_2O$	6.69	6.40	−0.29
			6.69	6.52	−0.17
15	Naphthionic acid	$C_{10}H_9NO_3S$	6.27	5.98	−0.29
			6.27	6.16	−0.11
16	Azobenzene	$C_{12}H_{10}N_2$	15.37	15.24	−0.17
			15.37	15.36	−0.10

2. The analysis is conducted in ordinary glass test tubes. It lasts 1½ hours. The analyst can run from 5 to 6 determinations in the course of a working day. The accuracy of the analysis lies within the limits of 0.2-0.3%.

3. The data presented above provide a basis for expecting that the method described will be used in various laboratories.

LITERATURE CITED

[1] A. P. Lidov, *Analiz Gazov* [Gas Analysis] State Chemical Publishing House, 1928.

[2] A. P. Lidov, *Zhur. Russ. Fiz.-Khim. Obshchestva* **34**, 42 (1930).

[3] N. A. Menshutkin, *Analiticheskaya Khimiya* [Analytic Chemistry], State Publishers, 1928.

[4] A. R. Kizel', *Prakticheskoe rukovodstvo po biokhimi rastenii* [Practical Handbook on the Biochemistry of Plants], State Publishers of Biology and Medicine, 1934.

[5] M. O. Korshun and N. E. Gel'man, *Novye metody elementarnogo mikroanaliza* [New Methods of Elementary Microanalysis], State Chemical Publishing House, 1949.

[6] N. N. Ivanov, *Metody fiziologii i biokhimii rastenii* [Methods of Plant Physiology and Biochemistry], State Publishers of Biology and Medicine.

[7] V. S. Krym, *Uchebnik kolichestvennogo i tekhnicheskogo analiza* [Manual of Quantitative and Technical Analysis], State Chemical Publishing House, 1932.

[8] N. D. Pryanishnikov, *Praktikum po organicheskoi khimii* [Handbook of Organic Chemistry], State Chemical Publishing House, 1948.

[9] H. Meyer, *Analiz i predelenie stroeniya organicheskikh veshchestv* [Analyse und Konstitutionsermittlung organischer Verbindungen], State Chemical Publishing House, 1925.

[10] L. Gattermann and H. Wieland, *Prakticheskie raboty po organicheskoi khimii* [Die Praxis des organischen Chemikers], State Chemical Publishing House, 1948.

CHAPTER 49

A New Method for the Quantitative Determination of Sulfur, and for the Simultaneous Determination of Sulfur and Nitrogen in Organic Compounds*

P. N. Fedoseev and N. P. Ivashova

In our first paper [1] we showed that all the nitrogen in an organic compound is quantitatively converted on heating with metallic magnesium powder at 550-650° into magnesium nitride [2-10].

In this article a method is described for the determination of sulfur, and also for the determination of nitrogen and sulfur in one sample by the same principle. For the quantitative determination of sulfur in organic compounds, the compound is heated with magnesium powder at 550-650° for 30-40 minutes. All forms of organically bound sulfur are thus quantitatively converted into magnesium sulfide. The combustion of the organic sulfur compounds is carried out in an ordinary glass test tube. Nitrogen can also be determined simultaneously with sulfur [1]. On acid treatment an ammonium salt and hydrogen sulfide are formed. The hydrogen sulfide is determined iodometrically and, by addition of strong alkali to the flask after removal of the hydrogen sulfide, the ammonia can be distilled over into standard acid solution and determined by back titration of the excess acid with alkali.

Combustion (30-35 minutes) of a sample of the test compound is carried out as described in the previous article. The test tube and its content are ground up and transferred

*Zhur. Anal. Khim. 7, 116-19 (1952).

to a 500 ml round bottom long necked flask containing 50-80 ml of distilled water.

A stream of carbon dioxide is passed through the apparatus (see Fig. 1) in order to expel the air, in the presence of which, on addition of acid, an explosive mixture can be formed which can be ignited by the silicon hydride (which is inflammable in the presence of air). Air will also favor the oxidation of hydrogen sulfide. When the air has been expelled, a small excess (80-100 ml) of 12% hydrochloric acid is gradually added to the flask used for decomposition so as to dissolve the remainder of the magnesium and the magnesium sulfide. A large part of the hydrogen sulfide is expelled together with the hydrogen formed and is mainly absorbed in the first absorption flask, which contains 65 ml of a mixture of 10% $ZnSO_4$ and 10% CH_3COONa with 3-5 ml of concentrated acetic acid.

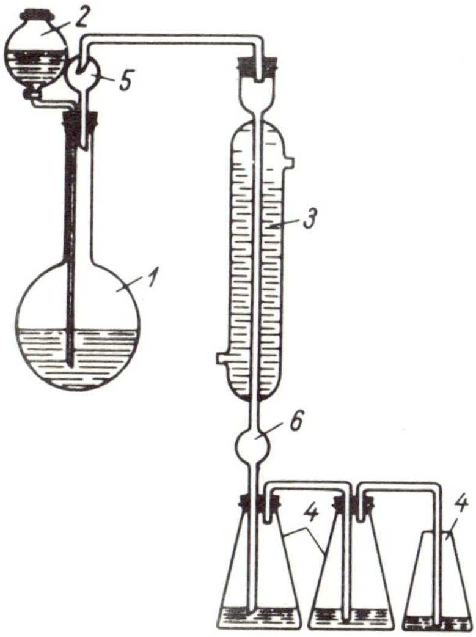

Fig. 1. Apparatus for expelling air with CO_2: 1) distillation flask; 2) dropping funnel; 3) condenser; 4) receivers; 5) splash head; 6) safety space.

TABLE 1

Molecular formula of compound and designation	S %			N %		
	Calc.	Found	Diff.	Calc.	Found	Diff.
Sulfosalicylic acid ($C_7H_6SO_6 \cdot 2H_2O$)	12.61	12.71	+0.10			
	12.61	12.69	+0.08			
Sodium alizarinsulfonate ($C_{14}H_7O_7SNa$)	9.37	9.20	−0.17			
	9.37	9.27	−0.10			
Sodium *p*-toluenesulfonate ($C_7H_7O_3SNa \cdot 2H_2O$)	13.92	13.65	−0.27			
	13.92	13.63	−0.29			
Bis(1,1-dimethylpenten-3-one) sulfide ($C_{14}H_{24}O_2S$)	12.60	12.50	−0.10			
	12.60	12.54	−0.06			
Acetothienone (C_6H_6OS)	25.25	25.28	+0.03			
	25.25	25.21	−0.04			
Ditolyl sulfide ($C_{14}H_{14}S$)	14.96	14.81	−0.15			
	14.96	14.69	−0.17			
Naphthenic acid ($C_{10}H_9NSO_3 \cdot \frac{1}{2}H_2O$)	13.81	14.04	+0.23	6.03	6.13	+0.10
	13.81	13.75	−0.06	6.03	6.16	+0.13
Sulfanilic acid ($C_6H_7O_3SN \cdot 2H_2O$)	15.32	15.30	−0.02	6.69	6.50	−0.19
	15.32	15.11	−0.21	6.69	6.73	+0.04

p-Toluenesulfamide ($C_7H_9SO_2N$)	18.72	17.51	-0.21	8.18	7.98	-0.20
	18.72	18.48	-0.24	8.18	7.89	-0.29
o-Toluenesulfamide ($C_7H_9SO_2N$)	18.72	18.56	-0.16	8.18	8.00	-0.18
	18.72	18.53	-0.19	8.18	8.01	-0.17
Diphenylthiourea ($C_{13}H_{12}N_2S$)	14.04	14.23	+0.19	12.27	11.98	-0.29
	14.04	14.15	+0.11	12.27	11.99	-0.28
p-Methoxyphenyl-Ψ-hydantoic acid	13.34	13.16	-0.18	11.66	11.86	+0.20
$C_{10}H_{12}O_3N_2S$)	13.34	13.39	+0.05	11.66	11.66	0.00
Sodium sulfanilate ($C_6H_6NSO_3Na \cdot 2H_2O$)	13.87	13.94	+0.07	6.06	5.90	-0.16
	13.87	13.80	-0.07	6.06	5.96	-0.10
N-Methylbenzanilide-*o*-methylsulfide	12.46	12.67	+0.21	5.44	5.30	-0.14
($C_{15}H_{15}SNO$)	12.46	12.23	-0.23	5.44	5.59	+0.15
2-(*p*-Aminobenzenesulfonamido)pyridine	12.86	12.66	-0.20	16.85	16.60	-0.25
$C_{11}H_{11}O_2N_3S$)	12.86	12.63	-0.23	16.85	16.68	-0.17
4-Sulfanilylsulfanilamide	19.59	19.48	-0.11	12.83	12.74	-0.09
$C_{12}H_{13}O_4N_3S_2$)	19.59	19.44	-0.15	12.83	12.67	-0.16
2-(*p*-Aminobenzenesulfonamido)thiazole	25.11	24.88	-0.23	16.46	16.26	-0.20
($C_9H_9O_2N_3S$)	25.11	25.02	-0.09	16.46	16.34	-0.12
2-Sulfanilamido-4,6-dimethylpyrimidine	11.52	11.29	-0.23	20.13	20.00	-0.13
$C_{12}H_{14}O_2N_4S$)	11.52	11.40	-0.12	20.13	19.93	-0.20

The second and third absorption flasks contain 50 and 25 ml of the same mixture respectively. After all (or almost all) the magnesium has dissolved, the flask is heated for 20-25 minutes to drive over any residual hydrogen sulfide with steam. In all, 70-80 ml of liquid is distilled over. When all the hydrogen sulfide has been driven over, 25 ml of 0.1 N iodine is added to the contents of the first two absorption flasks and acidified with a few drops of 12% hydrochloric acid. After standing for three minutes the excess iodine is titrated with thiosulfate.

Results of the determination of sulfur in a number of organic compounds are given in Table 1. They show that the analytical error lies within the range 0.2-0.3%.

For simultaneous determination of sulfur and nitrogen in organic compounds, combustion is carried out in the same way as for the separate determination of sulfur and nitrogen [1].

The residue after combustion of the organic compound with magnesium is transferred into a flask for decomposition. It is recommended that the sulfur be determined first by the method and in the apparatus already described (Fig. 1). When the hydrogen sulfide has been expelled, the receivers for the hydrogen sulfide are removed and flasks containing standard acid (hydrochloric or sulfuric acid) for absorption of the ammonia substituted instead. The condenser and Kjeldahl set-up are washed out while the two sets of flasks are being interchanged. Finally, by means of a separating funnel, 50-60 ml of 50% alkali is added to the contents of the flask and the ammonia distilled over.

Results for simultaneous determinations of sulfur and nitrogen are given in Table 1.

In addition we determined total sulfur in a number of coal samples, kindly given us by the Moscow All-Union Scientific Research Thermo-Technical Institute, for which we wish to express our thanks to the chief of the Heat Laboratory of the Institute, Comrade Zikeev. The results which are given in Table 2 show that our method is suitable for the quantitative determination of sulfur in coal.

TABLE 2

Coal used	% S found in the Labs of VTI[a]	% S found in the Labs of NKI[b]	Diff.	No. of VTI[a] analysis
Coal "T"	2.93	2.79	-0.14	5617
	2.93	2.65	-0.28	
Podmoskovni	3.48	3.30	-0.18	5771
	3.48	3.25	-0.23	
Silesian	2.23	2.04	-0.19	5753
	2.23	2.98	-0.25	
Coal "PS"	0.98	0.95	-0.03	5640
	0.98	0.86	-0.12	
Anthracite "ARSh"	1.93	1.64	-0.29	5644
	1.93	1.68	-0.24	
Komarovsk (Podmos.)	4.98	4.95	-0.03	6364
	4.98	4.94	-0.04	
Angrensk	2.70	2.73	+0.03	6694
	2.70	2.80	+0.10	
Donets	5.14	5.27	+0.13	6372
	5.14	5.02	-0.12	
Anthracite "ARSh"	4.35	4.46	+0.11	4907
	4.35	4.51	+0.16	
Brown Coal Dust	1.68	1.65	-0.05	6545
	1.68	1.60	-0.08	

a) F. E. Dzerzhinskii All-Union Scientific Research Thermotech. Inst.

b) Admiral S. O. Makarov Nikolaev Shipbuilding Institute

SUMMARY

1. A quantitative method for the determination of sulfur and for the simultaneous determination of nitrogen and sulfur in organic compounds is proposed.

2. The accuracy of the method lies in the range ±0.2-0.3%. The results of analyses give grounds for hoping that the given method may be used in a number of scientific-research and factory laboratories. Comparative results for the determination of total sulfur in different samples of coal carried out by the authors and the All-Union Scientific Research Institute, give good agreement.

3. The method of determination is simple, generally applicable, does not require complicated apparatus, and gives reliable results.

LITERATURE CITED

[1] P. N. Fedoseev and N. P Ivashova, *Zhur. Anal. Khim.* 7, 112, (1952).

[2] M. O. Korshun and N. E. Gel'man, *New Methods of Elemental Microanalysis*, 1949.

[3] N. E. Gel'man, Dissertation, Inst. Organic Chem., Acad. Sci. USSR, 1948.

[4] N. E. Gel'man, *Zavodskaya Lab.* **8**, 673 (1939).

[5] H. ter Meulen and J. Heslinga, *New Methods of Organic Chemical Analysis*, State Chem. Press, 1931.

[6] A. P. Shakhno, *Zavodskaya Lab.* **10**, 3 (1941).

[7] K. Burger, *Die Chemie* **54**, 479 (1941); **55**, 245 (1942).

[8] W. Zimmermann, *Mikrochemie* **31**, 15 (1943); **33**, 122 (1947).

[9] I. M. Kolthoff and E. B. Sandell, *Quantitative Analysis*, State Chem. Press, 1938.

[10] B. V. Nekrasov, *General Chemistry*, State Chem. Press, 1948.

CHAPTER 50

Micromethods for the Quantitative Determination of Carbon, Hydrogen, and Nitrogen in Organic Compounds With the Use of Vacuum*

P. N. Fedoseev and L. S. Ignatenko

The micromethod developed by V. A. Klimova [3, 4] is being widely used at the present time for the quantitative determination of carbon and hydrogen in organic compounds. The principle of this method consists in combustion of the sample of the organic compound in an empty tube at an oxygen flow rate of 35-50 ml/min under atmospheric pressure. Not only are the vapors of the substance subject to combustion but also, more importantly, the products of thermal decomposition of the sample.

The Pregl-Dumas [6] micromethod is used most frequently for the quantitative determination of nitrogen in organic substances. However, this method which is widely used in analytical practice has a number of drawbacks and gives unsatisfactory results for several classes of organic compounds. Thus, pyrazolines, hydrazines and their derivatives, diazo compounds, some methyl or ethyl esters and other substances burn poorly in the Dumas method. Running an analysis by this method requires special attention by the analyst and a whole series of precautions; viz., slow heating of the sample, scrupulous control of the carbon dioxide flow rate for displacement of the reaction products into the azotometer, etc. However, even when all these

Izvest. Akad. Nauk Turkmen.SSR* **1959, No. 1, 45-52.

precautions are observed, it is still possible to obtain unsatisfactory results. This is explained by various causes: cleavage of methane or ethane, leaks of undecomposed material, nitrogen dioxide, carbon monoxide, all of which give distorted analytical results.

The duration, the painstaking nature of the analysis and the fact that it is not always possible to displace the occluded air in the copper, especially in that part of the combustion tube which is not heated at the beginning of the analysis, must be counted among the shortcomings of the method which cause one to obtain high results almost all the time.

The difficulty of running the analysis according to the Dumas method had attracted the attention of many authors. Thus, N. E. Gel'man [2] points out that "everyone who must determine nitrogen by this method knows how dissimilar is the behavior of various substances during decomposition. The rate of decomposition of various substances cannot fit any standard conditions and may be regulated only by the hand and eye of the analyst. In some cases, during the nitrogen determination by the Dumas method, the decomposition of the sample is not accompanied, in general, by the evolution of gas bubbles in the azotometer. Then, the analyst runs the decomposition "blind", and often the first determinations for such substances give inaccurate results."

In our recent papers [7, 8, 9] we described a technique for the determination of C, H, and N working with samples of 30-40 mg. The semimicro methods developed by us were based on the principle of "vacuum combustion." Their essence consists in decomposition of the substance being analyzed under a vacuum (1-5 mm of mercury) in the combustion tube with heating in the presence of copper oxide, with a subsequent combustion of the reaction products in an oxygen stream, displacement of these into absorption tubes (determination of C and H), and displacement with a CO_2 stream of the gaseous reaction products into the azotometer (nitrogen determination). These methods were checked on a large number (about 600) of organic substances with various structures and compositions.

We noticed in this work that combustion under a vacuum greatly simplifies and facilitates the analysis. The combustion in a closed and evacuated tube guarantees the analyst freedom from the occurrence of flashes, explosions, passage of undecomposed sample material into the absorption tubes, and assures the complete reduction of nitrogen oxides over metallic copper, which is formed by reduction of the copper oxide by the decomposition products of the substance being analyzed.

The apparatus used by us is simple, available and dependable.

The difficult problem of the introduction of volatile liquids into the combustion tube without losses was also solved.

In the present paper we describe a micro modification of our methods for the quantitative determination of carbon, hydrogen and nitrogen with the use of vacuum, for samples weighing 3-7 mg.

DETERMINATION OF CARBON AND HYDROGEN

The Nature of the Method. The sample is subjected to thermal decomposition in an evacuated tube (1-5 mm of mercury) in the presence of copper oxide, with a subsequent combustion of the reaction products in an oxygen stream and their displacement into absorption tubes containing Anhydrone (H determination) and Ascarite (C determination).

Nitrogenous substances (nitrogen in any form) are burned without the introduction of special absorbents or reducers for nitrogen oxides either in the combustion tube or outside of it. In the combustion of halogen- and sulfur-containing substances, the sulfur and the halogen are retained in the combustion tube by means of metallic silver that has been heated to the necessary temperature.

Preparation of Apparatus for Analysis and its Operation. The apparatus is assembled according to Fig. 1.

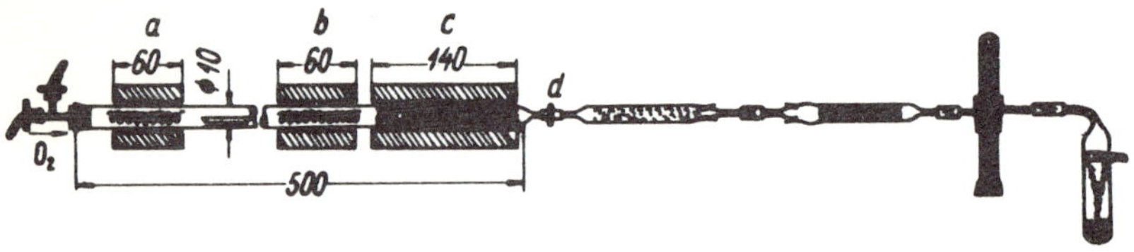

Fig. 1. Apparatus for the determination of carbon and hydrogen

The quartz combustion tube is connected to the drying system for oxygen through a glass Tee-adapter, one arm of which connects to a vacuum pump. Screw clamps A and B are placed on the rubber vacuum tubing leading from the adapter.

If one works with oxygen from an electrolytic source, the use of a copper oxide purifier and an electric furnace for heating it is necessary. Thus, the set-up in front of the combustion tube remains the same for the method being described as is accepted for C and H determinations [4].

A layer of copper oxide in the form of wire is introduced into the combustion tube after shaping into a plug 100-120 mm long, along with two oxidized copper spirals 60-70 mm long. The absorption tubes are connected to the drawn-out end of the combustion tube. The first tube, containing Anhydrone, is used for water absorption; the second one, containing Ascarite, for absorption of carbon dioxide. The second tube, besides Ascarite, is also filled with Anhydrone in the form of a layer about 30 mm thick (at the exit from the tube) to capture moisture that is contained in the Ascarite. The ground joints are sealed with a special cement [4].

After the tube with Ascarite come the Abrahamczik tube containing Anhydrone and the guard tube with concentrated sulfuric acid for the prevention of the entry of water vapor from the atmosphere into the system and for control of the speed of the gas sweeping. The rubber vacuum tubing which connects the drawn-out tip of the combustion tube to the absorption tube is selected to have such a length (50 mm) as to facilitate the placement of the screw clamp D onto it. Three hinged electric furnaces *a*, *b* and *c* are necessary for heating the combustion tube.

The assembled apparatus is checked for leaks, then the combustion tube is heated over its entire length in an oxygen stream with the furnaces *a*, *b* and *c* heated to 800-850° for 30-50 min. A blank run is made followed by a check combustion of a known pure substance (*p*-aminobenzoic acid, azobenzene, or sucrose). If these experiments indicate that the apparatus is ready for analysis, one begins the combustion of the substance given for the investigation.

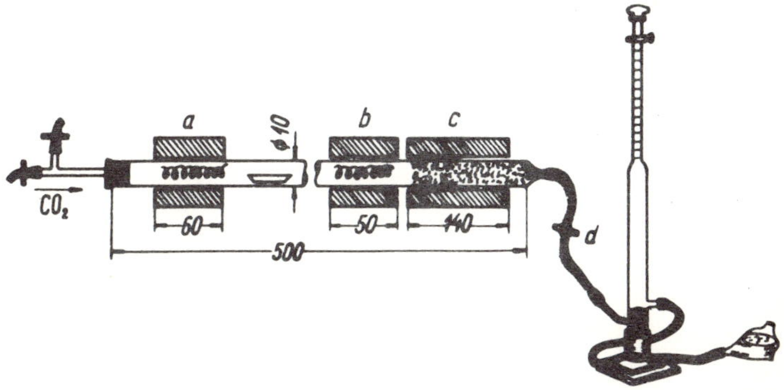

Fig. 2. Set-up for nitrogen determination with the use of vacuum.

The sample of the solid (3-7 mm) is placed in a boat. Taking out the first spiral, one moves the boat with the sample into the tube to a distance of 45-50 mm from the furnace and replaces the spiral. By this time furnaces *a* and *b* have been heated to 800-850° and furnace *c* to 850-900°. Furnace *a* remains in the heated condition next to the combustion tube during the introduction of the sample.

Following the introduction of the sample, furnace *a* is brought over the tube. With clamps A and D closed, air is pumped from the combustion tube (20-30 sec) by means of the vacuum pump (the sample during this period remains in the cold zone). Closing clamp C, furnace *b* is rapidly brought over the sample. The substance is subjected to thermal decomposition and part of it burns with the oxygen from the copper oxide. In the course of 3-5 min the combustion tube is heated over its entire length by movement of furnace *c*. If the sample contains nitrogen in oxidized form, the reaction products are kept in the closed tube for 10 min to complete the reduction of nitrogen oxides by metallic copper formed from the copper oxide, after which clamp A is opened (since the pressure inside the tube is lowered if the sample is not more than 8 mg), after which clamp D is opened and the reaction products are displaced into the absorption tubes by a stream of oxygen, which requires 5-8 min with a simultaneous heating of the combustion tube over its entire length. Gas sweeping is run at the rate of 30-35 ml/min (for the first 3-4 min), after which it is raised to 50-55 ml/min for removal of water from the tip of the combustion tube and the capillary of the absorption apparatus with Anhydrone.

The absorption tubes are disconnected, wiped, passed over a burner flame, transferred to the weighing room and, after cooling, weighed. The percentage content of carbon and hydrogen is calculated from the well-known formulas

[4]. The error in the hydrogen determination is ±0.15-0.20%, that for carbon amounts to ±0.15-0.30%.

For combustion of halogen- and sulfur-containing substances, one introduces into the combustion tube some metallic silver in addition to the copper oxide.

Inside the first spiral (zone of furnace *a*), which is made of copper oxide, another spiral is placed, this being of smaller diameter and made of silver wire. Instead of the second spiral (zone of furnace *b*) one uses a silver spiral 60-70 mm long. The absorption of sulfur oxides and halogens is carried out at 700-800°.

Combustion of Volatile Substances. For the combustion of substances like naphthalene, the sample is placed in the boat and pure, ignited anhydrous powdered copper oxide is added, this being carefully tamped down. The boat is placed inside the combustion tube, as usual.

Our studies showed that during the pumping out (20-30 sec) no appreciable loss of the substance is observed as far as the analysis is concerned (as in analysis of naphthalene, for example).

The sample of a liquid volatile substance (benzene, octane, pentane, etc.) is taken in a sealed ampule and placed in a special copper container which consists of two parts. The breaking of the ampule occurs in the evacuated but already closed tube. The movement of the furnace over the sample (furnace *b*) is made not at once but some 2-3 minutes following the breaking of the ampule.† An electromagnetic coil [1] may be used to break the ampule.

† The technique of combustion of such substances is described in the paper "A New Method for the Quantitative Determination of Carbon and Hydrogen in Organic Substances with the Use of Vacuum," *Izvest. Akad. Nauk Turkmen. SSR* 1957, No. 3, 24. The size of the container must be reduced to fit the microcombustion tube.

A sample of a nonvolatile liquid is taken in an unsealed ampule. This is placed in the boat and the analysis is run as usual.

The data from the analyses of various substances which were run by the method described here are shown in Table 1. It is evident that the accuracy of the carbon and hydrogen determination is quite satisfactory.

NITROGEN DETERMINATION

Nature of the Method. The organic substance is subjected to thermal decomposition in an evacuated tube in the presence of copper oxide. The products of thermal decomposition are burned with the oxygen from the copper oxide, after which they are displaced by a CO_2 stream into the azotometer with alkali.

Assembly of the Apparatus and Tests for its Tightness. The apparatus is assembled according to Fig. 2.

The combustion tube is connected with a 3-way glass adapter to a source of carbon dioxide (a Kipp generator) and to the vacuum pump. The tube filling is the same as used for the C and H determination. The azotometer filled with 50% potassium hydroxide solution is attached to the drawn-out end of the combustion tube by a piece of rubber vacuum tubing (18-20 cm).

The assembled apparatus is tested for tightness: with clamps A and D shut, air is pumped out of the tube, after which a stream of carbon dioxide is passed through it for 1-2 min, after which the tube is again pumped out. The apparatus is left for 15-20 min (clamps A, C and D closed) after which carbon dioxide is introduced again through the tube. If the system is tight, no air should be found in the azotometer.

Performance of the Analysis. Being satisfied with the tightness of the apparatus, one begins the analysis.

The sample of the solid (3-7 mg) is placed in the boat and covered by a layer of powdered copper oxide. Liquid substances are weighed in an ampule, and volatile substances in a sealed ampule.

The sample is introduced into the tube and air is pumped out (20-30 sec) of the combustion system with clamps A and D closed. Opening clamps A and D, the system is flushed (1-2 min) with a stream of CO_2 and is again pumped out. The two pumpings of the gas from the tube guarantee the absence of any possible residue of air in the system.

The above operation is conducted with the furnaces kept hot: *a* and *c* at 800-850°, *b* at 850-900° (furnace *a*, hot, is outside the system; the sample is located in the cold zone of the tube).

After the second removal of gas from the system, furnace *a* is brought over the tube and, when the combustion tube has been heated in the heating zone of this furnace, furnace *b* is brought over the sample. The substance pyrolyzes and burns with the oxygen from the copper oxide. For 5-6 min the combustion tube is heated by movement of furnace *b*. The other furnaces stay in place: their purpose is to prevent passage of completely undecomposed material past the rubber stopper or the azotometer.

Nitrogen oxides are reduced by the resulting metallic copper. In the combustion of a substance from which much nitrogen oxide may be expected, the reaction products are kept in the closed tube, while the latter is continuously heated for some time (7-8 min), after which clamp D is cautiously opened and the behavior of the mercury column in the take-off leg of the azotometer is observed. If the pressure in the combustion tube is low, clamp D is closed, after which clamp A is opened (followed by D) and the gaseous reaction products are displaced with the CO_2

TABLE 1
Determination of Carbon and Hydrogen

Substance and Empirical Formula	mg Sample	mg H_2O	mg CO_2	% H Calc.	% H Found	% H Diff.	% C Calc.	% C Found	% C Diff.
p-Aminobenzoic acid	5.556	2.600	12.502	5.14	5.23	+0.09	61.31	61.40	+0.09
$C_7H_7NO_2$	5.905	2.785	13.290	5.14	5.27	+0.13	61.31	61.40	+0.10
Diazoaminobenzene	7.950	4.100	21.219	5.62	5.77	+0.15	73.07	72.83	-0.24
$C_{12}H_{11}N_3$	6.370	3.329	17.022	5.62	5.84	+0.22	73.07	72.92	-0.15
m-Nitrobenzaldehyde	8.633	2.777	17.653	3.35	3.59	+0.24	55.63	55.80	+0.17
$C_7H_5NO_3$	5.580	1.769	11.386	3.35	3.54	+0.19	55.63	55.68	+0.05
Anthracene	5.898	3.101	20.379	5.66	5.88	+0.22	94.34	94.29	-0.05
$C_{14}H_{10}$	6.001	2.989	20.746	5.66	5.57	-0.09	94.34	94.34	0.00
Phenanthrene	5.250	2.659	18.150	5.66	5.66	0.00	94.34	94.34	0.00
$C_{14}H_{10}$	5.135	2.608	17.699	5.66	5.68	+0.02	94.34	94.06	-0.28
β-Methylnaphthalene	6.750	4.350	22.973	7.08	7.21	+0.13	92.91	92.87	-0.04
$C_{11}H_{10}$	8.149	5.211	27.702	7.08	7.15	+0.07	92.91	92.77	-0.14

Azobenzene	6.350	3.258	18.468	5.53	5.74	+0.21	79.09	79.36	+0.27
$C_{12}H_{10}N_2$	6.578	3.315	19.122	5.53	5.63	+0.10	79.09	79.33	+0.24
β-Naphthylamine	7.950	4.501	24.350	6.33	6.33	0.00	83.88	83.58	-0.30
$C_{10}H_9N$	5.380	3.020	16.495	6.33	6.28	-0.05	83.88	83.67	-0.21
Diphenylamine	6.593	3.899	20.585	6.55	6.61	+0.06	85.16	85.20	+0.04
$C_{12}H_{11}N$	7.030	4.115	21.899	6.55	6.55	0.00	85.16	85.01	-0.15
Picric acid	6.455	0.100	7.498	1.31	1.73	+0.43	31.45	31.69	+0.26
$C_6H_2O_7N_3$	6.389	0.081	7.367	1.31	1.41	+0.10	31.45	31.46	+0.01
Benzene	6.875	4.851	23.205	7.75	7.89	+0.14	92.25	92.11	-0.14
C_6H_6	4.346	2.998	14.701	7.75	7.71	-0.04	92.25	92.31	+0.06
Isooctane	5.848	8.096	18.000	15.88	15.49	-0.39	84.11	83.99	-0.12
C_8H_{18}	5.342	7.559	16.497	15.88	15.83	-0.05	84.11	84.27	+0.16
Trithioformaldehyde	7.548	3.117	7.216	4.37	4.62	+0.25	26.05	26.08	+0.03
$C_2H_6S_3$	8.465	3.475	8.129	4.37	4.59	+0.22	26.05	26.20	+0.15
Dinitrochlorotoluene	5.342	1.131	7.597	2.32	2.36	+0.04	38.81	38.80	-0.01
$C_7H_5O_4N_2Cl$	3.414	0.705	4.869	2.32	2.31	-0.01	38.81	38.92	+0.11

TABLE 2

Nitrogen Determination

Name of substance and Empirical formula	mg Sample	T°	P_{mm}	ml Nitrogen	% N Calc.	% N Found	Diff.
p-Aminobenzoic acid $C_7H_7NO_2$	4.450	21°	763	0.397	10.21	10.39	+0.18
	5.000	22°	763	0.452	10.21	10.49	+0.28
Azobenzene $C_{12}H_{10}N_2$	7.340	22°	763	0.980	15.37	15.50	+0.13
	4.650	22°	763	0.621	15.37	15.50	+0.13
β-Naphthylamine $C_{10}H_9N$	5.250	22°	763	0.443	9.78	9.79	+0.01
	4.250	23°	763	0.370	9.78	10.08	+0.30
Diazoaminobenzene $C_{12}H_{11}N_3$	4.600	23°	763	0.854	21.30	21.49	+0.19
	4.850	23°	763	0.902	21.30	21.53	+0.23
m-Nitrobenzaldehyde $C_7H_5NO_3$	6.750	19.5°	770	0.537	9.26	9.42	+0.16
	5.361	19.5°	770	0.423	9.26	9.35	+0.09
3.4-Tetramethylene-5.5-pentamethylenepyrazole picrate $C_{18}H_{21}N_5O_7$	4.658	20°	771	0.659	16.69	16.72	+0.03
	4.865	20°	771	0.685	16.69	16.64	-0.05
1.3.5-Triphenylpyrazole $C_{21}H_{16}N_2$	4.752	20°	771	0.379	9.45	9.42	-0.03
	5.135	20°	771	0.412	9.45	9.48	+0.03

Sodium salt of 1-dithiocarboxy-							
3,5,5-trimethylpyrazoline	3.231	20°	771	0.364	13.32	13.31	−0.01
$C_7H_{11}N_2S_2Na$	4.225	20°	771	0.478	13.32	13.37	+0.05
3-Methyl-5-α-furylpyrazole	5.758	20°	771	0.930	18.90	19.09	+0.19
$C_8H_8N_2O_7$	4.385	20°	771	0.708	18.90	19.08	+0.18
4-Ethyl-5-propylpyrazole picrate	5.522	21°	767	0.901	19.06	19.12	+0.06
$C_{14}H_{17}N_5O_7$	5.065	21°	767	0.823	19.06	19.04	−0.02
3-*tert*-Butylpyrazole picrate	7.285	21°	767	1.228	19.82	19.75	−0.07
$C_{13}H_{15}N_5O_7$	4.915	21°	767	0.831	19.82	19.81	−0.01
3,4,5-Trimethylpyrazole picrate	7.350	22°	764	1.313	20.64	20.77	+0.13
$C_{12}H_{13}N_5O_7$	5.665	22°	764	1.001	20.64	20.56	−0.08
3,5-Diphenylpyrazole picrate	5.350	20°	770	0.706	15.48	15.57	+0.09
$C_{21}H_{15}N_5O_7$	5.296	20°	770	0.705	15.48	15.70	+0.22
1-Phenyl-3-aminopyrazole picrate	5.230	21°	770	0.980	21.64	22.03	+0.39
$C_{15}H_{12}N_6O_7$	6.590	20°	775	1.202	21.64	21.66	+0.02
3-Methyl-5-α-furylpyrazole picrate	5.455	20.5°	774	0.862	18.56	18.72	+0.16
$C_{14}H_{11}N_5O_8$	6.130	20.5°	774	0.971	18.56	18.77	+0.21
Piperidine carbonate	5.145	23°	770	0.531	12.05	12.05	0.00
$C_{11}H_{24}N_2O_3$	5.217	23°	770	0.532	12.05	11.91	−0.14

stream into the azotometer with 50% KOH solution at the rate of 3-4 bubbles per second (toward the end of the displacement the passage rate may be increased to 5-6 bubbles, and it is permissible to turn off the furnaces). The duration of the displacement (5-8 min) depends on the size of the sample used and on the content of nitrogen in the compound being analyzed. The end of the displacement is determined by the appearance of microbubbles in the azotometer.

After the completion of the displacement, clamp D is closed and the azotometer leveling bulb is raised to the level of the potassium hydroxide solution, and is then fixed in that position. The azotometer is not disconnected from the combustion tube during this operation. The potassium hydroxide solution is allowed to flow down the inner walls of the azotometer (10 min) and the volume of the collected nitrogen is read with an accuracy of 0.001 ml. Noting the barometric pressure and temperature, one calculates the percentage content of nitrogen using the well-known formula [5].

The error of the method is ±0.1-0.25%.

Some of the analytical results in nitrogen determinations are shown in Table 2.

Various Notes. Prior to the beginning of the displacement process the filling of the azotometer should be checked. The end of the displacement of nitrogen is best done with the furnaces cooling off. After the completion of the displacement, furnace *a* is removed from the tube and pushed back so that the front part of the tube will cool more rapidly.

In the analysis of substances with a high percentage of nitrogen (over 30%) the graduated part of the azotometer may be insufficient, since the standard microazotometers usually have a graduated part of only 1.5 ml. For the nitrogen determinations in such compounds it is possible to use an azotometer without the upper funnel and to transfer the nitrogen into a eudiometer of appropriate capacity.

With the accumulation of a large amount of metallic copper in the combustion tube, the tube is heated in a stream of oxygen. The replacement of the alkali in the azotometer is done after 10-15 combustions.

In taking the sample it is best to avoid samples larger than 10 mg since in such cases a considerable pressure may develop in the combustion tube which might lead to the pushing out of the rubber stopper.

In the analysis of volatile and unstable substances the pumping out of the gas is done but once. In this case a correction is made for air, the traces of which may remain after the first pumping-out. This volume of air is established by a separate experiment prior to the combustion of the sample. The size of the correction is usually not over 0.01-0.005 ml.

Discussion of the Method. An entire series of organic compounds was analyzed by the above-described method for quantitative nitrogen determination, these compounds having given unsatisfactory results by the Dumas method.

The microtechnique of nitrogen determination with the use of vacuum was checked in Professor A. P. Terent'ev's laboratory at the Moscow State University and gave satisfactory results.

The ease of combustion of difficultly combustible and volatile substances is explained by us on the basis of the presence of the substance being analyzed in a closed system until the complete termination of the process of its decomposition and reduction of nitrogen oxides.

SUMMARY

1. New micromethods are described for the quantitative determination of carbon, hydrogen and nitrogen in organic substances. These methods are based on the decomposition of such substances in an evacuated tube in the presence of copper oxide with heating and a subsequent displacement of the reaction products into absorption tubes.

2. It was shown feasible to run the analysis without special reducing agents or absorbents for nitrogen oxides in the combustion of any nitrogenous substances.

3. The methods are simple to carry out, allow one to analyze rapidly, conveniently, dependably and without appreciable losses, various volatile and unstable substances as well as polynitro compounds. The conditions of formation of explosive mixtures are excluded under the conditions for these methods.

4. Working with a single apparatus in one working day one may run 10-12 combustions for nitrogen and 8-10 for carbon and hydrogen.

5. The methods of analysis in closed evacuated tubes may be quite useful and find a wide area of application in analytical practice.

LITERATURE CITED

[1] S. Baibaeva, *Microanalysis of the Elemental Composition of Low-Boiling Petroleum Products*, *Novosti Neftyanoi Tekh.*, *Neftepererabotki 1954*, No. 2.

[2] N. E. Gel'man, *Microdetermination of Sulfur and Nitrogen by Hydrogenization of Organic Compounds*, Dissertation, Inst. Org. Chem. Acad. Sci. USSR, 1948.

[3] V. A. Klimova, *Micro- and Semimicromethod for the Quantitative Determination of C and H in Organic Substances*, Dissertation, Inst. Org. Chem. Acad. Sci. USSR, 1948.

[4] M. O. Korshun and N. E. Gel'man, *Novye metody elementarnogo mikroanaliza* [New Methods of Elemental Microanalysis], Moscow, 1949.

[5] J. Niederl and W. Niederl, *Micromethods of Quantitative Organic Analysis* [Russian translation], Goskhimizdat [State Publishing House of Chemical Literature], 1949.

[6] F. Pregl, *Quantitative Organic Microanalysis* [Russian translation] Moscow-Leningrad, 1934.

[7] P. N. Fedoseev and L. S. Ignatenko, "A New Method for the Quantitative Determination of Carbon and Hydrogen in Organic Compounds with the Use of Vacuum," *Izvest. Akad. Nauk Turkmen. SSR* **1957**, No. 3, 24.

[8] P. N. Fedoseev and L. S. Ignatenko, "A Method for the Quantitative Determination of Nitrogen in Organic Substances with the Use of Vacuum," *ibid.*, **1957**, No. 4, 13.

[9] P. N. Fedoseev and L. S. Ignatenko, "A Method for the Quantitative Determination of Carbon and Hydrogen in Organic Compounds Containing Sulfur and Halogens, with the Use of Vacuum," *ibid.*, **1957**, No. 6, 84.

CHAPTER 51

Micromethods for Quantitative Determination of Carbon, Hydrogen, and Nitrogen in Organic Substances in Small Tubes Under Vacuum*

P. N. Fedoseev and T. E. Chernysheva

One of the serious shortcomings of all the methods that have been proposed for the quantitative determination of carbon, hydrogen, and nitrogen in organic substances is the fact that combustion is conducted in an open-ended flow system. This frequently leads to explosions and breakthrough of incompletely oxidized substance into the absorption apparatus, with detrimental results to analytical accuracy. The difficulty is present even in the combustion of volatile liquids and nitro compounds, which may form explosive mixtures with the oxygen in which the combustion is conducted.

We have set out to simplify the combustion of organic substances and to develop a method which would not require the analyst to resort to different combustion techniques for different types of substances.

The principle of the micromethods that we have developed for the determination of C, H, and N consists of a pyrolytic decomposition of the sample (3-5 mg) at 800-900° in a closed evacuated tube (1-5 mm Hg) in the presence of cupric oxide, followed by holding the products of the pyrolysis of the substance in the closed tube for 3-5 minutes, during which pyrolysis is completed and oxides of nitrogen are decomposed. Finally, oxygen (to complete

Izvest. Vysshikh Ucheb. Zavedeniĭ, Khim. i Khim. Tekhnol.* **2, 899-903 (1959).

the combustion of the pyrolysis products in the determination of C and H) or carbon dioxide (in the determination of nitrogen) is admitted into the tube and the CO_2 and H_2O formed are displaced into the absorption apparatus (or, in the case of N_2, into a eudiometer tube or nitrometer) [1, 2, 3].

Conducting the analysis in a closed tube under vacuum has the following advantages.

1. The most difficult part of the analysis, vaporization and pyrolysis, takes place smoothly, without special attention of the analyst, and in the same manner for any substance. Vacuum creates ideal conditions for pyrolysis.

2. Any substance containing C, H, O, and N under such conditions is completely converted into carbon, carbon dioxide, water, and elemental nitrogen.

3. The vacuum method makes it possible to reduce to a minimum or even to suppress completely the formation, during analysis, of products such as CO, CH_4, and oxides of nitrogen, which invariably are detrimental to analytical results.

4. The method of determining nitrogen offers a possibility of analyzing those groups of substances (picrates, pyrazolines, hydrazines, nitriles, and others) which in other known methods are analyzed with difficulty and give unreliable results.

In developing the "vacuum microcombustion" method, we have made use of the ordinary apparatus used in microanalysis. However, during the course of this work we came to the conclusion that for sample sizes of 3-5 mg it is desirable to use smaller combustion tubes, 350-400 mm long and 3.5-4 mm in diameter (resulting in about a tenfold decrease in the volume of the tube compared to tubes customarily used in microanalysis). Such tubes are cheaper and more readily available.

The reduction of the size of the combustion tube is of special importance in the determination of nitrogen, since the removal of air from such a small tube before the decomposition of the sample, and the displacement of nitrogen into the nitrometer after combustion, require far less time than in work with ordinary tubes. Thus, the time required for one determination of nitrogen is reduced to 3-5 minutes.

Combustion of more than two hundred organic substances in small tubes showed that quantitative determination of C, H, and N under such conditions gives more consistent and reproducible results.

It is known that the weights of the samples used in the analysis by the macro-, semimicro-, and micro-method are roughly in a ratio of 100 : 10 : 1. While the size of the tubes used in macroanalysis, tested by more than 125 years of use, is not open to any special objections, tubes used in microanalysis are much larger than is justified, approximately by a factor of ten [4]. In microanalysis, use should be made of tubes having a size in the same proportion to the size of the sample as in macroanalysis.

The present study was carried out to confirm this assumption.

EXPERIMENTAL PART

Determination of Carbon and Hydrogen

The construction of the apparatus is shown in Fig. 1. Into the quartz combustion tube is placed cupric oxide wire (5-6 g) and an oxidized copper spiral 50-60 mm long. In the analysis of halogen- and sulfur-containing substances the combustion tube is provided also with two silver spirals, each 30 mm long (Fig. 2). The vacuum rubber tubing from the T-tube, which connects the purifying system with the vacuum pump, is pushed directly over the end of the combustion tube. Such a connection is reliable, strong,

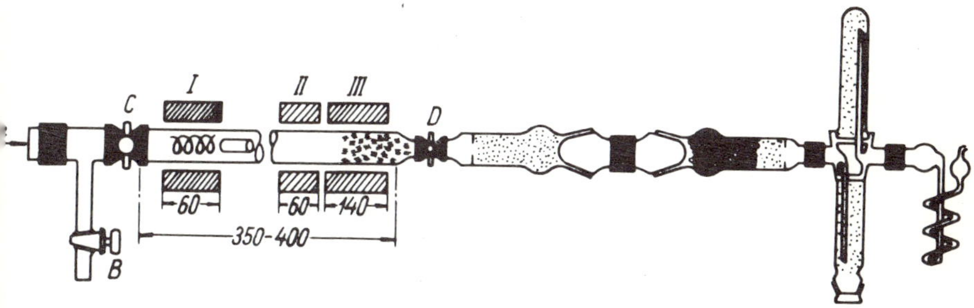

Fig. 1. Apparatus for determination of carbon and hydrogen: I, II, III) hinged electric furnaces; B) stopcock; C, D) pinch clamps.

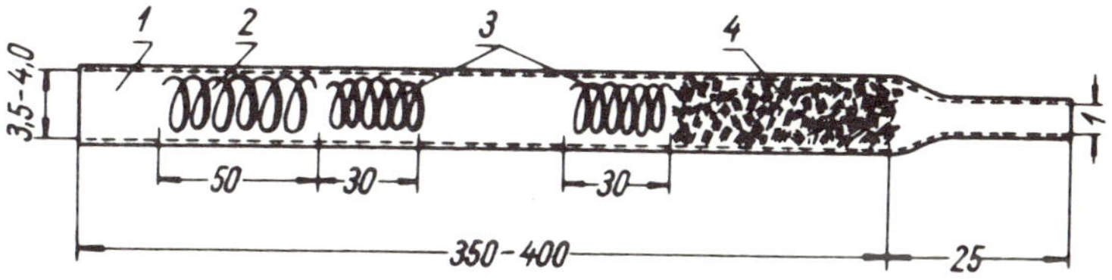

Fig. 2. Combustion tube and filling: 1) quartz tube; 2) oxidized copper spiral; 3) silver spiral; 4) cupric oxide wire (5 g).

and convenient. The absorption tubes are connected to the narrow end of the tube. The absorption tubes are filled in the usual manner [5]. Over the vacuum tubing between the narrow end of the combustion tube and the absorption tubes is slipped the pinch clamp D. The combustion tube is heated by three hinged electric furnaces, I, II, III. The assembled apparatus is checked for leaks [1, 2]. The combustion tube is then heated over its entire length in a stream of oxygen for 30 minutes at a furnace temperature of 750-800°. A blank determination is run, after which the actual analysis is carried out.

The sample of the substance (3-5 mg) is placed in a quartz capillary 20-25 mm long and 2.5-3 mm in diameter. The spiral is taken out, the capillary is placed in the combustion tube, and the spiral is replaced. At this point the furnaces I and III should be at 750-800° and furnace II at 850-900°. After introducing the sample, furnace I (which stands alongside) is moved under the combustion tube. Air is pumped (to 1-5 mm Hg) from the combustion tube for 2-3 seconds with the pinch clamp D closed. After closing stopcock B and pinch clamp C, furnace II is rapidly moved over the sample; the substance decomposes.

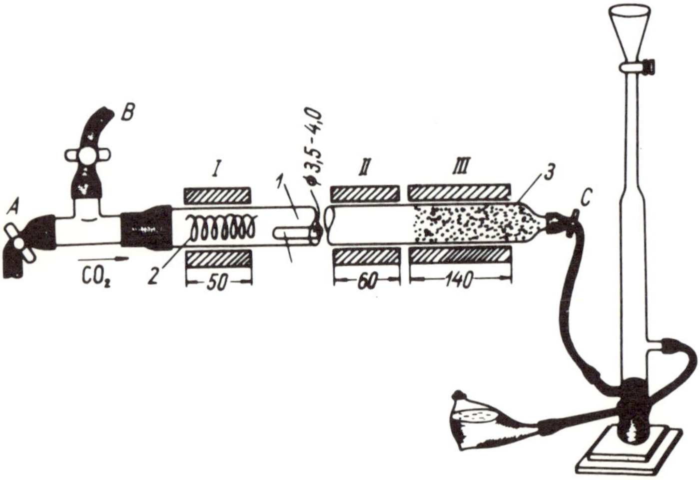

Fig. 3. Apparatus for determination of nitrogen: 1) quartz tube; 2) oxidized copper spiral; 3) cupric oxide wire; 4) quartz capillary; A, B, C) pinch clamps; I, II, III) hinged electric furnaces.

Moving furnace II, the combustion tube is heated for 3-5 minutes. When the substance contains nitrogen in an oxidized form, the time of heating is increased to 8 minutes until the oxides of nitrogen have completely decomposed under action of the copper formed in the combustion tube. The products of the reaction are swept out with a stream of oxygen at a rate of 25 ml/min (3-5 min), after which the rate is increased to 50 ml/min (1-2 min) to remove water from the combustion tube. Twelve combustions can be carried out in one working day.

Liquids are burned in sealed glass ampules placed in a copper cartridge [2]. Solid volatile substances are analyzed in a quartz capillary, covered with a layer of cupric oxide.

Results of the analysis of various substances are listed in Table 1.

DETERMINATION OF NITROGEN

The construction of the apparatus is shown in Fig. 3. The dimensions of the combustion tube and its filling are the same as in the determination of carbon and hydrogen. The combustion tube is connected through a glass T-tube to a Kipp apparatus and a vacuum pump.

The sample of the substance (3-5 mg) is placed into a quartz capillary and covered with a layer of cupric oxide; the capillary is then placed in the tube, from which air is exhausted with the pump for 2-3 seconds with the pinch clamps A and C closed. The pinch clamp B is then closed, clamps A and C are opened, a stream of CO_2 is passed through (0.5-1 min), and air is again pumped out. Two evacuations of the tube ensure the absence of traces of air in the system. The operation is carried out while furnaces I and III are at 750-800°, with the sample in the cold portion of the tube. Furnace II, which stands aside and is heated to 850-900°, is then drawn over the sample.

TABLE 1
Determination of Carbon and Hydrogen

Name and Formula of Sample	% H Calc.	% H Found	% H Diff.	%C Calc.	%C Found	%C Diff.
Cyclohexylamine benzoate,	8.65	8.64	−0.01	70.55	70.36	−0.19
$C_{13}H_{19}NO_2$	8.65	8.79	+0.14	70.55	70.26	−0.28
Morpholine β-phenylpropionate,	8.07	8.36	+0.29	65.79	66.02	+0.23
$C_{13}H_{19}NO_3$	8.07	8.32	+0.25	65.79	65.99	+0.20
N-Butyrylhexamethylenimine,	11.31	11.52	+0.21	70.95	70.82	−0.13
$C_{10}H_{19}NO$	11.31	11.56	+0.25	70.95	70.78	−0.17
1-Benzoyl-2-(N,N-hexamethyleneamino)	8.28	8.38	+0.10	67.27	66.99	−0.28
ethane hydrochloride, $C_{15}H_{22}ClNO$	8.28	8.10	−0.18	67.27	66.98	−0.29
1-(*p*-Bromobenzoyl)-2-(N,N-hexamethylene	6.10	6.14	+0.04	51.96	51.99	+0.03
amino) ethane hydrochloride, $C_{15}H_{21}BrClNO$	6.10	6.34	+0.24	51.96	51.79	−0.17
Cyclohexylamine - β-phenylpropionate,	9.29	9.27	−0.02	72.25	71.95	−0.30
$C_{15}H_{23}NO_2$	9.29	9.43	+0.14	72.25	71.96	−0.29
1-(*p*-Ethoxybenzoyl)-2-(N,N-hexamethylene	8.40	8.59	+0.19	65.47	65.19	−0.28
amino)ethane hydrochloride, $C_{16}H_{27}ClNO_2$	8.40	8.66	+0.26	65.47	65.26	−0.21

The substance is pyrolyzed and burned by the oxygen of the cupric oxide.

With substances whose combustion does not give a large amount of oxides of nitrogen, the tube is heated for not more than 3 minutes, while in the analysis of nitro compounds the time of heating is increased to 3-5 minutes. The gaseous reaction products are swept out with a stream of carbon dioxide at a rate of 3-4 bubbles per second for not more than 2-3 minutes, with the pinch clamps A and C open.

In the combustion of substances with a high nitrogen content, one nitrometer is frequently not sufficient for the entire volume of the nitrogen produced. In these cases two nitrometers are connected to the combustion tube by means of a T-tube.

The results of analyses of this series of experiments are listed in Table 2.

TABLE 2
Determination of Nitrogen

Name and Formula of Sample	Per cent nitrogen		
	Calcd.	Found	Diff.
3,5-Diethyl-5-methylpyrazoline, $C_8H_{16}N_2$	19.97	19.96	-0.01
		20.03	-0.06
Isoamyl hexamethylenedithiocarbamate, $C_{12}H_{23}NS_2$	5.70	5.65	-0.05
		5.61	-0.09
2-(p-Toluyl)-1-(N.N-hexamethyleneamino) propane hydrochloride, $C_1.H_2.ClNO$	4.73	4.49	-0.24
		4.61	-0.12
3.4-Diethyl-5-methylpyrazole picrate, $C_{14}H_1.N_5O.$	19.06	19.04	-0.02
		19.08	+0.02
Benzylamine -β-phenylpropionate, $C_1.H_{19}NO_2$	5.44	5.38	-0.06
		5.41	-0.03
3,5-Diethyl-4-methylpyrazole, $C.H_{14}N_2$	20.26	20.30	-0.04
		20.32	+0.04

SUMMARY

1. The possibility of using in the quantitative microdetermination of carbon, hydrogen, and nitrogen, combustion tubes reduced in size to about one-tenth of those used in organic elemental analysis has been studied.

2. It has been shown that quantitative determination of carbon, hydrogen, and nitrogen in such tubes gives more consistent and reproducible results. The error in the determination of carbon is ±0.15-0.30%, in the determination of hydrogen ±0.10-0.20%, and in the determination of nitrogen ±0.10-0.25%.

3. The method may be recommended for extensive application in various scientific-research and educational laboratories.

LITERATURE CITED

[1] P. N. Fedoseev and L. S. Ignatenko, *Izvest. Akad. Nauk Turkmen. SSR 1957*, No. 3, 23; *1957*, No. 4, 13; *1957*, No. 6, 28; *1959*, No. 1, 45.

[2] P. N. Fedoseev, L. S. Ignatenko and T. E. Chernysheva *Izvest. Vysshikh Ucheb. Zavedenii MVO SSSR*, Khim. i Khim. Tekhnol. *1958*, No. 5, 42.

[3] P. N. Fedoseev, L. S. Ignatenko and T. E. Chernysheva, *Referaty VIII Mendeleevskogo S'ezda* 1959, No. 3, 32.

[4] J. Körbl, *Collection Czechoslov. Chem. Communs.*, 20, 948, 953, 948, 953, 1026 (1955).

[5] M. O. Korshun and N. E. Gel'man, *Novye metody elementarnogo mikroanaliza* [New Methods of Elemental Microanalysis], Goskhimizdat [State Publishing House of Chemical Literature], Moscow-Leningrad, 1949.

CHAPTER 52

New Method for Rapid Automatic Combustion of Organic Substances, Micro determination of Carbon and Hydrogen*

M. Vecera, F. Vojtech and L. Synek

In the presence of active combustion catalysts which give off their own oxygen for combustion the necessary time of 3 to 5 minutes has been established for the quantitative combustion of organic substances in the case of carrying out of microelementary analyses [1]. With the earlier known automatic apparatuses, the combustion of organic substances cannot be carried out in a short enough time. The apparatuses calling for the automatic displacement of the combustion furnace [2] require observation in the case of short combustion times and regulation of the displacement velocity according to the course of the combustion. Earlier known complete automatic apparatuses [3,4] make possible combustion of the substance within 20 to 30 minutes; the reported apparatuses are complicated and require high frequency heating or manual ignition of the combustion tube after ending the combustion.

It was therefore our effort to find a method for the rapid combustion of organic substances with which means one could burn even volatile substances with high vapor tension as well as those decomposing explosively in the heating. We based our work on our good experience with the zone combustion of organic substances in an inert atmosphere as in the case of nitrogen combustion [5] and on the

Collection of Czechoslovak Chemical Communications* **25, 93-100 (1960).

experiments of the course of the rapid combustion of organic substances in oxygen atmosphere with the application of cobalt (II, III) oxide [6]. The problem of the rapid combustion is solved through distribution of the combustion space into several zones, for which heating is switched in step-wise and regulated automatically. The combustion in the case of insufficient oxygen supply which makes possible the rapid combustion of substances decomposing explosively in oxygen (for example, picrates) is solved by using a quartz capsule. This principle was proposed by Marek [7] for the combustion of macro samples of organic substances by manual displacement of the burner and was applied in analagous arrangement by Korshun [8] and Kuck [9] in the case of the combustion of micro samples in an empty combustion tube in the rapid stream of oxygen.

In the experimental part of this work the method for the micro determination of C and H in organic substances is described; analytical results with test compounds of different structure and their behavior upon combustion are presented. The automatic apparatus with the zone-wise combustion of the sample is described.

EXPERIMENTAL PART

Reagents. As combustion catalyst cobalt (II, III) oxide on corundum was used. Manganese (IV) oxide for collecting oxides of nitrogen was prepared from potassium permanganate and manganese (II) sulfate on the basis of the procedure described by us earlier. In view of the fact that manganese (IV) oxide upon storage takes on moisture, it was dried at 130°. Suitable preparations contained 16 to 18% water. Carbon dioxide is collected with sodium hydroxide on asbestos, water with magnesium perchlorate, halogen and oxides of sulfur with silver wool.

Apparatus. An apparatus illustrated in Figure 1 was employed. Oxygen from the pressure cylinder flows over the regulation valve *1* and the flowmeter *2* into the purifying apparatus *3* heated to a temperature of 600-650° and from here over the absorption tube *4* filled with magnesium perchlorate and sodium hydroxide asbestos into the quartz combustion tube *5*. The combustion of the organic substance is carried out by means of step-wise switching on of the combustion zones of the furnace *6*; the substance is placed in the platinum boat *7*, which is inserted in the quartz capsule *8*. In the quartz combustion tube are located the combustion catalyst *9* and the layer of silver wool *10*. Water is absorbed in apparatus *11* which is heated in its first third to a temperature of 120°C. by means of the furnace.

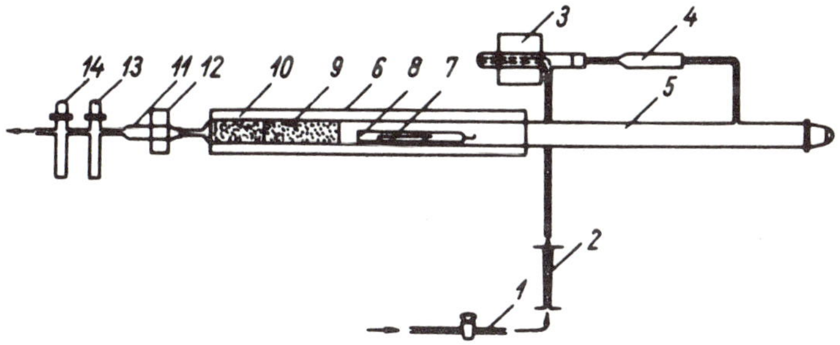

Fig. 1. Scheme of the apparatus for microdetermination of C and H with zone combustion. 1) regulator valve, 2) rotameter, 3) apparatus for oxygen purification, 4) absorption tube, 5) combustion tube, 6) electric furnace, 7) platinum boat, 8) quartz tube, 9) combustion catalyst, 10) silver wool, 11) water apparatus, 12) heating furnace, 13) apparatus for absorption of nitrogen oxides, 14) carbon dioxide apparatus.

Combustion Tube and Absorption Apparatus. The overall dimensions of the combustion tube and its filling are evident in Figure 2. The sample is weighed in the platinum boat (for dimensions see Figure 3) which is pushed into the quartz capsule illustrated in Figure 4. The absorption apparatuses are of the same type and similar dimensions as in our earlier work [1] and are joined to one another as well as to the combustion tube by means of ground joints.

Oxygen Purification and Rotameter. An apparatus illustrated in Figure 5 serves for the purification of the oxygen. The quartz test tube which is filled with the

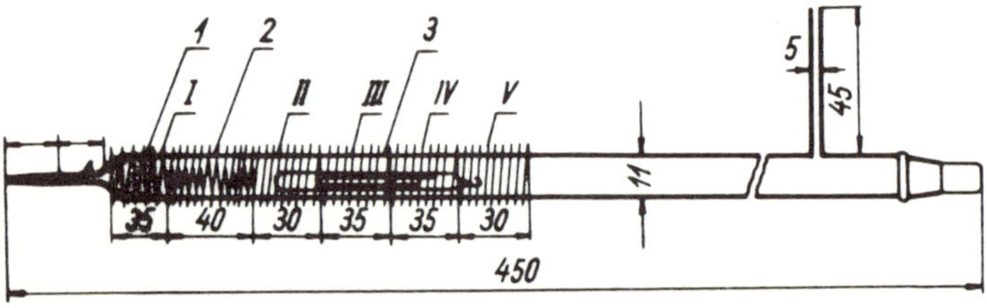

Fig. 2. Quartz Combustion Tube and Electric Combustion Furnace with the zones 1 to 5. 1) Silver wool, 2) cobalt (II, III) oxide on corundum, 3) platinum boat in the quartz capsule (dimensions in mm.).

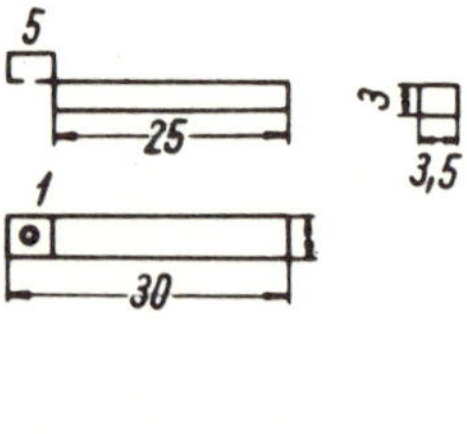

Fig. 3. Platinum Boat (Dimensions are given in mm.)

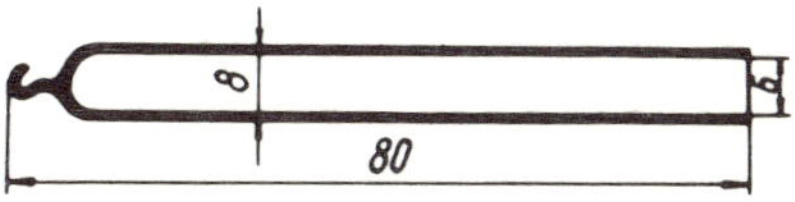

Fig. 4. Quartz Capsule (Dimensions are given in mm.)

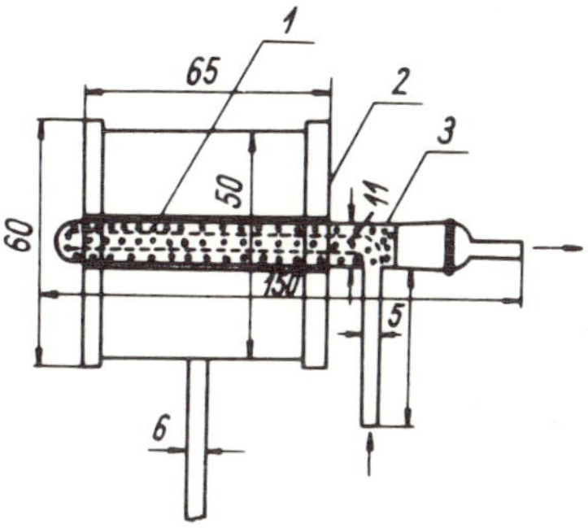

Fig. 5. The Oxygen Purification Apparatus. 1) cobalt (II, III) oxide on corundum, 2) electric furnace, 3) quartz tube.

combustion catalyst is placed in an electric furnace with a working temperature of 600-650°. A small rotameter serves to measure the flow velocity of the oxygen; with it, the velocity of 7-50 ml. of oxygen per minute can be measured. A needle valve serves for fine regulation of the oxygen stream whose cone is adjusted by filing.

Combustion Apparatus. For the combustion furnace the dimensions of the individual zones as well as the arrangement of the combustion tube of the quartz capsule and the platinum boat are visible in Figures 2-4. In carrying out analyses in series, zone (1), in the places in which the combustion catalyst is placed, is heated to a temperature of 700-750° and in the places with silver wool, is heated to 600°. After insertion in the combustion tube of the capsule with the sample, individual zones are automatically switched on in the sequence 5, 2, 3 and 4. The time interval of switching on is equal and can be set from a minute to a few hours. As soon as all zones are turned on, the combustion period is ended and the combustion products are rinsed from the tube. The rinsing period is the same as that of the combustion period. After ending of the rinsing, zones 2 to 4 are automatically switched off and the apparatus gives a light or a sound signal that the analysis is over. However, the switching in and out of the individual zones for the automatic combustion and ignition can also be regulated by hand. The temperature is adjustable in the region from 500-800°. The combustion apparatus is so constructed that the combustion tube cools quickly after switching off. The heating coil is wound on the quartz tube, the individual zones are so dimensioned and the winding distance so selected that the furnace works under the required working conditions, so that, upon switching on the zones, uniform temperature distribution prevails, the temperature between the individual zones does not fall, and the constant time of heating and cooling is low.

Mechanical Arrangement of Apparatus (Figures 6-7). After switching on the apparatus, the current is led into transformers *1* and *2* and into the electronic timer *3*, which provides the impulses for the control currents to the relay from the anode current of the electronic control tube. The time constant of the impulse potential of the electronic tube grid is regulated by a variable resistance; through changes in resistance, the time interval of the impulses may be varied from 2 minutes to several hours. By pressing knob *5* the current impulse passes over into the stepping relay *6*. The rotor arm of the stepping relay springs up to a contact segment and switches in the circuit through relay *7*. The contacts of this relay close the circuit to heater No. II. On the stepping relay *6* the contact *a* is switched in simultaneously, which closes the circuit *6* to relay *10* and whose contacts switch in the circuit to heater No. V. The stepping relay *6* receives an uninterrupted impulse from the time switch *3* by means of the closed contact *c* existing on relay *11*. When the rotor arm has advanced to another contact segment, relay *8* is switched out and relay *9* is switched in, which with its contacts closes the circuit to zone (or heater) III. Now zone II and zone III are heating. After further travel of the rotor arm to another contact segment, relay *8* is switched out and relay *9* is switched in, which with its contacts closes the circuit to zone IV so that now all furnaces are switched in. Finally the rotor arm touches the last automatic segment contact, relay *11* is switched in, and through the current flow through contact *b* the control lamp *12* lights up or another signal arrangement of the apparatus is brought into operation. Simultaneously the normally closed contact *c* which brings the circuit of the impulse to selector *6* is opened. The time switch continues further with the work circuit switch out. With commutation of the selector to the last automatic contact, contact *a* is also opened which also breaks relay *10* and thereby breaks the current flow to heater V. The apparatus is prepared for the next work cycle and is put into operation by pressure on knob *5*.

Carrying out the Analyses. Unless otherwise mentioned, the work procedure is the same as described earlier [1].

Solid Sample. After weighing a 3 to 5 milligram sample in the platinum boat, this is introduced into the quartz capsule and is pushed into the combustion tube as shown in Figure 2. The equipment with the weighed absorption

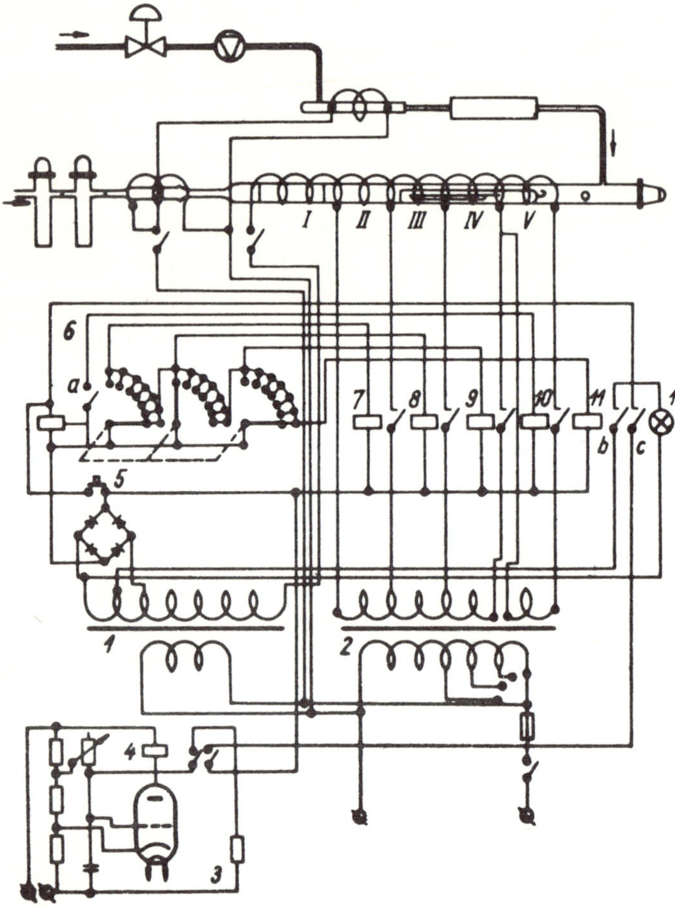

Fig. 6. Mechanical and Electrical Arrangement of the Automatic Apparatus. For description see text.

apparatus closed is switched on and after opening the oxygen velocity is set at 15 millimeters per minute. Heater 1 is switched on; the combustion and rinsing time is set at 12 minutes by means of the time switch, and by pressing knob 5 the apparatus is put in operation. The combustion proceeds automatically. As soon as the light signal lights the combustion is ended. The apparatus is closed, disconnected, and wiped with a dry linen rag; then the water apparatus is weighed on the tenth and the carbon dioxide apparatus on the fifteenth minute.

Liquid Samples. These are weighed in a glass capillary; the installation of the capillary and the quartz capsule is visible in Figure 8. The combustion time including the rinsing is set at 15 minutes. The ordinary procedure is the same as in the foregoing case.

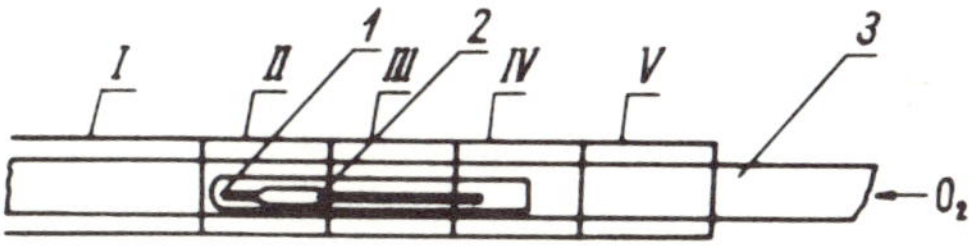

Fig. 8. Apparatus for the Combustion of Liquids. 1) quartz capsule, 2) capillary with sample, 3) quartz tube, I-V — electrically heated zones.

RESULTS AND DISCUSSION

The described arrangement for burning organic substances is simple. It eliminates displacement of the combustion furnace and no arrangement is necessary to regulate the combustion furnace according to the behavior of the sample under analysis. On the basis of our briefly made investigation, cobalt (II, III) oxide makes possible the combustion of micro samples of readily burnable substances in less than a minute. It is apparent that the diminution in time of the whole determination is to be attributed to the true behavior of the sample upon combustion in oxygen. The necessary samples of substances decomposing explosively and volatile substances cannot, however, be burned with the same velocity. Therefore we have put the sample in a long boat with a quartz capsule, divided the combustion space into 3 zones, and switched in these manually or automatically in the opposite sense to the oxygen stream. The last zone (see Figure 2, zone V) prevents a back diffusion of combustion products into the forepart of the tube and also heats the oxygen in the case of the combustion of liquids. The quartz capsule prevents the strong diffusion of the oxygen to the sample being burned so that upon raising the temperature pyrolysis of the sample takes place and its decomposition does not go explosively. In our recent practice we have burned all solid samples including violently decomposing compounds (picrates) within 7 minutes. Samples behaving completely abnormally can be burned more slowly. If necessary the individual zones can be switched in by hand. In the case of combustion of liquids weighed out in a capillary this is inserted in the combustion tube in a reversed position in contrast to our earlier custom (i.e. the capsule now faces the entering O_2 stream). The mouth of the capillary is placed in the zone whose heating is switched in last; rapid evaporation of the liquid is facilitated by the heated oxygen. In this way benzene is burned by us without difficulty.

The combustion procedure in the case of solid and liquid samples is completely automatic, requires no supervision, and is completely switched off after the time determined for rinsing the apparatus. During the combustion and rinsing, other jobs can be carried out. The principle of combustion can also be employed in the determination of other elements, for example, for the nitrogen determination, which was described in our earlier communication. Also it finds application in the halogen and sulfur determination. These possibilities are being further investigated by us.

The proposed method of automatic combustion was brought into this work for the micro determination of C and H. The results collected in Table I show that the apparatus works reliably. For these analyses we have employed substances of different composition and of different combustion behavior. It must be emphasized that by means of the described method even volatile substances with high vapor pressure whose combustion could be carried out until now only slowly under supervision and with great precaution can now be burned easily and quickly.

On the basis of the knowledge obtained by carrying out elementary analysis it is established that the working procedure may be accelerated with a minimum amount of supervision and may be simplified. Combustion of the sample is carried out under constant conditions and the personal factor of the individual people carrying out the analysis is removed so that a further improvement in precision of the method can be expected. In contrast to the earlier known types of automatic installation the construction of the apparatus is simpler and the individual parts can be mass produced.

SUMMARY

A new method for automatic combustion of organic substances has been worked out, which makes possible the quantitative combustion of solid and liquid compounds in 3 to 8 minutes, and calls for the step-wise switching on of combustion zones; the substance is placed in a long platinum boat with a quartz capsule. Liquids are burned in a capillary. In the case of micro determination of carbon and hydrogen in the presence of cobalt oxide (II, III) it has yielded very good results for solid and liquid samples of various composition, including difficultly combustible, explosive, and readily volatile compounds.

TABLE I

Summary of Results

Sample Empirical Formula	Mol. Wt.	Calc. %C	Calc. %H	Found %C	Found %H	Difference %C	Difference %H
Anthraquinone $C_{14}H_8O_2$	208.2	80.76	3.87	81.06	3.60	+0.30	-0.27
Azobenzene $C_{12}H_{10}N_2$	182.2	79.09	5.53	79.02	5.61	-0.07	+0.08
2,3,5,6-Tetrachloro-4-nitrobenzene $C_6HCl_4NO_2$	260.9	27.62	0.39	27.91	0.44	+0.29	+0.05
p-Chlorobenzoic acid $C_7H_5ClO_2$	156.6	53.68	3.21	53.72	3.22	+0.04	+0.01
Bicetyl $C_{32}H_{66}$	450.8	85.24	14.76	84.96	14.39	-0.28	-0.37
Naphthalene $C_{10}H_8$	128.2	93.07	6.29	93.64	6.22	-0.06	-0.07
Anthracene $C_{14}H_{10}$	178.2	94.34	5.66	94.19	5.66	-0.15	—
Salicylic Acid $C_7H_6O_3$	138.1	60.87	4.38	60.79	4.26	-0.08	-0.12
Vanillin $C_8H_8O_3$	152.1	63.15	5.30	63.20	5.29	-0.05	+0.01
Maltol $C_6H_6M_8$	126.1	57.17	4.80	57.17	4.80	+0.03	—
Thiourea CH_4N_2S	76.07	15.79	5.30	16.09	5.32	+0.30	+0.02
Piperonal $C_8H_6O_3$	150.1	64.00	4.03	63.81	4.10	-0.19	+0.07
Tribromophenol $C_6H_3Br_3$	314.8	21.72	0.89	22.03	0.85	+0.31	-0.04
Cholesteryl benzoate $C_{34}H_{50}O_2$	490.7	83.21	10.27	83.29	10.23	+0.08	-0.04
Piperidine picrate $C_{11}H_{14}N_4O_7$	314.2	42.04	4.49	42.09	4.59	+0.05	+0.10
Camphor $C_{10}H_{16}O$	152.2	78.89	10.59	79.1	10.71	+0.22	+0.12
Saccharose, $C_{12}H_{22}O_{11}$	342.3	42.10	6.48	42.20	6.40	+0.10	-0.08
Carbazole, $C_{12}H_9N$	167.2	86.20	5.43	86.18	5.44	-0.02	+0.01
Theobromine $C_7H_8N_4O_2$	180.2	46.66	4.48	46.53	4.38	-0.13	-0.10
2-Ethoxynaphthalene $C_{12}H_{12}O$	172.2	83.69	7.02	84.06	7.11	+0.37	+0.09
p-Acetotoluide $C_9H_{11}NO$	149.2	72.45	7.43	72.31	7.45	-0.14	+0.02
Benzylthiuronium	202.7	47.40	5.47	47.22	5.52	-0.18	+0.05
chloride, $C_8H_{11}ClN_2$				47.55	5.36	+0.15	-0.11
Xylene C_8H_{10}	106.2	90.50	9.50	90.50	9.72	—	+0.22
		.	.	90.34	9.72	-0.16	+0.22
Benzene C_6H_6	78.1	92.26	7.74	92.37	7.82	+0.11	+0.08
		.	.	92.44	7.85	+0.18	+0.11
		.	.	92.23	7.70	-0.03	-0.04

LITERATURE CITED

[1] M. Vecera and L. Synek, *Collection Czechoslov. Chem. Communs.* **23**, 1202 (1958); *Chem. listy* **51**, 2266 (1957).

[2] M. Vecera and L. Synek, *Chem. listy* **49**, 1891 (1955).

[3] F. Fischer, *Anal. Chem.* **21**, 827 (1949).

[4] B. Bobranski, *Mikrochim. Acta* **1956**, 1735.

[5] M. Vecera, *Chem. listy* **47**, 1090(1953.

[6] M. Vecera, *Chem. listy* (in press).

[7] I. Marek, *Arhiv. Hem. Farm.* **1**, 188 (1927); *Chem. Abstr.* **29**, 1933 (1928).

[8] M. O. Korshun and V. A. Klimova, *Zhur. Anal. Khim.* **2**, 274 (1947).

[9] J. A. Kuck and P. L. Altieri, *Mikrochim. Acta.* **1956**, 1550.

[10] M. Vecera and D. Snobl. *Collection Czechoslov. Chem. Communs.* **23**, 636 (1958); *Chem. listy* **51**, 1482 (1957).

[11] M. Vecera and L. Synek, *Ibid.* **24**, 3402 (1959).

Fig. 7. Automatic apparatus for microelectrical analysis.

CHAPTER 53

Electric Heating Element for Elementary Analysis*

M. Vecera, F. Vojtech, and B. Sekera

Electric furnaces for elementary analysis, usually used for heating combustion tubes, are not always suitable for a rapid method of organic elementary analysis, since they have a large time factor for heating and cooling. In order to achieve rapid cooling, there are used removable lid furnaces whose construction is expensive and requires special ceramic elements.

We aimed to build an electric heating unit which would be inexpensive, easily made, with a small time factor for heating and cooling and good heat transfer across the combustion space, and which would possess an easily adjustable heat characteristic along the length of the combustion space inside the tube.

The required properties of the furnace were attained in this project by using a ceramic shell in which there is

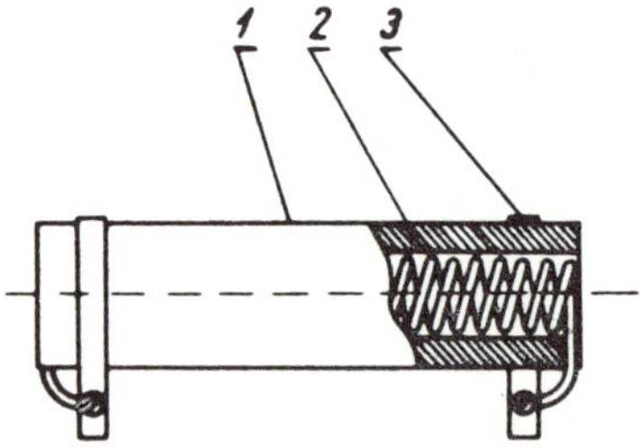

Fig. 1. Electric heating furnace. 1—ceramic shell; 2—Kanthal resistance wire having a 0.9mm diameter; 3—rustproof sheet sleeve.

*Chem. listy **53**, 1054-57 (1959)

freely placed a heating coil which directly transmits the heat to the heated tube. The required heat characteristics in the combustion space are suitably regulated by an adjustable increase in the turns of the heating coil.

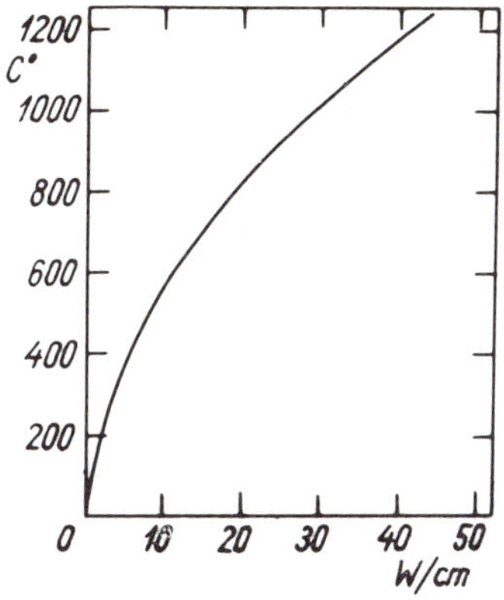

Fig. 2. Temperature variation inside the quartz tube depending on power input per 1 cm of length of the furnace.

DESCRIPTION OF THE APPARATUS

Insulating shell (Fig. 1) There are used tubes made of any non-combustible material having good insulating properties (various sorts of ceramic having a wall thickness 3-5mm, and an inside diameter which is 1-2mm larger than the diameter of the heating coil; the tube length is set according to need.

Sleeves (Fig. 1) are made from rustproof ribbon having a thickness of 1 mm, and a width of 6 mm.

Heating Coil (Fig. 1) is wound of a resistant wire (e.g. Kanthal) having a diameter of at least 0.75 mm; in order to be strong enough mechanically. In the preparation of ordinary elements, turns are wound tightly side by side, and after winding, the coil is stretched so that its length is about 100% longer than before stretching. The interior diameter of the coil is made 1 mm larger than the diameter of the heated tube. In order to obtain various temperatures in the same furnace, a part of the coil is further stretched as needed (Table I and Fig. 3).

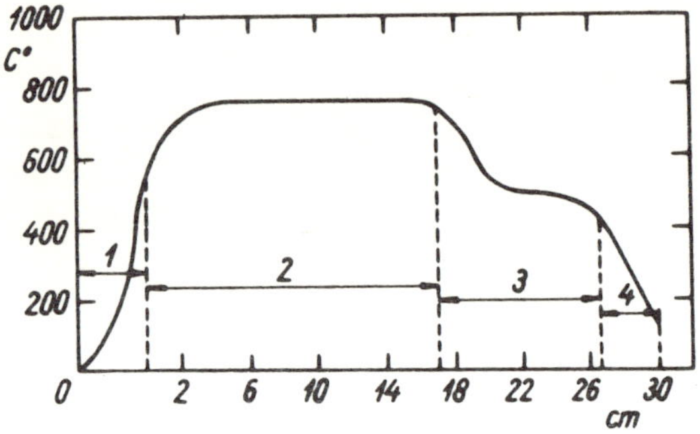

Fig. 3. Temperature variation in the quartz combustion tube at various pitches of the heating coil.

Area	Condition
1, 4	outside of furnace
2	coil pitch of 1.8 mm
3	coil pitch of 5.5 mm

TABLE 1

Length mm.	Temperature in the combustion tube, °C.	Input W	Voltage V	Current A
80	part 750[a]	126	23	5.5
	part 500[a]			
150	800	275	50	5.5
100	800	182	33	5.5
100	500	126	23	5.5

[a] In this furnace the coil was wound with turns next to each other and stretched in such a way that one part of the coil with a pitch of 1.8 mm heated to 750°; the other with a pitch of 5.5 mm heated to 500°.

RESULTS

The apparatus described was used for microdetermination of carbon, hydrogen [1, 2], sulfur [3], halogens [4, 5], and oxygen [6]; it is also suitable for semimicro methods. Some examples of the use of the elements for elementary analysis are shown in Table I.

Furnaces made from Kanthal resistance wire (1150°), with a diameter of 0.9 mm, suitable for heating a quartz tube with an outside diameter of 11 mm, and an inside diameter of 9 mm, show a variation of the power input calculated per cm of length of the furnace with the temperature inside the tube, according to Fig. 2.

The furnaces are self-contained and are used in elementary analysis one after another, and only between ends of neighboring furnaces is there a temperature drop. This is insignificant. In Fig. 3, there are shown heat characteristics of the furnace used in the carbon determination (6).

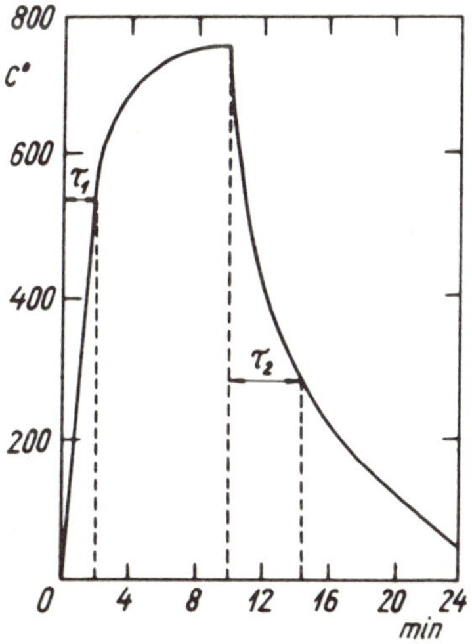

Fig. 4. Time lag in heating and cooling the inside space of the combustion tube. T_1 and T_2 are times of delay.

Part of the furnace heats the tube to 750°C., and another part to 500°C.; Fig. 4 shows the time lag in heating and cooling of the same furnace.

The furnaces are used also for other purposes; for example, for heating tubes in catalytic processes, in gas chromatography devices, and also for heating metal tubes, if the heating coil is suitably insulated from the metal tube by means of inner ceramic pieces.

The use of the furnaces described is not restricted to the examples cited above.

SUMMARY

In the foregoing work, an electric furnace with a small time lag in heating and cooling, with good heat transfer into the combustion space, and with readily calibrated temperature along the combustion zone is described. The individual parts of the combustion furnace are: a ceramic insulating shell (Fig. 1) of wall thickness 3-5 mm., a stainless steel skin, a heating coil of Kanthal resistance wire of 0.75 mm. minimum diameter. By stretching the coil, different temperatures are attainable. The arrangement described is employed for the microdetermination of carbon, hydrogen, sulfur, halogen and nitrogen and is also suitable for semi-micro analytical methods.

LITERATURE CITED

[1] M. Vecera, L. Synek, *Chem. listy* 52, 2266 (1958).

[2] M. Vecera, D. Snobl, L. Synek, *Mikrochim. Acta* 1958, 9.

[3] M. Vecera, D. Snobl, *Mikrochim. Acta* 1958, 28.

[4] M. Vecera, J. Bulusek, *Chem. listy* 51, 1475 (1957).

[5] M. Vecera, J. Bulusek, *Mikrochim. Acta* 1958, 41.

[6] M. Vecera, L. Synek, *Collection, Czechoslov. Chem. Communs.* (in press).

CHAPTER 54

Automatic Microdetermination of Carbon and Hydrogen in Organic Substances*

M. Večeřa

In the automation of the microdetermination of carbon and hydrogen, various procedures for the ignition of the sample have been drawn upon [1-5], but in the final phase of the operation weighing of the absorption tubes is ordinarily necessary. Only recently articles have been published according to which other principles of the final determination have found application (measurements of changes of electrical conductivity in solution [6, 7] or thermal conductivity in the gas phase [8, 9] coulometry [10]) and present in this manner possibilities for automating the procedure. The methods mentioned are mainly in the evolutionary stage and it is difficult to give an appraisal in regard to their suitability.

One can say, however, that the methods have the prospect for spreading, with gradual displacement of the gravimetric microdetermination, through which the entire time of the determination, which at the present amounts to 20-25 min. in the use of the gravimetric determination [5], is essentially shortened, the operation simplified, and the personal factors which at present have to be reckoned with in the case of micro methods, eliminated.

Collection Czechoslov. Chem. Communs.* **26, 2298-2307 (1961).

The methods [6, 7] based on measurement of changes in electrical conductivity after absorption of carbon dioxide in dilute alkali require a relatively complicated apparatus to ensure quantitative absorption and, with regard to the extent of time, are more demanding than the rapid gravimetric methods. The methods based on the principle of gas chromatography [8,9] are either time-consuming [9] or require so many complicated steps [8] that it is difficult to think of complete automation.

After all this, it can be said that a completely automatic method for the micro determination of carbon and hydrogen in organic substances is still unknown.

The investigation of the kinetics and the mechanism of the complete oxidation of organic substances [11] on active combustion catalysts enabled us to develop a simple method which permits the quantitative ignition of micro samples of organic substances in the course of some ten seconds, without danger of explosion, both in an inert atmosphere as well as in a stream of oxygen [11, 12]. In this manner, as we reported elsewhere [12], the procedures for the gravi-

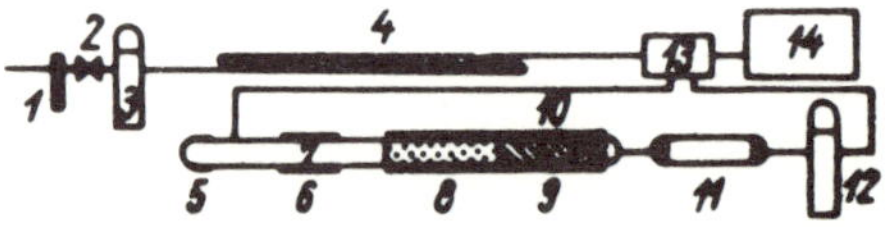

Fig. 1. Apparatus for the microdetermination of carbon in organic substances.
1—Tube with catalyst for oxygen purification, 2—fine gas reduction valve, 3—flow meter, 4—glass capillary, 0.5 mm. i.d., 5—quartz combustion tube, 6—movable electric furnace, 7—platinum boat with sample, 8—layer of combustion catalyst, 9—silver wool layer, 10—electric furnace, 11—absorption tube with Anhydrone, 12—absorption tube with MnO_2, 13—conductivity sensing element, 14—recorder.

metric carbon and hydrogen determination and for the gas-volumetric nitrogen determination in organic substances are extremely simplified and considerably speeded up; moreover, we were able to proceed to the direct determination of the combustion products by measurement of thermal conductivity and to contemplate development of simple, completely automatic micro methods.

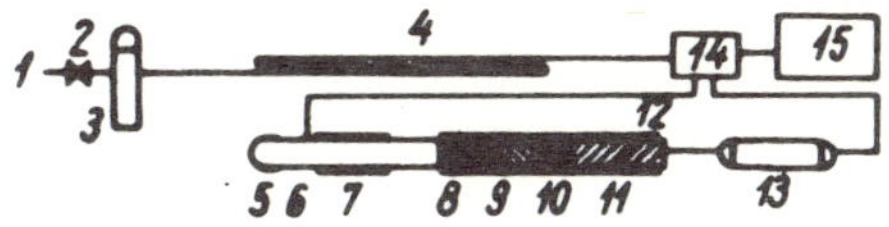

Fig. 2. Apparatus for the microdetermination of hydrogen in organic substances.
1—Nitrogen inlet, 2—fine gas reducing valve, 3—flow meter, 4—glass capillary, 5 mm. i.d., 5—quartz combustion tube, 6—movable electric burner, 7—platinum boat with sample, 8—copper oxide layer, 9—Co_3O_4-layer, 10—copper layer, 11—layer of iron filings, 12—electric furnace, 13—absorption tube with Anhydrone and sodium hydroxide on asbestos, 14—conductivity cell, 15—recorder.

In the case of the carbon determination, the weighed sample is burned in a stream of oxygen in the apparatus illustrated in Figure 1 and, after absorption of halogen, water, and nitrogen oxides, the carbon dioxide is determined by measurement of the changes in thermal conductivity indicated by means of a recorder. In the case of the hydrogen determination, the sample is burned in a stream of nitrogen in an apparatus illustrated in Figure 2, the water being converted into hydrogen by conduction over iron filings at 650°, which—analogously to the case of the carbon determination — is determined by measurement of thermal conductivity.

EXPERIMENTAL PART

REAGENTS

Combustion catalyst—Cobaltous-cobaltic oxide on corundum is prepared in a manner described in an earlier work [11]. Cobalt (II, III) oxide is prepared by thermal decomposition of cobaltous oxalate [11]. For the hydrogen determination, the quantity of substance required for the daily series of analyses was ignited for a short time and stored in a desiccator over Anhydrone. Silver wool was used for the absorption of halogen; water was absorbed on magnesium perchlorate, carbon dioxide on sodium hydroxide on asbestos, and nitrogen oxides on active manganese dioxide. [13]. For the conversion of water to hydrogen, about 3 mm. lengths of iron filings were employed. Copper wire prepared by reduction of copper oxide wire serves for removal of oxygen.

APPARATUS, EQUIPMENT AND COMBUSTION FURNACES

Stationary and movable electric furnaces described in our earlier communications [5, 14] were used for the carbon and hydrogen determination.

The platinum combustion boat was 5 mm long, 5 mm wide, and 4 mm high.

Absorption Tubes—Magnesium perchlorate was used for the absorption of water in the carbon determination, in a 100 mm long, Pregl-type absorption tube described in detail previously [16]. A tube of similar type serves for the water and CO_2 absorption in the hydrogen determination. The length of the sodium hydroxide-asbestos layer was 45 mm; that of the magnesium perchlorate layer 40 mm. The filling of the rest of the apparatus and the length of the individual layers has already been mentioned previously [16].

For the determination of the combustion products, a semi-diffusion conductivity cell was used, whose bridge was energized with a current of 120-150 milliamperes calibrated to an accuracy of ±2%.

The graphic trace-out was accomplished by means of an automatic Honeywell recorder with a scale range of 10 millivolts; the speed of the paper (chart) displacement was 25 mm/min.

COMBUSTION TUBES

Combustion Tube for the Carbon Determination: Silver wool to a length of 50 mm was introduced into a 9 mm i.d. quartz tube, 320 mm long, with an oxygen inlet side-arm. Then cobalt (II, III) oxide on corundum is shaken into the tube and a layer 50 mm long is prepared. The layer of combustion catalyst is kept in place by means of a stopper of 0.2 mm platinum wire approximately 5 mm long. After a 30-minute continuous ignition in a stream of oxygen, the tube is ready for analysis.

Combustion Tube for the Hydrogen Determination: A tuft of asbestos is introduced into a 9-mm i.d. quartz tube, 450 mm long with a nitrogen inlet side arm, lightly compressed, and iron filings to a layer length of 120 mm are put into the tube. Then wire-form copper is shaken into the tube and a 30-mm layer prepared. There follows a 50-mm layer of cobalt (II, III) oxide on corundum and a 20-mm layer of cupric oxide. The filling of the combustion tube is kept in place with a 2-mm long stopper of 0.2 mm platinum wire. After a 30-min long ignition in a stream of nitrogen, the tube is ready for analysis.

ANALYTICAL PROCEDURE

Carbon Determination (Figure 1):

From 1.0-1.6 mg of sample is weighed out in a platinum boat and the apparatus is assembled according to Figure 1. The layer of silver in the combustion tube is heated to about

600°C by means of an electric furnace; the layer of combustion catalyst to about 700°C. The sample in the boat is covered with about 30-40 mg of cobalt (II, III) oxide and is inserted to about 10 mm distance from the tube filling. The movable ignition furnace is set 10 mm from the platinum boat and, after switching on the heating (700-800°C), the displacement mechanism is set in operation. The displacement speed is set so that the furnace passes over 20 mm of the combustion path in 60 seconds. After the sample ignition is complete, the movable furnace is left in operation another 2 minutes, whereupon it is switched off and brought back to the original position. For rapid cooling of the tube and the movable furnace, a stream of air can be led into the ignition furnace. The oxygen stream in the combustion tube is set at 15 ml per minute; the bridge of the conductivity sensing element is energized with a current of 15 milliamperes. As soon as the apparatus has been rinsed (in 4-6 min), another sample can be inserted into the combustion tube.

Hydrogen Determination (Figure 2):

Only the parts of the determination are described which differ from the foregoing determination. The combustion tube filling including the iron filings is heated to a temperature of 650°C. To the sample in the boat is added 30-40 mg of previously ignited cobalt (II, III) oxide, stored in a desiccator over magnesium perchlorate. The stream of nitrogen in the combustion tube is set permanently at 20 ml/min; the bridge of the conductivity cell is energized with a current of 120 milliamperes. If the hydrogen content in the organic sample is less than 3%, the sensitivity of the conductivity sensing element can be increased by raising the current strength to 140 milliamperes.

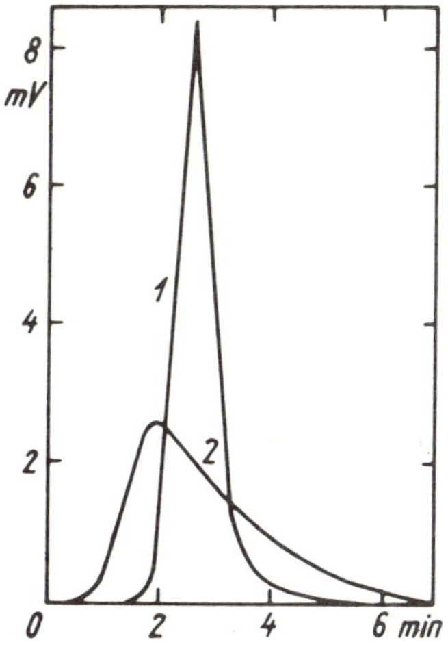

Fig. 3.

Trace-out of CO_2 concentration in the combustion of acetanilide (curve 1).

Trace-out of H_2 concentration in the combustion of 1-chloroanthraquinone (curve 2).

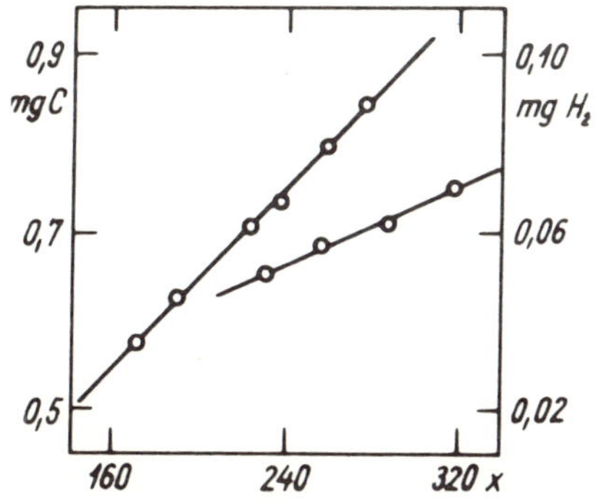

Fig. 4.

Calibration curve for the carbon and hydrogen determination
X = weight of paper in mg.

Evaluation and Calculation of Results:

After carrying out a series of 20 determinations with standard compounds, the areas traced out by the recorder were cut out and weighed on a semimicro balance to an accuracy of ± 0.1 mg. The weighings of the paper varied within the limits of 150-300 mg. A typical trace pattern obtained in the carbon-hydrogen determination is seen in Fig. 3. It was further established from the experiments carried out that the weight of the paper area that carries the recorder traces can be reproduced with an accuracy of ± 1 mg.

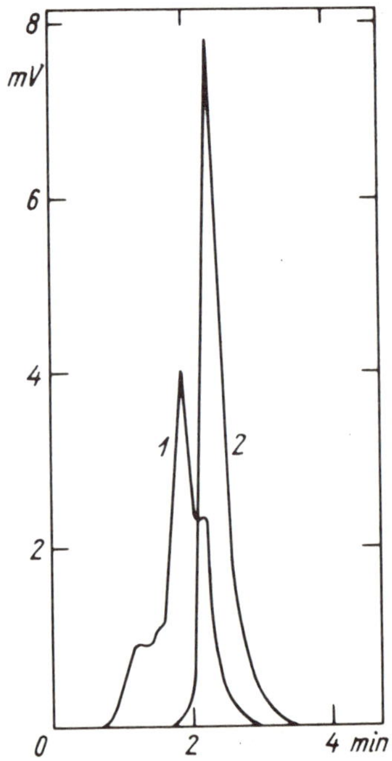

Fig. 5. Ignition of terephthalic acid in an oxygen stream. Curve 1: without Co_3O_4; Curve 2: with Co_3O_4.

The results obtained were collected after calculation into a calibration curve, in which the weight of the paper for the given experiment design is plotted on the abscissa and the corresponding quantity of carbon on the ordinate (Figure 4). The calibration curve for the hydrogen determination was obtained in the same way (Figure 4).

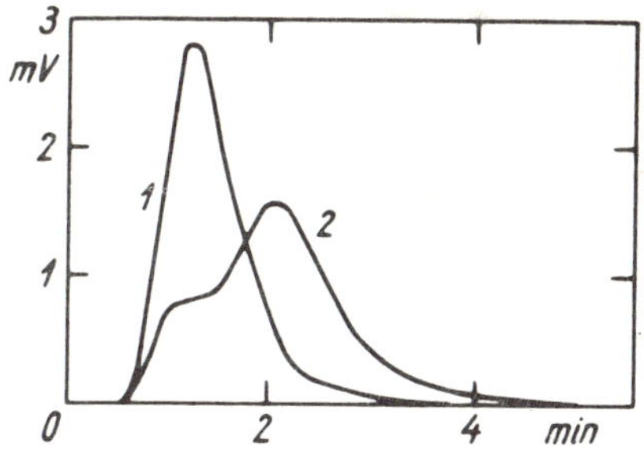

Fig. 6. Ignition of antipyrine in a CO_2 stream. Curve 1: with Co_3O_4; Curve 2: with Co_3O_4.

RESULTS AND DISCUSSION

The problem of flash ignition of organic substances was solved by covering the analytical sample in the platinum boat with 20-30 mg of active cobalt (II, III) oxide prepared by thermal decomposition of cobalt (II) oxalate. Thus, a complete oxidation of organic samples can be carried out practically with any desired speed, a short layer of the said ignition catalyst in the combustion tube sufficing to guarantee a quantitative reaction course. (See also [11]). The products of the complete oxidation of the organic compound leave the combustion tube in a single concentration wave, even in the case of ignition of substances of different constitution (Figures 5, 6; see above all [12]). Herein exists a fundamental difference from the ordinary method of combustion, in which the greatest part of the sample is so burned that it forms several concentration waves (Figures 5, 6; see also [12]), wherein the micro sample of most types of organic compound must be burned at least 3-5 minutes to eliminate the danger of explosion.

The flash ignition method made possible application of the simplest procedure for the concentration determination of the combustion products, namely measurement of thermal conductivity, which belongs to the most accurate procedures of the quantitative evaluation of gas mixtures [17]. The decisive condition for the application of thermal conductivity consists in the carrying out of the measurement at constant gas flow through the measuring cell. In the application of the flash ignition method, pressure changes take place which are produced by combustion in the brief time interval and even earlier, as the front of the combustion products reaches the measuring cell, so that the measurement is not influenced. In the case of the continuously used combustion method, direct measurement of the combustion products cannot be utilized, since the changes in flow rate produced by pressure changes in the combustion influence the measurement of thermal conductivity. Therefore, in this case, the combustion products must first be condensed [8,9], whereby the determination and the automation of the entire process are made very difficult.

It must also be pointed out that the direct measurement of thermal conductivity is quicker and more simple than measurement of the conductivity change of solutions after the absorption of carbon dioxide in dilute caustic, which requires a complicated measuring apparatus and a further apparatus to ensure quantitative absorption in the solution. Especially in the case of the carbon dioxide determination, this problem brings a series of complications with it which makes complete automation of the entire process very difficult.

The results of the carbon determination in some compounds are listed in Table 1.

TABLE 1

Results of the Carbon Determination

Compound (Mol. Wt.)	%C Calc.	mg. Sample	%C Found	Deviation %C (abs.)
Sucrose	$C_{12}H_{22}O_{11}$	1.787	42.32	+0.21
(342.3)	(42.11%)	1.506	42.60	+0.49
		1.397	42.71	+0.60
		1.451	42.18	+0.07
		1.431	42.13	+0.02
		1.373	41.61	+0.49
		1.545	42.66	+0.55
		1.404	42.44	+0.34
		1.545	41.82	−0.28
		1.488	42.42	+0.32
		1.501	42.08	−0.03
		1.516	41.57	−0.54
Benzoic acid	$C_7H_6O_2$	1.080	68.03	−0.81
(122.1)	(68.84%)	1.203	68.06	−0.78
		1.095	69.64	+0.80
Picric acid	$C_6H_3O_7N_3$	1.649	31.14	−0.31
(229.11)	(31.45%)	1.653	32.06	+0.61
		1.707	31.18	−0.27
Acetanilide	C_8H_9ON	1.144	71.47	+0.38
(135.2)	(71.09%)	1.150	71.70	+0.61
		1.193	71.32	+0.23
		1.225	71.03	−0.06
		1.289	71.06	−0.03

The calculated standard deviation s = 0.45% is less than that given by the authors [8,9], who also make use of thermal conductivity. The determined standard deviation is also comparable to the standard deviations of gravimetric microdetermination, [5, 18], considering that the determination is carried out with one half the sample and that accuracy in gravimetric determinations falls off considerably in the case of samples under 2.5 mg.

The results of the hydrogen determination are collected in Table 2. The calculated standard deviation s = 0.16% show that the method in this set-up gives as accurate results as the earlier known methods. In the method described, the great difference in the thermal conductivity of hydrogen in comparison to nitrogen as the carrier gas is used to advantage; the water formed in the combustion is converted to hydrogen by passage over iron filings. In the stationary state, the equilibrium of the reaction

$$H_2O + Fe \rightleftharpoons FeO + H_2$$

is known to be dependent on temperature, and the reported equilibrium constants [19] show that in the temperature range 500-700°C about two-thirds of the water in the reaction mixture is converted to hydrogen. However, our experiments have shown [20] that in passing water vapor, in amounts which come into consideration in elemental microanalysis, over a 120 mm long layer of iron filings in a quartz tube with an i.d. of 9 mm and a flow rate of 15-20 ml/min, conversion of water to hydrogen takes place to an extent of 96-98%. This finding shows on the one hand that the reaction mentioned can be used advantageously for quantitative purposes; on the other hand it must be taken into consideration that by passing water vapor over the layer of iron filings a continuous disturbance of the reaction equilibrium takes place, so that under the conditions employed the conversion proceeds practically to completion.

TABLE 2

Results of the Hydrogen Determination

Compound (Mol. wt.)	%H Calc.	mg Sample	%H Found	Deviation %H (abs)
1-Chloroanthraquinone	$C_{14}H_7O_2Cl$	1.542	2.89	-0.02
(242.6)	(2.91)	1.577	2.93	+0.02
		1.607	2.98	+0.06
Sucrose	$C_{12}H_{22}O_{11}$	1.078	6.22	-0.26
(342.2)	(6.48)	1.080	6.66	+0.18
		1.083	6.62	+0.14
		1.144	6.66	+0.18
		1.179	6.26	-0.23
Cholesteryl acetate	$C_{29}H_{48}O_2$	1.034	11.44	+0.15
(428.7)	(11.29)	1.056	11.33	+0.04
		1.078	11.10	-0.19

The greatest advantage of the method is its speed. Forty to fifty determinations are carried out with ease within 6 hours, but not involving final values. By use of the optimum dimensions of the combustion tube and the absorption tube, the elution time can be further appreciably shortened. Thus, for example, by shortening to 20 mm the sodium hydroxide-asbestos layer in the absorption tube and omitting the magnesium perchlorate layer, a trace-out is obtained (Figure 7) from which it is evident that the sample combustion, including the rinsing of the system, is complete within 4 minutes.

Finally, the prospects must be cited with regard to the methods based on catalytic flash ignition using active cobalt (II, III) oxide. Above all, a significantly greater precision for both carbon and hydrogen determination can be expected. For the experiments described in this work, a serial conductivity cell was employed which was not thermostated, nor was any special attention given to the thermostating of the gas flowing into the cell. The maintenance of constant temperatures should be readily practicable, and, furthermore, the establishment of constant gas flow rate through the apparatus can be rendered much more precise by the use of thermostated capillaries, by which in our opinion the greatest source of error is removed.

According to expectation, the methods can be made even more precise and faster, for the purpose of quantitatively interpreting changes in conductivity, by the use of an integrator of a special type such as is employed in coulometers with reading of integrated values by means of electronic calculators, possibly with automatic recording, whereby this type of quantitative evaluation also makes possible work in wider concentration ranges.

It must also be emphasized that the micro methods described have been worked out for samples of 1.0-1.7 mg., i.e., for half the sample size used in the classical micro methods. In fact, there exists the possibility of reducing the sample to the range of ultramicro elementary analysis, by using more sensitive recording methods. By using an integrator and the principles of gas chromatography, possibly after previous conversion of water to acetylene, one can also proceed to the development of a method for the simultaneous determination of carbon and hydrogen.

The method described opens up for the first time a real way to the complete automation of the entire process, since

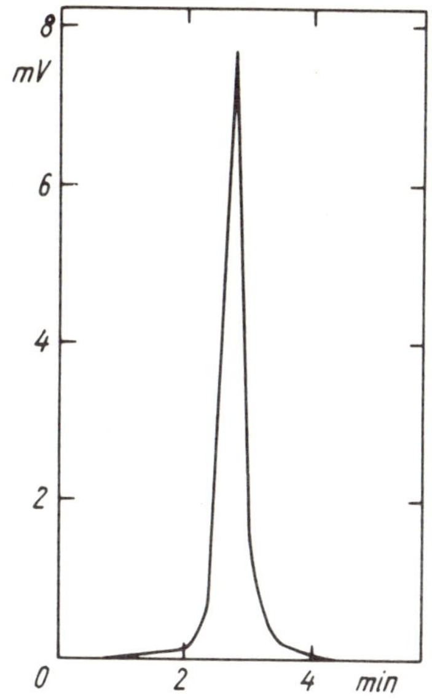

Fig. 7. Combustion of 1-chloroanthraquinone in a nitrogen stream, using a short layer of sodium hydroxide-asbestos for CO_2 absorption.

the problems with regard to the individual procedure-steps still to be automated have been already solved in our laboratories. We are presently working on the construction of the completely automatic apparatus.

The flash ignition process for organic substances also makes possible the development of a simple process for automatic nitrogen determination. Some of our further experiments are also aimed in this direction.

Finally, it must be pointed out that the method has for the present been developed for the analysis of solid and nonvolatile liquid compounds, and that in analysis of nitrogen compounds the possibility of influence on the results through the presence of nitrogen in the products that reach the conductivity cell has not been taken into account. An investigation of this point can be carried out in the event that it is possible to increase the accuracy of the method developed.

SUMMARY

A new automatic carbon and hydrogen determination depending upon the flash ignition of samples in the presence of cobaltous-cobaltic oxide has been developed. Carbon dioxide and hydrogen are determined by direct measurement of thermal conductivity. Standard deviation of the carbon determination $s = 0.45\%$, for the hydrogen determination $s = 0.16\%$.

LITERATURE CITED

[1] T. Kono, *Mikrochim. Acta* **1958**, 461.

[2] B. Bobránski, *ibid.* **1956**, 1753.

[3] M. Vecera and L. Synek, *Chem. listy* **49**, 1891 (1955).

[4] M. Vecera, et al., *Collection Czechoslov. Chem. Communs.* **25**, 93 (1960).

[5] M. Vecera, et al., *Mikrochim. Acta* **1958**, 9.

[6] H. Malissa, *ibid.* **1957**, 553.

[7] S. Greenfield, *Analyst* **85**, 486 (1960).

[8] A. A. Duswalt and W. W. Brandt, *Anal. Chem.* **32**, 272 (1960).

[9] O. E. Sundberg and C. Maresh, *ibid.* **32**, 274 (1960).

[10] W. Olensen, et al., *Angew. Chem.* **63**, 557 (1951) ; **64**, 24 (1952).

[11] M. Vecera, D. Snobl, and L. Synek, *Mikrochim. Acta* (in press).

[12] M. Vecera, V. Vojtech, and L. Synek, *Collection Czechoslov. Chem. Communs.* (in press).

[13] M. Vecera, et al., *ibid.* **25**, 2013 (1960).

[14] M. Vecera, et al., *Chem. listy* **53**, 1054 (1959).

[15] M. Spala, Unpublished results.

[16] M. Vecera and L. Synek, *Collection Czechoslov. Chem. Communs.* **23**, 1202 (1958).

[17] M. Wurst, *Chem. listy* **54**, 1042 (1960).

[18] M. Vecera and D. Snobl, *Collection Czechoslov. Chem. Communs.* **23**, 636 (1958).

[19] J. Chipman and S. Marshall, *J. Am. Chem. Soc.* **62**, 299 (1950).

[20] M. Vecera, J. Petránek, and V. Rusek, Hitherto unpublished results.

CHAPTER 55

Automatic Determination of Nitrogen in Organic Substances*

M. Večeřa

Gasometric nitrogen determination according to Dumas has not undergone many fundamental changes in the course of its further development. The modifications of the original procedure can be divided into three groups. For automating the combustion process, Zimmermann's solution was an especial contribution [1]; in this, combustion is carried out in a closed system. Since the combustion of the substance is carried out in an inert atmosphere, the danger of an explosive course of oxidation is practically excluded, and the combustion process can therefore be automated in a very simple manner, namely by application of what is called zone combustion [2]. The automation of the combustion process used in the classical method of Pregl was likewise solved by automatic displacement of the ignition furnace [3]. The problem of carrying out a more rapid combustion than in the case of the classical micromethod of Pregl (CO_2 stream 4 ml/min.) was partially solved on the one hand by Zimmermann's procedure [1]; on the other by more active ignition catalysts [4], if necessary by increasing their surface area [5].

Ways have also been sought to speed up and increase the precision of the end phase of the procedure, namely the determination of the amount of nitrogen. All modifica-

Collection Czechoslov. Chem. Communs.* **26, 2308-2314 (1961).

tions consist in changes of the apparatus for the gasometric determination of the nitrogen collected over 50% KOH. Until now there has been no other method known for the final determination of the amount of nitrogen in the gases emerging from the combustion tube.

The flash ignition of organic substances under application of active cobalt (II, III) oxide [11, 12, 13], and its high activity as an oxidation catalyst, made it possible for us to enter upon experiments for developing a method for the microdetermination of nitrogen in organic substances by direct determination of the nitrogen content in the carrier gas through conductivity changes, and thus to provide the prerequisites for complete automation of the process.

In the method for nitrogen determination described in this paper, the weighed sample is burned in a stream of carbon dioxide and, after reduction of the nitrogen oxides to nitrogen by copper, and after absorption of halogen and water in the carrier gas, the nitrogen is determined by measurement of the changes in thermal conductivity noted by means of a recorder. In another modification of this procedure, the substance is first burned in a closed system, so that nitrogen can also be determined in volatile liquids.

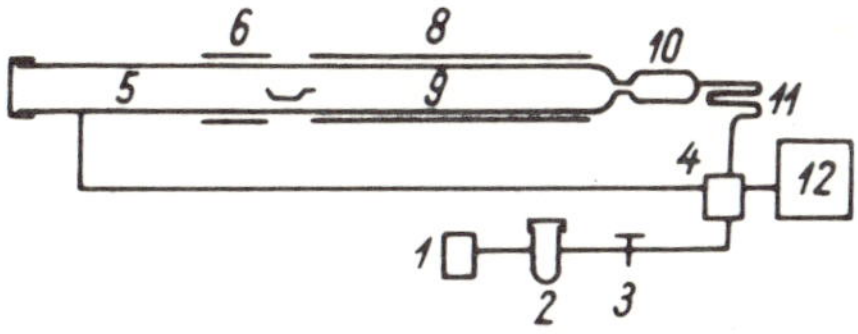

Fig. 1. Flow scheme for the automatic micro nitrogen determination by thermal conductivity measurement. 1—CO_2 source; 2—flow meter; 3—stopcock; 4—thermal conductivity sensing element; 5—quartz tube; 6—movable ignition furnace; 7—combustion boat; 8—heating furnace; 9—oxidation filling; 10—absorption tube with magnesium perchlorate; 11—capillary; 12—recorder.

EXPERIMENTAL PART

Ignition catalysts—Wire-form copper oxide, cobaltous-cobaltic oxide on corundum [4], and powdered cobaltous-cobaltic oxide prepared by thermal decomposition of cobaltous oxalate [4] were employed.

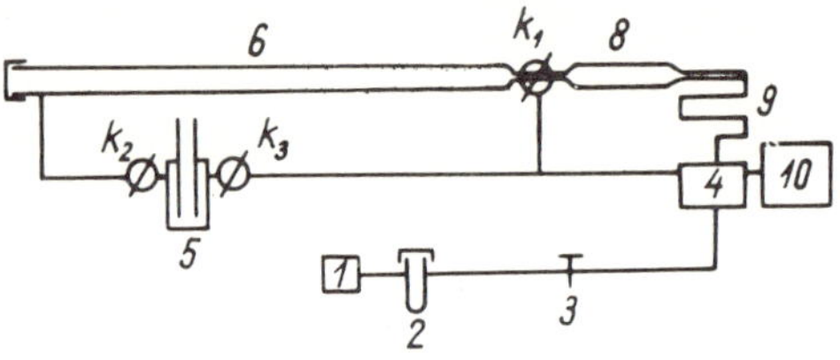

Fig. 2. Flow scheme of apparatus for automatic micro nitrogen determination by combustion in a closed system. 1—CO_2 source; 2—flow meter; 3—stopcock; 4—thermal conductivity sensing element; 5—pressure equalization vessel with stopcocks K_2 and K_3; 6—quartz combustion tube with oxidation filling and electric heating furnaces; 7—three-way stopcock K_1; 8—absorption tube with magnesium perchlorate; 9—capillary; 10—recorder.

Reagents. Wire-form copper was prepared by reduction of copper (II) oxide with hydrogen; silver wool was used for uptake of halogen; water is absorbed in magnesium perchlorate. The mixture of finely ground copper oxide and cobaltous-cobaltic oxide was prepared in this way: wire-form copper oxide was crushed into small pieces about 1-2 mm long and freed of dust on a fine sieve; 5 parts of the finely pulverized oxide was then mixed with one part Co_3O_4.

Apparatus, Set-up and Combustion Furnaces. For the combustion we used an apparatus with automatic displacement described in an earlier paper [14]. Combustion was carried out in boats 4 × 4 × 30 mm and 3 × 3 × 5 mm; quartz boats 4 × 4 × 30 mm were also used. A 30-mm long magnesium perchlorate layer in a 60-mm long glass tube with an i.d. of 4 mm was used for absorption of water. A semi-diffusion conductivity sensing element [12] was used for the determination of combustion products; its bridge was energized with a 200 milliampere current adjusted with an accuracy of ±0.2%. The graphic trace outs were made with the aid of a Honeywell automatic linear recorder with a scale range of 10 millivolts; the speed of chart displacement was 50 mm min.

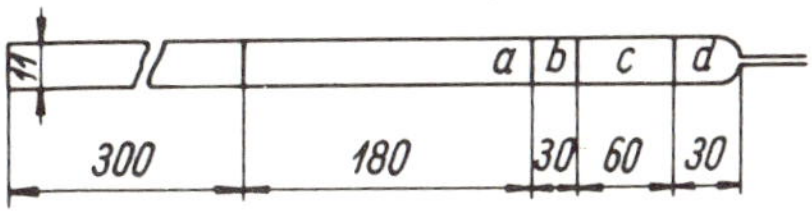

Fig. 3. Tube filling for direct combustion
Dimensions in mm.
a = CuO, b = Co_3O_4, c = Cu, d = Ag.

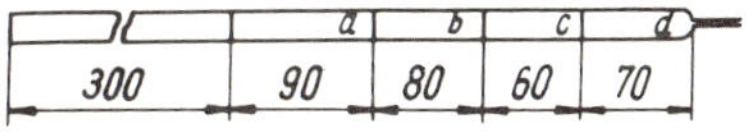

Fig. 4. Tube filling for combustion in a closed system
Dimensions in mm.
a = CuO, b = Co_3O_4, c = Cu, d = Ag.

COMBUSTION TUBES

Direct Combustion. An 11-mm i.d., 12-mm o.d. quartz combustion tube, 600 mm long, was used. The tube was filled in the usual manner; the composition of the filling and the length of the individual layers are seen from Fig. 3.

Combustion in a Closed System. A 6-mm i.d., 9-mm o.d. quartz combustion tube, 600 mm long, was used; the composition of the permanent tube filling is seen from Fig. 4.

Pressure Equalization Apparatus: This is made of heavy walled glass with capillary stopcocks and is filled with mercury. (Fig. 5).

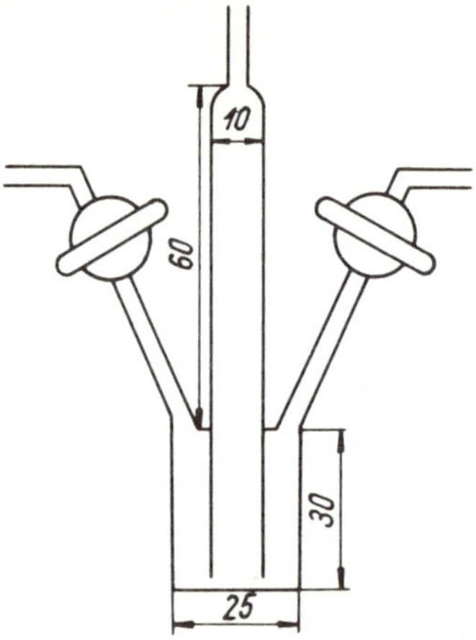

Fig. 5. Pressure equalization apparatus with mercury. Dimensions in mm.

ANALYTICAL PROCEDURE

Nitrogen determination by direct combustion. (Fig. 1)

From 1.5-5 mg of sample, depending on nitrogen content, is weighed out in a platinum boat and the apparatus is assembled according to Figs. 1 and 3. The silver and copper layers are heated to about 500° by means of the electric furnace; the layer of combustion catalyst is heated to 700°. The sample in the boat is covered with about 30-40 mg of powdered Co_3O_4 and the boat is inserted into the combustion tube to 10 mm distance from the tube filling. A quartz

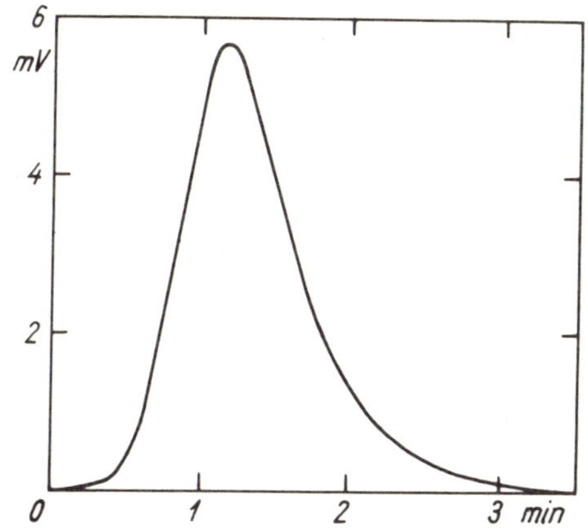

Fig. 6. Trace-out of nitrogen concentration in the analysis of antipyrine.

boat with approximately 50 mg Co_3O_4 is inserted at a distance of 10 mm from the sample. The movable ignition furnace is set at a distance of 20 mm from the stationary furnace. As soon as the apparatus has been rinsed with CO_2 (after 3-5 minutes), heat is turned on in the movable ignition furnace (700-800°), and after 30 seconds the displacement mechanism is set in motion. The displacement speed is adjusted so that the furnace completes the 20 mm long combustion path in 30 seconds. After combustion of the sample the movable furnace is left turned on at the left outer edge of the stationary furnace. As soon as the pen of the recorder returns to its original position, the heating of the movable burner is switched off and the latter is brought back to its starting position. For rapid cooling

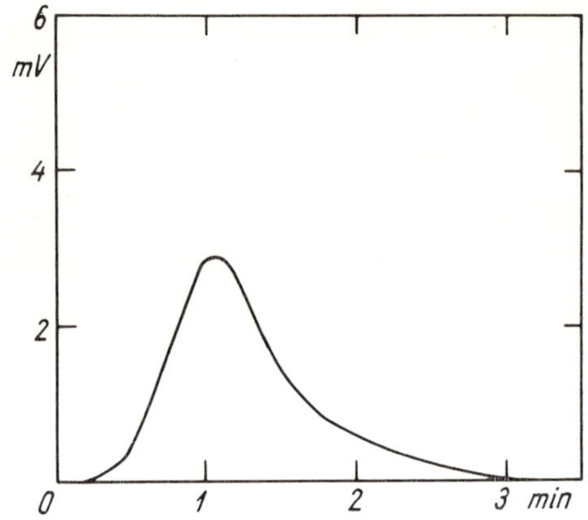

Fig. 7. Trace-out of nitrogen concentration in the analysis of diethylaniline

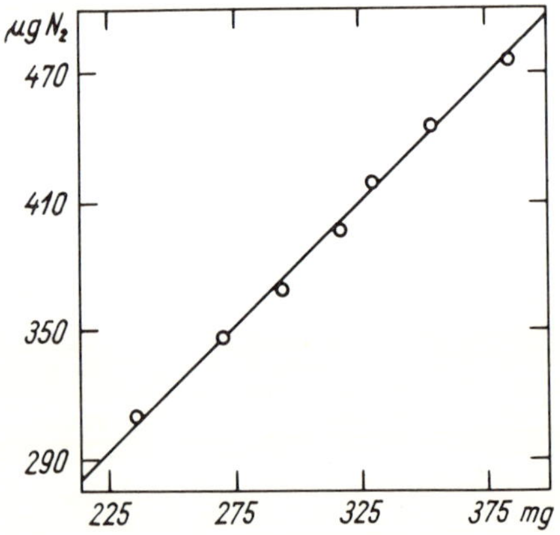

Fig. 8. Calibration curve

of the tube, an air jet can be directed into the space. The CO_2 flow during analysis is set at 10 ml per min; for rinsing, the flow rate can be raised to 20-30 ml per min so that the rinsing time can be correspondingly reduced. The bridge of the conductivity sensing device is energized with a current of 200 milliamperes.

Nitrogen determination by combustion in a closed system (Figs. 2 and 4). From 1.5-5mg of sample is weighed out in a small platinum boat; liquid samples are weighed into capillaries. A 15-mm layer of CuO is first poured on to the permanent filling of the combustion tube, then the sample boat (or capillary) is inserted into the combustion tube, and a mixture of CuO and Co_3O_4 is placed on the boat (or capillary) in such a way that a 100-mm long layer is formed. The movable ignition furnace is placed 30 mm from the stationary furnace, the tube is connected to the rest of the apparatus, and this is rinsed with a rapid stream of CO_2 (20 to 30 ml per min). After rinsing for 3 minutes, the CO_2 flow rate is adjusted to 8 ml per min and stopcock K_3 is closed. Stopcock K_2 is opened, and stopcock K_1 is adjusted so that a connection of the sensing element with both the CO_2 source and the combustion tube is made possible. The movable furnace is switched on and, after 30 seconds, connection of the combustion tube with the conductivity cell is made by turning K_1. CO_2 now flows from the CO_2 source, over the comparison channel of the conductivity sensing element; through K_1 into the measuring channel of the conductivity sensing element. At the same time, travel of the ignition furnace is started by adjusting K_1. The travelling speed is adjusted so that the movable furnace completes the combustion path in 120 seconds. As soon as combustion is complete, K_1 and K_3 are turned to make the connection between CO_2 source, combustion tube, and sensing element, and 60 seconds later the displacement of the recorder is set in motion. The rest of the procedure is as before.

EVALUATION AND CALCULATION OF RESULTS

The calibration curve for the calculation was obtained in the same manner as in the previous paper [12]. A typical trace-out for the determination of nitrogen in antipyrine is shown in Fig. 6; a trace-out for the combustion of diethylaniline in a closed system is seen in Fig. 7. The calibration curve for the experimental set-up described is shown in Fig. 8.

RESULTS AND DISCUSSION

The results, summarized in Tables 1 and 2, of N determination by measurement of thermal conductivity after flash ignition of samples with the use of active Co_3O_4 again confirm the results that we obtained in the study of the kinetics and mechanism of the complete oxidation of organic substances [11]. It was proved that the flash ignition of organic substances, including ones hard to burn (antipyrine), can be carried out even in an inert atmosphere in the absence of gaseous oxygen. The flash method of complete oxidation of organic substances in an inert atmosphere by means of chemically combined oxygen, as is shown by another of our communications, greatly simplifies and accelerates the procedure for volumetric N determination, and also permits the development of a micro method, which permits a direct determination of the N concentration in the carrier gas by measurement of thermal conductivity. Using the same measuring apparatus as in C and H determination [12], we achieved an accuracy approaching that obtained with the classical Pregl micro method. The standard deviations found can probably be decreased by the use of more accurate methods for evaluating the trace-outs.

TABLE 1.

Nitrogen Determination by Direct Combustion

Sample and Formula (Mol. wt.)	mg Sample	% N Calc.	% N Found	% N Diff.
Antipyrine	2.914	14.88	14.67	—0.21
$C_{11}H_{12}N_2O$ (188.2)	2.936	14.88	14.70	—0.18
	2.941	14.88	15.36	+0.48
	2.954	14.88	14.79	—0.09
	3.157	14.88	14.88	0
	3.200	14.88	15.28	+0.40
	3.246	14.88	14.76	—0.12
	3.370	14.88	15.14	+0.26
	3.397	14.88	14.89	+0.01
	3.527	14.88	14.70	—0.18
Acetanilide	4.214	10.36	10.15	—0.21
C_8H_9NO (135.2)	4.095	10.36	10.55	+0.19
	4.229	10.36	10.61	+0.25
	4.271	10.36	10.15	—0.21
	4.525	10.36	10.12	—0.24
Piperidine picrate	2.219	17.83	17.54	—0.29
$C_{11}H_{14}N_4O_7$ (314.3)	2.548	17.83	18.41	+0.59
	2.592	17.83	18.14	+0.31
Phenacetin	4.114	7.82	7.60	—0.22
$C_{10}H_{13}NO_2$ (179.2)	4.390	7.82	7.69	—0.13
	4.579	7.82	8.04	+0.22
	4.839	7.82	8.03	+0.21
	4.928	7.82	8.03	+0.21
Cyclohexanone thiosemicarbazone	2.228	24.55	25.10	+0.55
$C_7H_{13}N_3S$ (171.3)	2.373	24.55	24.33	—0.22
	2.348	24.55	24.69	+0.14
	2.439	24.55	24.70	+0.15
	2.615	24.55	24.34	—0.21
Aceto-*p*-toluidide $C_9H_{11}NO$ (149.2)	3.966	9.39	9.17	—0.22
Dibenzylsulfilimine $C_{21}H_{21}NO_2S_2$ (383.4)	5.116	3.65	3.39	—0.26
Picric acid $C_6H_3N_3O_7$ (229.11)	3.081	18.34	18.58	+0.24

TABLE 2.

Nitrogen Determination by Combustion in a Closed System

Sample and Formula (Mol. wt.)	mg Sample	% N Calc.	% N Found	Diff.
Antipyrine	3.121	14.88	14.84	—0.04
$C_{11}H_{12}N_2O$ (188.2)	3.205	14.88	14.55	—0.33
	3.465	14.88	15.18	+0.30
Cyclohexanone thiosemicarbazone	1.131	24.55	24.77	+0.22
$C_7H_{13}N_3S$ (171.3)	1.171	24.55	24.80	+0.25
	1.237	24.55	24.29	—0.26
Aceto-*p*-toluidide	1.058	9.39	9.06	—0.33
$C_9H_{11}NO$ (149.2)	1.303	9.39	9.57	+0.18
	1.342	9.39	9.42	+0.03
	1.348	9.39	9.53	+0.14
Diethylaniline	1.290	9.39	9.45	+0.06
$C_{10}H_{15}N$ (149.2)	1.487	9.39	9.29	—0.10
	1.893	9.39	9.48	+0.09
	2.238	9.39	9.44	+0.05
	2,200	9.39	9.53	+0.14
	1.762	9.39	9.56	+0.17

Further possibilities for increasing the precision are generally the same as in C and H determination; also, it will be advantageous to decrease the sample size and also to use more sensitive recording apparatus.

Combustion in a closed system was tested along with combustion in a CO_2 stream. A simple device for pressure equalization (Fig. 5) and a combustion tube with smaller dimensions were used, so that more rapid rinsing was obtained. The N concentration wave reaching the sensing element is actually narrower than that observed in direct combustion (cf Figs. 6 and 7). More accurate results

($s = \pm 0.202\%$) were obtained with combustion in a closed system, but it would be premature to draw any conclusions from this fact. The advantage of combustion in a closed system is that liquid compounds can also be analysed, as is shown by the results for N determination in diethylaniline summarized in Table 2. This step in the procedure will possibly also find application in a C determination in liquid compounds.

The method described is the first to show a real way towards complete automation of the total process. The prospects for further development of the method described are the same as those stated in the previous paper [12]; the accuracy achieved thus far is already sufficient to permit use of the method for serial analyses.

SUMMARY

A new method for automatic nitrogen determination, based on flash ignition of samples in the presence of cobalt (II, III) oxide in a stream of CO_2, has been developed. Nitrogen in the carrier gas is determined by direct measurement of thermal conductivity changes. The standard deviation of the nitrogen determination is $s = \pm 0.27\%$. In a second modification of the method, which is also suitable for the analysis of liquid compounds, combustion is carried out in a closed system.

LITERATURE CITED

[1] W. Zimmermann, *Mikrochemie ver. Mikrochim. Acta* **31**, 42 (1943).

[2] M. Vecera, *Chem. listy* **47**, 1090 (1953).

[3] G. Kainz and L. Heinberger, *Z. anal. Chem.* **169**, 27 (1959).

[4] M. Vecera and L. Synek, *Mikrochim. Acta* **1960**, 208.

[5] G. Kainz and H. Horwatitsch, *Z. anal. Chem.* **175**, 166 (1960).

[6] C. Weygand, *Chem. Tech. (Berlin)* **16**, 15 (1943).

[7] C. W. Koch, T. R. Simonson and V. H. Tashinian, *Anal. Chem.* **21**, 1133 (1949).

[8] J. W. Miller and D. D. DeFord, *ibid.* **30**, 295 (1958).

[9] T. Mitsui, *Mikrochim. Acta* **1960**, 150.

[10] E. Stehr, *ibid.* **1954**, 213.

[11] M. Vecera, D. Snobl and L. Synek, *ibid.* **1961**, 370.

[12] M. Vecera, *Collection Czechoslov. Chem. Communs.* **26**, 2298 (1961).

[13] M. Vecera, *Talanta* **8**, 446 (1961).

[14] M. Vecera and J. Bulusek, *Mikrochim. Acta* **1958**, 41.

CHAPTER 56

Advances in Organic Elemental Analysis

A. P. Terent'Ev and E. A. Terent'Eva

The intensive development of studies in the field of organic chemistry, leading to the preparation of new organic compounds, produces increased demands for elemental organic analysis. The time has passed long ago when, even in large chemical institutes, each organic chemist also had to be an analyst. There are now groups of analysts to perform elemental organic analysis. Their task is to increase the productivity of their work by using rapid methods of analysis, working with small quantities of material, and developing methods of analysis for organic compounds containing "uncommon" as well as "common" elements. Methods have been developed, and are finding increasing use, that permit the determination of 3, 4, or more elements in one sample, rather than 1 or 2.

The greatest number of new papers in the field of organic analysis deals with micro (i.e., milligram) samples, but equal attention is also given to methods of semimicro (centigram) analysis. The emphasis on analysis with small samples does not mean that analyses with macro (decigram) samples have completely lost their significance. The latter can be used where it is difficult to provide the conditions required for microchemistry, and when analyses have to be conducted on mixtures or on substances that are not chemically pure. In these cases it is very difficult to take a correct average sample or, if necessary, to determine the content of small amounts of impurities.

**Khim. Nauka i Prom.* 4, 242-249 (1959).

CARBON AND HYDROGEN

The determination of C and H is based, as before, on the quantitative conversion of these elements to CO_2 and water. But, since the time of the classical work of Pregl, the efforts of investigators have been directed towards obtaining conditions for decomposition of the substance that will ensure the most complete, most rapid, and easiest course of the decomposition and oxidation of the sample, and also towards developing rational and accurate methods for quantitative determination of the final analytical form.

Investigators have followed several routes towards an easier decomposition and oxidation of the substance. A so-called "universal" filling of the combustion tube has been proposed, designed not only to ensure the processes mentioned, but also to absorb all the interfering combustion products: oxides of N and S [1], and halogens. However, such a multi-layer filling, consisting of various catalysts, oxidants, and adsorbents, is complicated and inconvenient in operation. It often proves to be generally non-universal. Changes in composition of the filling [2-4], and improved methods for preparing its individual components—PbO_2, Cu on quartz [5, 6] have been proposed. Instead of a multi-layer filling, some authors recommended the use of substances (silver vanadate, MnO_2) capable not only of catalyzing oxidation and of lowering the decomposition temperature, but also of absorbing interfering products [7].

A series of papers on the use of the thermal decomposition product of silver permanganate [8] is of great interest.

A special place is occupied by studies searching for decomposition catalysts and absorbents for the C & H determination in organic compounds that contain metals, fluorine, phosphorus, silicon, etc. [9].

The chief new direction of interest has been toward combustion in a tube without filling [10]. Investigation has followed two lines. One school of workers attempted to compensate for the absence of oxidants in the tube with a large excess of gaseous oxygen provided by a considerable increase in the rate of supply to the system. Studies showed that this likewise requires stronger heating [11, 12]. But even these changes did not ensure complete decomposition and oxidation, and great complications had to be introduced into the construction of the combustion tube to increase contact between the vapors of the substance and oxygen [13, 14, 15]. An oxidation catalyst (platinum) was still required. Since the method involves complex apparatus design, it cannot be considered universal.

Korshun [16, 17] and coworkers succeeded in avoiding the difficulties of combustion in an "empty" tube because of the fact that she, in contrast to foreign investigators, departed from the classical method of distilling the organic substance in a stream of oxygen and used the principle of preliminary pyrolysis of the sample in a limited space in the absence of oxygen (i.e. in a quartz capsule). The pyrolysis leads to the formation of active decomposition products, which readily undergo the further chemical conversions required (oxidation and absorption). These conditions eliminate the need for oxidation catalysts, higher temperature, and increase in heating zone. The process takes place in a few minutes in a short, empty combustion tube in a rapid stream of oxygen at 900°. Pyrolytic combustion has not only made possible the determination of C and H in hard-to- analyze, complex organic compounds, but has also permitted development of a new trend in elemental analysis—the determination of 3 to 5 elements in one sample [18]: C, H, S (or halogen), and one of the elements such as P [167], Si [168], Hg, and a number of other metals. The different combustion products are retained in the decomposition zone or beyond it by suitable absorbents and are weighed.

Pyrolytic combustion methods have recently found widespread use both in the Soviet Union and abroad.

The problem of removing nitrogen oxides from the combustion gases has received great attention in many investigations. The use of PbO_2 in the combustion tube filling for this purpose has always received deserved criticism [15, 25]; methods of removing nitrogen oxides by means of other compounds have been suggested [26, 27]. At the present time, absorption of oxides often takes place outside the tube—between the H and C apparatus—by agents such as $K_2Cr_2O_7$ and conc. H_2SO_4 on silica gel [28], MnO_2 [15].

There are interesting papers on the formation of nitrogen oxides during pyrolytic combustion of substances containing nitrogen in a different form [29, 30]. To remove nitrogen oxides from the combustion products, solutions of potassium permanganate and chromate in concentrated H_2SO_4 have been used; the absorbents are located outside the combustion tube [21, 22]. A method for simultaneous determination of C, H, and N has been developed [170]. It has been shown that higher oxides of Mn and Cr, at high temperature, have significant volatility; they can be retained by a solution of iron vitriol in 36% H_2SO_4.

There is an interesting suggestion for using alkaline earth silicates to absorb the sulfur oxides obtained on combustion of sulfur-containing compounds [31, 32].

Let us note papers devoted to the determination of C and H in small percentage. Thus, Unterzaucher [33] by using the method he developed, determined oxygen in the CO_2 and H_2O formed [33]. Holt [34] determines the quantity of CO_2 in a precision manometer [34]. Malissa [35, 36] absorbs CO_2 in an alkali solution and determines CO_2 from the change of conductivity of the solution.

An original method for analysis of C and H has been suggested [37-41], by combustion with CuO in a closed, evacuated tube, which permits rapid and equal distillation and pyrolysis of the sample. Only CO_2, H_2O, N_2, and C are obtained. After roasting in the tube, a stream of oxygen is passed through to burn up the carbon and remove the combustion products. The method has also proved suitable for semimicro and micro samples of compounds containing N, S, and halogens as well [40]. Vacuum-pyrolytic combustion greatly simplifies the most difficult phase of analysis and is a transition towards semiautomatic operation.

Recently papers have been published dealing with the use of statistics for examining the advantage of various methods for determining C and H and the use of different apparatus [15, 19, 42, 43].

In developing methods for C and H determination, there is a trend towards automating the process. Many suggestions, however, deal only with the use of automatic furnaces [44, 45]. Others are based on a more efficient principle of automation: control of the degree of heating as a function of the decomposition of the substance [46, 47]. The very lengthy process of weighing the absorption apparatus has as yet not been automated. Up to the present the only attempts to save time have been through more intensive cooling of the apparatus before weighing [48].

OXYGEN

Of the methods for direct determination of O, the most obvious were those based on the reaction of the O of an organic compound with free metals, H, and, finally, C. At the beginning of the 1940's, Korshun in the USSR [49-51] and Unterzaucher in Germany [52] simultaneously and independently formulated the principle of a micro method for direct determination of O, based upon decomposition of the sample in a stream of N_2, quantitative conversion of all O-containing decomposition products to CO by means of carbon (in the form of carbon black) heated to 1150°, and then to CO_2 over I_2O_5. Unlike Unterzaucher, Korshun suggested determining O by three factors: increase in weight of CO_2, amount of iodine liberated, and loss in weight of I_2O_5. Another advantage of Korshun's method was the minimal significance of experimental blanks by virtue of the ingenious principle for preliminary removal of trace oxygen in the nitrogen: the gas was passed through a purifying system consisting of the same apparatus as that used for collecting the gases after decomposition of the organic compound.

Since 1942, papers by many authors have been devoted to a solution of the following problems: improvement in purification methods for nitrogen [53-57, 162], study of the decomposition process of the sample [58-60], and a search for better methods of determining CO [53, 61-64]. The principle of pyrolytic decomposition of the analytical sample has been used [16, 18], permitting a significant decrease in length of the contact layer of carbon black [65] and decreasing the possibility of errors due to seepage of atmospheric oxygen into the system. To avoid harmful interaction between C and quartz, the temperature of the contact layer was lowered to 900° by using platinized carbon black [59, 60]. I_2O_5, which was a source of errors, was replaced by CuO (at 100-150°). The authors succeeded in developing a simultaneous determination of halogen and oxygen. The halogen was converted to hydrogen halide by hydrogen from pure paraffin added to the sample, and the halide was determined by weighing (where N was absent from the molecule) or by titration [66].

In Unterzaucher's method absorption of the liberated I_2 by crystalline sodium thiosulfate has been suggested [68]. The use of metallic Cu [60] has been advocated for removing S in the analysis of a S-containing compound, and of magnesium nitride for removing F [67]. The use of porcelain combustion tubes [57] has been suggested to prevent reaction between carbon and quartz. The use of H_2 [68] instead of N_2 as carrier gas has been recommended.

The method of hydrogenation has also been modified; the water formed has been caught in the absorption apparatus and weighed [69]. A new method for decomposing the sample with carbon and strontium oxide in a nickel bomb has been suggested; oxygen is converted to carbonate and determined by a volumetric method [70]. There are directions for a manometric determination of the oxygen liberated by reaction between the organic compound and a mixture of BF_3 and SbF_5 at 500° in vacuum [71].

It should, however, be noted that the direct determination of O, despite the numerous papers, still remains a trouble spot in organic analysis and chemists as before are still satisfied with determining O "by difference".

NITROGEN

The determination of N by the Dumas method has been the subject of many studies [72-76] in which their authors attempted to shorten the pyrolysis time or to modify slightly the usual apparatus. Wurzschmitt [77] showed that, by using dry CO_2, hydrogen can also be determined in the Dumas analysis.

Among the modifications of this method we may mention some that permit combustion of compounds difficult to analyze by the classical method, such as nitrogen-containing heterocycles, compounds with a large number of nitro groups, with angular methyl groups, etc. [26, 33, 78-80]. However, these do not always lead to correct results.

Some papers [81, 82] were aimed at a change in the decomposition method of the sample, in order to establish conditions for easy combustion of the compound.

Within the range of this classical method, a principally new and original method of vacuum pyrolysis has been developed [31-47]. The combustion tube is filled, as usual, with copper oxide; after introducing the sample, the tube is flushed with CO_2 and evacuated. In this state it is ignited for 10-15 minutes. Since the pyrolysis products are passed over copper oxide and reduced copper, carbon, CO_2, H_2O, and N_2 remain in the tube. N_2 is then expelled into an azotometer by means of a rapid stream of CO_2.

An original method for determining N was proposed by Ingram [83]. The sample is burned in a rapid stream of a mixture of CO_2 and O_2, excess O_2 is removed by heated Cu, which also reduces nitrogen oxides; N is measured in an azotometer. The subject of a study of Kjeldahl methods was the composition of sulfuric acid mixtures—catalysts, reducing agents in the analysis of nitro compounds [84]. The use of α-naphthylpyrogallol [85] has been suggested for this purpose; this gives better results than the generally used salicylic acid. For analysis of substances containing N in an oxidized form, preliminary reduction of the sample by zinc and HCl in organic solvents was used [86].

Determination of ammonia (by the Kolthoff method) without distillation has been used [87]; however, this requires maintaining a series of these conditions of preliminary combustion with H_2SO_4, so that this method is not likely to receive widespread use.

Very simple to carry out and reliable is the method of simultaneous determination of N, S, and halogens [88]. A sample of the compound is heated with powdered Mg metal without access of air. Nitrogen in any form is thereby converted to magnesium nitride. Picric acid and picrates are first reduced in the same tube with Mg powder by adding acetic acid. Preliminary removal of air can be carried out by adding a few drops of ether to the tube and heating. On cooling, the tube is broken, the contents are dissolved in acid, and, after adding alkali, distillation and determination of ammonia are carried out by the usual methods. The use of metallic calcium in this method has a number of advantages over the Mg method.

The Mg method is used [90] for microdetermination of N and for small samples.

An original and practical modification of the method of hydrogenation for determination of N has been described [91]. . The authors used an Fe catalyst and preliminary pyrolytic decomposition of the sample in a stream of hydrogen. A more rapid reaction is thus achieved; moreover, the Fe catalyst can be used repeatedly for determination of N in S-containing compounds.

SULFUR

The methods for determining S in organic compounds can be subdivided into oxidative and reductive methods according to the manner of decomposing the sample. Oxidative decomposition leads to the formation of a mixture of S oxides, which, by further oxidation with the same or another agent,are converted to SO_4 ion. This includes the absorption of gaseous combustion products by a solution of H_2O_2 according to Pregl [92-94]. Further oxidation of S oxides and their absorption in the form of SO_4 ion is also carried out by heating with metallic silver [95-99]. The mechanism of this process has been covered in a series of papers [99]; in the most plausible variant of the mechanism it was shown [17,100] that metallic silver in this case acts not only as a chemical absorbent but also as a catalyst for the conversion of SO_2 to SO_3.

The authors of individual methods recommend different temperatures for heating the silver. It was shown by the studies of Korshun and coworkers that the use of relatively low temperatures (~400-450°) leads to rapid exhaustion of the silver. This is because at these temperatures only the surface of the silver takes part in the absorption process. On raising the temperature to 650°, the silver, as it were, breaks up and begins to react with its whole mass, which considerably increases its capacity for S oxides. Pure silver, prepared electrolytically [101], is characterized by its high efficiency. The highest adsorptive capacity is shown by a preparation of silver on granulated, purified pumice [102].

Sulfate ion, fixed in the form of Ag_2SO_4, is determined by the increase in weight of the metallic silver, if the compound does not contain halogens, or in the wash water, after washing it out from the silver halide [15]. The accuracy of the gravimetric determination depends in many respects on the design of the apparatus. The most efficient variation of the gravimetric determination of S is its collection by silver in a special apparatus attached by a ground joint to the end of the combustion tube, or, even better, in a quartz cartridge inserted in the combustion tube in the method for simultaneous determination of C, H, and S [100].

In serial analyses, the old Carius method is still frequently used: decomposition of the compound by prolonged heating with HNO_3 in sealed tubes [103]. In the S determination (particularly for biological objects) wet combustion by various oxidizing media is frequently used for decomposition of the samples [104, 105]. Oxidation with Na_2O_2 in a Parr bomb is a more rapid method, but this leads to a large salt number in the melt solution, which interferes with further titration [106]. Preliminary removal of excess Na ions by ion-exchangers has been suggested [107]. The above difficulty can be avoided by using combustion in an oxygen bomb [108].

The method of burning the sample in an oxygen atmosphere [90, 109, 110] has recently received wide recognition.

For determination of the SO_4 ion obtained in all oxidative methods, any of the common methods can be used. It is proper to mention here gravimetric methods of determination in the form of various insoluble inorganic or organic bases [111, 112]. Various direct and reverse volumetric methods have been used successfully: high-frequency or conductometric titration [113, 114], titration with barium salts, rhodizonic acid and its salts [115, 116]. An indirect determination, by titration or colorimetric determination of the excess precipitant, is also used. The quantity of precipitate can be determined turbidimetrically [117] and spectrophotometrically [118].

Of the reductive methods, we might mention decomposition in a stream of H_2 with formation of H_2S [119], and in the presence of metallic K [120, 121] or Ni catalyst [122] with formation of metal sulfide. $SnCl_2$ is also used as a reducing agent [123]. Reductive methods are terminated by determination of the sulfide ion by iodometric titration or precipitation as silver sulfide and determination of excess silver [124]. Zimmermann uses a more complex route of liberating H_2S from the sulfide, oxidizing it with iodate, and determining excess iodate; the method gives sufficiently accurate results due to compensation of errors in its individual procedures. The H_2S liberated from the sulfide is also absorbed by H_2O_2 and determined polarographically [122].

The literature contains methods of indirect determination of S in hydrocarbons by means of X-ray absorption [125].

Use of various metal oxides and silicates for absorption of S oxides has been suggested [31, 32, 164], which would permit dispensing with the use of silver. This makes possible the determination of C, H, and S in one sample. The method is inapplicable in the presence of halogens.

For combustion of a S-containing compound according to Körbl [8], in a stream of O_2 in a tube in the presence of the decomposition product of silver permanganate, the S is retained as a mixture of Mn and Ag sulfates; these salts are extracted with hot water, the silver ions are oxidized with MnO_2 to form an equivalent of Mn (II) ions, which is determined complexometrically.

HALOGENS

Halogens are frequently determined simultaneously with C and H, by fixation with metallic silver. S can also be determined simultaneously with halogen, by washing out Ag_2SO_4 with hot water [12, 15, 96, 97].

Of interest are studies in which it was shown that, on pyrolytic combustion of halogen-containing organic compounds, the pyrolysis products contain gaseous halogen derivatives, which inhibit the oxidation of CO to CO_2, thus leading to a low C value. The author therefore recommends using a Pt catalyst in the heated zone [127] or catching halogens on Ag in a special weighed quartz cartridge placed before the oxidation zone.

Reductive methods for halogen cleavage are found quite widely in the literature. In the Stepanov method, the use as solvent of a 15% solution of ethylene glycol in isobutanol is suggested [128]. The halides are then titrated with 0.05N mercury nitrate.

Alkali metal biphenyls are also active mineralizing agents for halogen. Liggett [129] used sodium and lithium biphenyl in glycol or [its] methyl ether, and Sezarat [130] worked with sodium biphenyl in boiling toluene.

There are indications [131] that in some chloro compounds (CCl_4, $CHCl_3$) Cl can be split off by action of chromous oxide solution, prepared from $K_2Cr_2O_7$ by reduction with amalgamated Zn.

A universal means for splitting off halogen is the direct action of metals at high temperature. We mention here fusion of a sample with metallic K at 800° in a steel microbomb. A systematic analytical procedure has been developed, permitting determination of 2 to 3 halogens, including F, in one sample [132, 165].

Fedoseev and Sobko suggested an original macro method for analysis of halogen, using the "displacement" principle [133, 134]. A sample of the compound is burned in a stream of oxygen, and the combustion products are passed through a KBr tube (at 700°) and a KI tube (at 500°). Successive displacement of heavier by lighter halogens takes place. For determination, either a gravimetric or a volumetric iodometric method can be used .

The Fedoseev-Sobko displacement method has been used for micro samples of ordinary and polymeric compounds [135].

A special analytical task is the determination of a halogen such as fluorine which forms reactive products upon decomposition of organic fluorine compounds. Good results are obtained by fusion with alkali metals in a bomb. In a solution of the melt, the F ion can be determined either directly, by thorimetric titration [132], or by titration after distillation of F from $HClO_4$ solution as H_2SiF_6 [136, 166]. A gravimetric method for determining F as PbFCl has also been developed [137]. For substances composed of C, H, O, and F, a convenient method comprises combustion in a quartz tube at 900° in the presence of Pt, absorption of the combustion products by water, and titration with alkali [138]. For highly fluorinated compounds, combustion at 1150° in moist oxygen is necessary.

The simultaneous determination of F, C, and H in one sample is of great interest. During pyrolytic combustion, the F is retained by MgO, placed in a special quartz cartridge in the combustion tube. C and H are determined in the usual way, and the MgF_2 formed is pyro-hydrolyzed by steam and the F determined thorimetrically [139].

The same principle has been used for the simultaneous determination of F and N [169].

"WET" COMBUSTION METHODS

This is the name given to methods of organic elementary analysis in which complete or partial oxidation of the organic sample is carried out in a solution of strong oxidants ($KMnO_4$, CrO_3, etc.). Although, as a rule, H cannot be determined by these methods, in a series of cases they have a significant advantage over the usual direct combustion methods and can be used in less well equipped laboratories. Some of these methods are particularly suitable for the analysis of organic compounds containing metals; the latter are mineralized and can be determined by the usual analytical methods.

In 1954 there was published [104] a thorough review of "wet" combustion methods, devoted mainly to the use of CrO_3 in H_2SO_4 solution. The most recent studies are papers on the use of $HClO_4$, having a more individual character [141, 142], and use of an oxidizing mixture of KIO_3 and H_3PO_4, described in two papers [143]. Determination of CO_2 was carried out by the usual method of precipitation as $BaCO_3$. Oxidation of organic substances by means of potassium or ammonium persulfate in H_2SO_4 or H_3PO_4 solution has been suggested for the analysis of drinking water and sewage. The process is generally carried out in the presence of silver ions. In the papers by Gertner [144] and Katz [145], CO_2 is determined by a gravimetric, and in that by Battley [146] by a manometric method.

Oxidation by Chromic Acid

Of great practical importance are the methods in which the oxidants are chromic acid and its salts in H_2SO_4 and H_3PO_4 solutions. In the main, three methods have been suggested for using CrO_3 in the analysis of organic compounds.

1. *Titrimetric Determination of Excess Oxidant*

This method has been used successfully for a long time in many individual cases [140]. Excess oxidant is usually determined iodometrically.

2. *Wet Oxidation without Combustion*

Oxidation by chromic mixture with determination of the CO_2 formed has been used by a number of authors; better oxidation is promoted by adding Ag_2CrO_4 to the CrO_3-H_2SO_4 mixture. However, this method was far from applicable in all cases. A number of compounds form CO and acetic acid, which volatilize without oxidation. Introduction of KIO_3 into the chromic mixture was suggested in 1951 [147]. This results in complete oxidation without CO formation. The process is carried out at the boiling point of the mixture and is complete within a few minutes. CO_2 is absorbed by an alkali solution containing hydrazine sulfate, and is then determined manometrically in a Van

Slyke-Neill apparatus. The method is used widely in biochemistry; it is not applicable to volatile substances. There are, however, indications that it also cannot be used for several nonvolatile substances (acetates, polystyrene, choline sulfate, etc.).

Many papers have been published on the development of an analytical technique and various methods for gravimetric and titrimetric determination of CO_2 [148-152]. We may note that wet oxidation by a chromic acid mixture is used successfully in the analysis of compounds containing C^{14} [153, 154].

3. *Oxidation by Chromic Acid with Combustion*

The main desire of the practicing analyst is to have an accurate, reliable, simple method suitable for the analysis of the majority of organic compounds. This demand is particularly keen in the case of "new" organic compounds complicated by metals or, like many polymeric substances, by caking and by difficulty in burning under ordinary conditions. Oxidation by chromic acid mixture in different variations permits easier combustion of such substances, during which the metals are mineralized, and can then be determined by the usual methods of inorganic analysis, both chemical and physico-chemical (polarography, colorimetry, electrolysis). Of course, in this case, one must dispense with the determination of S and H.

Oxidation by chromic acid mixture can be adopted in analytical laboratories without competing with the usual methods of "dry" combustion. One great disadvantage is that, in a number of cases, the combustion is incomplete. The numerous attempts to find an ideal oxidizing mixture sometimes appear to be a kind of "chemical sport." Actually, the problem is solved much more simply by subjecting the products of wet oxidation to additional combustion in a stream of oxygen by passage over catalysts in a hot tube. Thus it becomes possible, using one sample varying widely in size, to determine, in addition to C, halogen and frequently N, Si, As, and metal.

Let us point out a number of more recent papers. For combustion, a multilayer tube filling has been used, consisting of Pt, CuO, $PbCrO_4$, and Ag [155]. A Van Slyke-Folch oxidizing mixture is used, and combustion is carried out in a tube with CuO and Ag spiral (in the case of halogen-containing compounds). The results, however, are not satisfactory in all cases [156].

Use of the same oxidizing mixture in conjunction with combustion of volatile products in a stream of oxygen in a tube filled with copper oxide, Pt, and Ag has been suggested [157]. This method was used for the analysis of C^{14} compounds. The method proved unsatisfactory for the analysis of polyhalide compounds of benzene. In general, for Cl derivatives, the wet combustion method is not completely satisfactory if copper oxide is used in the combustion tube. In this case, Cl forms the volatile chromyl chloride, and the copper halides formed also have appreciable volatility, clog the tube, and are collected with great difficulty. Various methods and absorbents for halogen have been proposed: potassium and antimony iodide [158], stannous chloride and metallic tin [159]. The most successful fixation of halogen was achieved by Houghton [160], who used a hydrazine solution for this purpose. This author did not use combustion, which is why the method suggested cannot, on the whole, be considered universal.

A practical wet-dry method for the analysis of organic compounds with complex qualitative composition has recently been developed [161]. The essential innovation here is the use, for combustion in an oxygen stream, of a quartz or porcelain tube with chromium sesquioxide. This catalyst, introduced into analysis by Fedoseev, makes possible the rapid oxidation of a great variety of substances. Along with carbon, one can determine Cl and Br, which are absorbed by hydrazine solution and titrated in the Volhard method. The method permits the analysis of organosilicon compounds, metal complexes, etc.; in addition to C and Cl, one can determine in the same sample those elements that remain in the reaction flask in the form of nonvolatile products. Chromium oxide on pumice has proved a very "enduring" catalyst, operating over a wide temperature range and not losing activity on prolonged operation with various compounds.

* * *

In the present review of methods of elemental organic analysis, it was not our task to cover exhaustively all variations of the different methods, but rather to note mainly those papers reporting principally new methods or significant improvements of existing methods. Particular attention was paid to the work of Soviet scientists, especially since reviews published in foreign journals report this work after a considerable delay. Frequently, serious misinterpretations are also tolerated.

This review contains no section on the analysis of organosilicon compounds, since this subject was treated exhaustively by A. P. Kreshkov in the transactions of the conference on "The Chemistry and Practical Use of Organosilicon Compounds," held in Leningrad in September 1958.

LITERATURE CITED

1) J. B. Niederl and V. Niederl: "Micromethods of Quantitative Organic Analysis" [Russian Edition], Goskhimizdat (1949), p. 111.
2) R. Belcher and A. L. Godbert: "Semi-Micro Quantitative Organic Analysis," London: Longmans, Green & Co. (1954).
3) C. L. Ogg, C. O. Willits, C. Ricciuti, and J. A. Connelly, *Anal. Chem.* **23**, 911-14 (1951).
4) O. G. Backeberg and S. S. Israelstam, *Anal. Chem.* **24**, 1209-14 (1952).
5) G. Kainz, A. Resch and F. Schöller, *Mikrochim. Acta* **1956**, 850-7.
6) F. R. Cropper, *Mikrochim Acta* **1954**, 25-48.
7) G. Ingram, *J. Soc. Chem. Ind.* **62**, 175-6 (1943).

8) J. Körbl and R. Pribil, *Collection Czechoslov. Chem. Communs.* **21,** 322-6 (1956).
9) E. L. Head and C. E. Holley, *Anal. Chem.* **28,** 1172-4 (1956).
10) R. Belcher and C. E. Spooner, *Fuel* **20,** 130-4 (1941).
11) R. Belcher and G. Ingram, *Anal. Chim. Acta* **4,** 118-29 (1950).
12) G. Ingram, *Mikrochemie ver. Mikrochim. Acta* **36/37,** 690-705 (1951).
13) A. F. Colson, *Analyst* **73,** 541-7 (1948).
14) R. Belcher and G. Ingram, *Anal. Chim. Acta* **7,** 319-23 (1952).
15) G. Ingram, *Mikrochim. Acta* **1956,** 877-98.
16) M. O. Korshun and V. A. Klimova, *Zhur. Anal. Khim.* **2,** 274 (1947).
17) M. O. Korshun, *ibid.* **7,** 10 (1952).
18) M. O. Korshun and N. E. Gel'man: "New Methods of Elemental Microanalysis" (in Russian), Goskhimizdat (1949).
19) J. A. Kuck and P. L. Altieri, *Mikrochim. Acta* **1956,** 1550-64.
20) P. N. Fedoseev and M. M. Pavlenko, *Zhur. Anal. Khim.* **5,** 296 (1950).
21) P. N. Fedoseev and M. M. Pavlenko, *ibid.* **6,** 317 (1951).
22) P. N. Fedoseev and M. M. Pavlenko, *ibid.* **8,** 158 (1953).
23) P. N. Fedoseev and M. M. Pavlenko, Author's Certificate 9,267 (1950).
24) M. M. Pavlenko, Dissertation, Institute of Organic Chemistry of the Academy of Sciences of the USSR (1952).
25) R. Levy, *Bull. soc. chim. France* **1952,** 672-87.
26) G. Ingram, *Mikrochim. Acta* **1953,** 71-8.
27) C. K. Cross and G. F. Wright, *Anal. Chem.* **27,** 1611 (1955).
28) M. O. Korshun and V. A. Klimova, *Zhur. Anal. Khim.* **6,** 228 (1951).
29) M. O. Korshun and V. A. Klimova, *ibid.* **3,** 176 (1948).
30) M. O. Korshun and V. A. Klimova, *ibid.* **4,** 292 (1949).
31) P. N. Fedoseev and R. M. Lagoshnaya, *ibid.* **9,** 224 (1954).
32) P. N. Fedoseev, R. M. Lagoshnaya, and N. P. Ivashova, *Trudy Nikolaevsk. Korablestroitel. Inst.* No. 9, (1957).
33) J. Unterzaucher, *Mikrochemie ver. Mikrochim. Acta* **36/37,** 706-26 (1951).
34) D. B. Holt, *Anal. Chem.* **27,** 1500 (1955).
35) W. Koch and H. Malissa, *Arch. Eisenhüttenw.* **27,** 695-9 (1956).
36) R. Boulin, *Chim. anal.* **40,** 72-6 (1958).
37) P. N. Fedoseev and L. S. Ignatenko, Author's Certificate 109,422 (1955).
38) P. N. Fedoseev and L. S. Ignatenko, *Izvest Akad. Nauk Turkmen. SSR* **1957,** No. 3, 24.
39) P. N. Fedoseev and L. S. Ignatenko, *ibid.* **1957,** No. 4, 13.
40) P. N. Fedoseev and L. S. Ignatenko, *ibid.* **1957,** No. 6, 84.
41) P. N. Fedoseev, L. S. Ignatenko, and T. E. Chernysheva, *Izvest. Vyssh. Shkoly* **5,** 42 (1958).
42) H Fujiwara and M. Kan, *J. Pharm. Soc. Japan* **75,** 1213 (1955), **76,** 883 (1956). *Anal. Chem.* **30,** 760-5 (1958).
43) G. Wernimont, *Anal. Chem.* **21,** 115-20 (1949).
44) G. L. Royer, A. R. Norton and O. E. Sundberg, *Ind. Eng. Chem. Anal. Ed.* **12,** 688-90 (1940).

45) K. Hozumi, K. Imaeda, and T. Tanaka, *J. Pharm. Soc. Japan*, **72**, 658 (1952).
46) T. Kono, *Mikrochim. Acta* **1958**, 461.
47) B. Bobranski, *Mikrochim. Acta* **1956**, 1735-46.
48) G. I. Robertson, L. M. Jett, and L. Dorfman, *Anal. Chem.* **30**, 132 (1958).
49) M. O. Korshun, *Zavodskaya Lab.* **10**, 241 (1941).
50) M. O. Korshun, Dissertation, Institute of Organic Chemistry of the Academy of Sciences of the USSR (1943).
51) M. O. Korshun, *Trudy Komissii Anal. Khim. Akad. Nauk SSSR* **1**, No. 4, 168 (1947).
52) J. Unterzaucher, *Ber.* **73B**, 391-404 (1940).
53) A. F. Colson, *Analyst* **79**, 784 (1954).
54) W. H. Jones, *Anal. Chem.* **25**, 1449 (1953).
55) M. Bastick, *Bull. soc. chim. France* **1954**, 775-80.
56) W. Kirsten, *Anal. Chem.* **25**, 75 (1953).
57) A. Hintermaier and R. Grützner, *Mikrochim. Acta* **1955**, 994.
58) J. Unterzaucher, *Bull. soc. chim. France* **1953**, C71-7.
59) F. H. Olivier, *Analyst* **50**, 935 (1956).
60) J. Oita and H. S. Conway, *Anal. Chem.* **26**, 600-2 (1954).
61) H. W. Deinum, *Brennstoff-Chem.* **35**, 334-7 (1954).
62) J. Unterzaucher, *Analyst* **77**, 584 (1952).
63) J. Unterzaucher, *Analyst* **79**, 784 (1954).
64) A. Lacourt, *Mikrochim. Acta* **1954**, 735.
65) M. O. Korshun and E. A. Bondarevskaya, *Trudy Nach. Issledovatel. Inst.*, Goskhimizdat (1958).
66) M. O. Korshun and E. A. Bondarevskaya, *Doklady Akad. Nauk SSSR* **110**, 220 (1956).
67) L. Mázor, *Mikrochim. Acta* **1956**, 1757.
68) K. Bürger, *Mikrochim. Acta* **1957**, 313.
69) R. N. Smith, J. Duffield, R. A. Pierotti, and J. Mooi, *Anal. Chem.* **28**, 1161 (1956).
70) T. S. Lee and R. Meyer, *Anal. Chim. Acta* **13**, 340 (1956).
71) I. Sheft and J. J. Katz, *Anal. Chem.* **29**, 1322 (1957).
72) W. Kirsten and S. Ehrlich-Rogozinsky, *Mikrochim. Acta* **1955**, 786.
73) G. Gorbach and H. Kogler, *Mikrochim. Acta* **1957**, 573.
74) G. D. Shah, V. S. Pansare, and V. N. Mulay, *Mikrochim. Acta* **1956**, 1140-3.
75) C. E. Childs, E. E. Meyers, C. K. Johnston and J. D. Mitulski, *Anal. Chem.* **28**, 1193 (1950).
76) M. Fukuda, *J. Pharm. Soc. Japan* **76**, 883 (1956).
77) A. B. Wurzschmitt, *Mikrochemie ver. Mikrochim. Acta* **36/37**, 614-27 (1951).
78) P. D. Sternglanz, R. C. Thompson, and W. L. Savell, *Anal. Chem.* **23**, 1027 (1951).
79) T. D. Parks, E. L. Bastin, E. J. Agazzi, and F. R. Brooks, *Anal. Chem.* **26**, 229 (1954).

80) W. Kirsten, *Mikrochemie ver. Mikrochim. Acta* **40,** 121 (1952).
81) V. A. Klimova and I. F. Dubinina, *Izvest. Akad. Nauk SSSR, Otdel. Khim. Nauk* **1958,** 129.
82) F. Martin, *Bull soc. chim. France* **1953,** C65-C70.
83) G. Ingram, *Mikrochim. Acta* **1953,** 131.
84) W. E. Dickinson, *Anal. Chem.* **30,** 992 (1958).
85) R. Bradstreet, *Anal. Chem.* **26,** 185, 235 (1954).
86) T. S. Ma, R. E. Lang, and J. D. McKinley, Jr., *Mikrochim. Acta* **1957,** 368.
87) R. Belcher and M. K. Bhatty, *Mikrochim. Acta* **1956,** 1183-6.
88) P. N. Fedoseev and N. P. Ivashova, *Zhur. Anal. Khim.* **7,** 116 (1952), **11,** 233 (1956), **13,** 230 (1958).
89) A. P. Terent'ev, P. N. Fedoseev, and N. P. Ivashova, *ibid.* **13,** 344 (1958).
90) W. Schöniger, *Mikrochim. Acta* **1956,** 869.
91) N. E. Gel'man and M. O. Korshun, *Zhur. Anal. Khim.* **12,** 12 (1957).
92) W. Grote and H. Krekeler, *Angew. Chem.* **46,** 106-109 (1933).
93) C. W. Beazley, *Ind. Eng. Chem., Anal. Ed.* **11,** 229-30 (1939).
94) O. Sundberg and G. L. Royer, *Ind. Eng. Chem., Anal. Ed.* **18,** 719-23 (1946).
95) G. L. Stragand and H. W. Safford, *Anal. Chem.* **21,** 625 (1949).
96) H. W. Safford and G. L. Stragand, *Anal. Chem.* **23,** 520 (1951).
97) E. W. Hoffman, *Ind. Eng. Chem., Anal. Ed.* **12,** 53 (1940).
98) M. Vecerá, *Chem. Listy* **48,** 613 (1954).
99) M. Vecerá, *Chem. Listy* **50,** 1941 (1956).
100) M. O. Korshun and N. S. Sheveleva, *Zhur. Anal. Khim.* **7,** 104 (1952).
101) T. Mitsui and H. Sato, *Mikrochim. Acta* **1956,** 1603.
102) N. V. Sokolova, *Zhur. Anal. Khim.* **11,** 728 (1956).
103) A. Steyermark: "Quantitative Organic Microanalysis," Philadelphia: Blakiston Co. (1951).
104) E. W. McChesney and W. F. Banks, *Anal. Chem.* **27,** 987 (1955).
105) P. B. Hawk and O. Bergeim: "Practical Physiological Chemistry," 10th Edition, Philadelphia: Blakiston Co. (1931), p. 872.
106) J. F. Alicino, *Ind. Eng. Chem., Anal. Ed.* **13,** 506 (1941).
107) J. S. Fritz and S. S. Yamamura, *Anal. Chem.* **27,** 1461 (1955).
108) E. C. Wagner and S. H. Miles, *Anal. Chem.* **19,** 274 (1947).
109) W. Schöniger, *Mikrochim. Acta* **1955,** 123.
110) W. Schöniger, "New Methods in Quantitative Organic Analysis," Regional Analytical Symposium, Philadelphia, 8, 11 (1957).
111) R. M. Lincoln, A. S. Carney, and E. C. Wagner, *Ind. Eng. Chem., Anal. Ed.* **13,** 358 (1941).
112) R. Belcher and A. J. Nutten, *Mikrochim. Acta* **1953,** 51.
113) W. H. S. Massie, *Analyst* **52,** 352 (1957).
114) I. Yamaji, *Repts. Govt. Chem. Ind. Research Inst. Tokyo* **50,** 203-6 (1955).
115) B. L. Donicz and T. Rosenquist, *Anal. Chem.* **24,** 404 (1952).
116) L. J. Anderson and R. R. Ravelle, *Anal. Chem.* **19,** 264-7 (1947).

117) G. Toennies and B. Bakay, *Anal. Chem.* **25**, 160 (1953).
118) B. Klein, *Ind. Eng. Chem., Anal. Ed.* **16**, 536 (1944).
119) H. Ter Meulen, *Rec. trav. chim.* **41**, 121-3 (1922).
120) W. Zimmermann, *Mikrochemie ver. Mikrochim. Acta* **33**, 122 (1947).
121) W. Kirsten and G. Carstens, *Anal. Chim. Acta* **5**, 272 (1951).
122) A. Trifonov, K. Ivanov, and D. Pavlov, *Izvest. Khim. Inst. Bolgar. Akad. Nauk* **3**, 3 (1955).
123) S. Ohashi, *Bull. Chem. Soc. Japan* **28**, 585 (1955).
124) I. Irimescu and E. Chirnoaga, *Z. anal. Chem.* **125**, 32-37 (1942).
125) B. H. Eccleston and M. L. Whisman, *Anal. Chem.* **28**, 545-8 (1956).
126) M. Vecerá, *Mikrochim. Acta* **1955**, 88.
127) M. O. Korshun and N. S. Sheveleva, *Zhur. Anal. Khim.* **11**, 376 (1956).
128) L. N. Lapin and R. K. Zamanov, *ibid.* **10**, 364 (1955).
129) L. M. Liggett, *Anal. Chem.* **26**, 738 (1954).
130) A. Sezerat, *Ann. pharm. franç.* **13**, 745-8 (1955).
131) T. Kiba and K. Terada, *J. Chem. Soc. Japan, Pure Chem. Sect.* **75**, 196 (1954).
132) M. O. Korshun and M. N. Chumachenko, *Doklady Akad. Nauk SSSR* **99**, 769 (1954).
133) P. N. Fedoseev and M. Ya. Sobko, Author's Certificate 100,355 (1954).
134) P. N. Fedoseev and M. Ya. Sobko, *Zhur. Anal. Khim.* **10**, 323 (1955).
135) N. V. Sokolova and K. T. Loseva, *ibid.* **13**, 349 (1958).
136) F. S. Ma and J. Gwirtsman, *Anal. Chem.* **29**, 140 (1957).
137) L. Mázor, *Mikrochim. Acta* **1957**, 113.
138) H. S. Clark, *Anal. Chem.* **23**, 659 (1951).
139) M. O. Korshun, N. E. Gel'man, and K. I. Glazova, *Doklady Akad. Nauk SSSR* **111**, 1255 (1956).
140) D. Van Slyke, *Anal. Chem.* **26**, 1706 (1954).
141) P. O. Bethge, *Anal. Chim. Acta* **10**, 317 (1954).
142) G. F. Smith, *Anal. Chim. Acta* **17**, 175 (1957).
143) H. W. Emerton, D. H. Page, and J. Watts, *Brit. Paper and Board Makers' Assoc., Proc. Tech. Sect.* **37**, 105-23 (1956).
144) A. Gertner and H. Ivekovic, *Z. anal. Chem.* **142**, 36 (1954).
145) J. Katz, S. Abraham, and N. Baker, *Anal. Chem.* **26**, 1503 (1954).
146) E. H. Battley, *J. Biol. Chem.* **226**, 237 (1957).
147) D. Van Slyke, J. Plazin, and J. Weisiger, *J. Biol. Chem.* **191**, 299 (1951).
148) P. J. Heald, *Biochem. J.* **49**, 684-5 (1951).
149) A. Lindenbaum, J. Schubert, and W. D. Armstrong, *Anal. Chem.* **20**, 1120 (1948).
150) J. A. Thorn and Ping Shu, *Can. J. Chem.* **29**, 558-62 (1951).
151) N. Baker, H. Feinberg, and R. Hill, *Anal. Chem.* **26**, 1504 (1954).
152) B. Duff and A. Knight, *Chem. & Ind. (London)* **1955**, 1469.
153) D. L. Buchanan and A. Nakao, *J. Am. Chem. Soc.* **74**, 2389 (1952).
154) J. G. Burr, Jr., *Anal. Chem.* **26**, 1395 (1954).
155) B. A. Skopintsev, *Zhur. Anal. Khim.* **4**, 192 (1949).
156) E. E. Archer, *Analyst* **79**, 30 (1954).

157) J. Rutschmann and W. Schöniger, *Helv. Chim. Acta* **40,** 428 (1957).
158) H. Kay, *Chem.-Ing.-Tech.* **26,** 156 (1954).
159) W. P. Pickhardt, A. N. Oemler, and J. Mitchell, Jr., *Anal. Chem.* **27,** 1784 (1956).
160) A. A. Houghton, *Analyst* **70,** 118 (1945).
161) A. P. Terent'ev and B. M. Luskina, *Zhur. Anal. Khim.* **14,** 112 (1959).
162) E. A. Druyan-Rempel and S. A. Babushkin, *Zavodskaya Lab.* **16,** 238 (1950).
163) P. N. Fedoseev, Dissertation, Moscow State University (1957).
164) P. N. Fedoseev, *Zhur. Anal. Khim.* **13,** 123 (1958).
165) M. O. Korshun, V. A. Klimova, and M. N. Chumachenko, *ibid.* **10,** 358 (1955).
166) T. S. Ma, *Microchem. J.* **2,** 91 (1958).
167) M. O. Korshun and E. A. Terent'eva, *Doklady Akad. Nauk SSSR* **100,** 707 (1955).
168) V. A. Klimova and E. G. Bereznitskaya, *Zhur. Anal. Khim.* **11,** 292 (1956); **12,** 424 (1957).
169) N. E. Gel'man, M. O. Korshun, M. N. Chumachenko, and N. I. Larina, *Doklady Akad. Nauk SSSR* **123,** 468 (1958).
170) V. A. Klimova and G. F. Anisimova, *Izvest. Akad. Nauk SSSR, Otdel. Khim. Nauk* **1958,** 791.

JOURNAL INDEX

AUTHOR INDEX

SUBJECT INDEX